AF394779

Linear Algebra with Applications in Machine Learning

Md. Jalil Piran

Linear Algebra with Applications in Machine Learning

From Intuitive Understanding to Python Coding

 Springer

Md. Jalil Piran (iD)
Department of Computer Science
and Engineering
Sejong University
Seoul, Korea (Republic of)

ISBN 978-981-95-5166-8   ISBN 978-981-95-5167-5   (eBook)
https://doi.org/10.1007/978-981-95-5167-5

This Springer imprint is published by the registered company Springer Nature Singapore Pte Ltd.
The registered company address is: 152 Beach Road, #21-01/04 Gateway East, Singapore 189721,
Singapore

If disposing of this product, please recycle the paper.

*To my beloved father and mother, whose
unwavering love, sacrifices, and guidance
have been the foundation of all my endeavors.*

Preface

Linear algebra serves as the foundational mathematical framework powering contemporary machine learning (ML). It underpins a wide array of applications, from image recognition and recommendation engines to Natural Language Processing (NLP) and deep neural networks, offering essential tools for representing, manipulating, and interpreting high-dimensional data. In the current Artificial Intelligence (AI)-centric landscape, mastering linear algebra transcends mere theory; it is an indispensable competency for architecting sophisticated intelligent systems.

This book, *Linear Algebra with Applications in Machine Learning: From Intuitive Understanding to Python Coding*, effectively connects abstract mathematical principles with tangible implementations. Tailored for students, researchers, and practitioners, it delivers an intuitive and comprehensive exploration of the discipline, firmly rooted in practical scenarios and interactive programming exercises.

The text seamlessly incorporates lucid expositions, graphical depictions, and Python-based demonstrations leveraging libraries like NumPy, SymPy, SciPy, Matplotlib, and scikit-learn. The objective is to render linear algebra approachable, replicable, and directly applicable to ML challenges.

Why This Book?

ML is reshaping industries across the board, with linear algebra at its nucleus providing robust mathematical instruments:

- Linear algebra constitutes the bedrock of data-driven learning algorithms.
- This book intertwines theoretical insights with practical utility, illustrating how conceptual abstractions drive authentic models.
- It prioritizes lucidity, conceptual intuition, and engagement via illustrative examples, problem sets, and computational implementations.

Who Is This Book For?

This book caters to:

- Undergraduate and graduate students in computer science, data science, AI, or applied mathematics;
- Researchers and engineers aiming to solidify their ML underpinnings;
- Independent learners bridging from basic programming to advanced ML.

Prerequisites are minimal: foundational algebra and introductory Python proficiency suffice. Complex topics are introduced progressively from fundamental concepts.

How the Book Works?

Chapters adhere to a methodical flow, advancing from conceptual intuition to real-world application, featuring:

- Geometric visualizations and explanatory diagrams;
- Sequential examples accompanied by Python code;
- Highlighted key concepts and chapter recaps;
- Exercises that fuse mathematical reasoning with programming practice.

All code accompanying the book is publicly available at [https://github.com/jalil-piran/Linear-Algebra-with-Applications-in-Machine-Learning].

What You'll Learn?

Upon completing this book, readers will possess a firm command of:

- Core elements including vectors, matrices, tensors, linear systems, transformations, determinants, eigenvalues and eigenvectors, vector spaces, orthogonality, matrix decompositions, and optimization;
- Essential ML methodologies such as regression, principal component analysis (PCA), and matrix factorization;
- Practical execution of these ideas using Python alongside visualization utilities.

A Flexible Learning Approach

Adaptable to sequential reading or targeted reference, the book accommodates:

- Visual-oriented learners through abundant illustrations;
- Practical enthusiasts via hands-on coding segments;
- Application-focused individuals with ML case studies.

Final Words

Far beyond a mere compilation of formulas, this book embarks on an exploration of data's intrinsic architecture and potential. By fusing mathematical theory with coding practice, you will cultivate both intuitive comprehension and applicative prowess essential for constructing advanced intelligent frameworks.

Welcome to the intersection of mathematics and ML.

Seoul, Korea (Republic of)   Prof. Md. Jalil Piran
September 2025

Competing Interests The author has no competing interests to declare that are relevant to the content of this manuscript.

Contents

About the Author

Md. Jalil Piran is an Associate Professor in the Department of Computer Science and Engineering at Sejong University, Seoul, South Korea. He received his Ph.D. in Electronics and Information Engineering from Kyung Hee University in 2016, followed by postdoctoral research at the Networking Laboratory of the same institution.

His expertise spans Artificial Intelligence, Data Science, Big Data, IoT, and Cybersecurity, with numerous publications in leading international journals and conferences. Prof. Piran serves on the editorial boards of the *IEEE Transactions on Intelligent Transportation Systems* and the *Elsevier Journal of Engineering Applications of Artificial Intelligence*, and has guest-edited special issues in several IEEE journals and magazines. He is also Vice-Chair of the IEEE Consumer Technology Society's MDA Subcommittee and has chaired sessions at major IEEE conferences, including ICCE 2024 and ICC 2022.

Professor Piran is a Senior Member of IEEE and represented South Korea in the MPEG standardization body (2014–2017). His international recognition includes the "Research Excellence Professor" Award 2025 from Sejong University, the "Young Scientist Medal" of the Year 2017 from IAAM (Sweden), and the "Outstanding Emerging Researcher" Award 2017 from the Iranian Ministry of Science and Technology. His Ph.D. dissertation was honored as "Dissertation of the Year 2016" by the Iranian Academic Center for Education, Culture, and Research.

Acronyms

AI	Artificial Intelligence
ANNs	Artificial Neural Networks
CNNs	Convolutional Neural Networks
CP	Canonical Polyadic
GANs	Generative Adversarial Networks
GD	Gradient Descent
LR	Linear Regression
LSA	Latent Semantic Analysis
LU	LU Factorization
ML	Machine Learning
MSE	Mean Squared Error
NLP	Natural Language Processing
PCA	Principal Component Analysis
QR	QR Decomposition
REF	Row Echelon Form
RMSE	Root Mean Square Error
RREF	Reduced Row Echelon Form
SGD	Stochastic Gradient Descent
SPD	Symmetric Positive Definite
SVD	Singular Value Decomposition
SVMs	Support Vector Machines
TT	Tensor-Train

Chapter 1
Introduction to Linear Algebra for ML

Introduction

Mathematics is the universal language of science and technology, providing tools to model natural phenomena, design algorithms, and extract insights from data. Among its branches, linear algebra stands out as a cornerstone for ML, enabling the representation, manipulation, and analysis of high-dimensional data. The term "linear" derives from the Latin *linearis*, meaning "pertaining to lines," and refers to relationships that preserve proportionality, such as straight lines in geometry or linear transformations in algebra. The term "algebra," from the Arabic *al-jabr* ("reunion of broken parts"), denotes the manipulation of symbols and equations to solve problems systematically. This chapter introduces linear algebra's role in ML, offering a high-level overview of key concepts, Python tools, and guidance on using this book. Designed for undergraduate and early graduate students, it sets the foundation for subsequent chapters that explore vectors, matrices, linear systems, linear transformations, determinants, eigenvalues, eigenvectors, vector spaces, orthogonality, matrix factorization, optimization, and advanced applications in detail.

Example 1.1 The function y = 2x is a linear function. It satisfies the properties of linearity:

- Additivity: $f(x_1 + x_2) = 2(x_1 + x_2) = 2x_1 + 2x_2 = f(x_1) + f(x_2)$
- Homogeneity: $f(cx) = 2(cx) = c(2x) = cf(x)$

In contrast, a nonlinear function like $f(x) = x^2$ introduces curvature, violating linearity.

Md. Jalil Piran, *Linear Algebra with Applications in Machine Learning*,
https://doi.org/10.1007/978-981-95-5167-5_1

Remark 1.1 A linear function in the algebraic sense must satisfy both additivity and homogeneity, which together imply $f(0) = 0$. Functions of the form $y = mx + b$ with $b \neq 0$ are called affine functions. Although they produce straight-line graphs, they are not linear maps because $f(0) = b \neq 0$ and the superposition property fails. For instance, $f(x) = 3x - 1$ gives $f(1) + f(2) = 2 + 5 = 7$, but $f(1 + 2) = f(3) = 8 \neq 7$. In applied contexts such as linear regression, "linear" often refers to linearity in the parameters rather than in the strict map sense used throughout this book.

Figure 1.1 illustrates the difference between a linear function $y = 2x$, shown as a straight blue line, and a nonlinear function $y = x^2$, shown as a red-dashed parabola. While the linear function has a constant rate of change and forms a straight line, the nonlinear function's rate of change varies with x, resulting in a curved graph. This visual contrast highlights the fundamental distinction between linear and nonlinear relationships in mathematical modeling and ML.

1.1 Why Linear Algebra Matters in ML

Linear algebra provides the mathematical framework for handling multidimensional data and performing efficient computations, which are central to ML. It enables the representation, manipulation, and analysis of large-scale datasets, allowing algorithms to uncover patterns, make predictions, and optimize models. Without linear algebra, many core ML techniques would be computationally infeasible or conceptually unclear. This branch of mathematics underpins everything from simple regression models to complex deep learning architectures, making it essential for understanding how ML systems process and learn from data. In ML, linear algebra is indispensable for several key aspects, as detailed throughout this book:

- **Data Representation**: Vectors represent individual data points, such as feature vectors in classification tasks or word embeddings in NLP, where each component captures a specific attribute or dimension (Chap. 2). Matrices organize collections of these vectors into datasets, like rows representing samples and columns features, while tensors extend this to higher dimensions for complex data like images (height, width, channels) or videos (adding time), preserving structural relationships for efficient processing (Chaps. 3 and 4). This structured representation allows algorithms to handle high-dimensional data scalably.
- **Model Training**: Matrix operations are the backbone of neural network computations, facilitating forward propagation where inputs are multiplied by weight matrices to produce outputs, and backward propagation for gradient calculations to update parameters (Chaps. 2 , 3, and 6). For instance, in deep learning, layers are represented as matrix multiplications, enabling efficient batch processing and leveraging hardware acceleration like GPUs for large-scale training.
- **Dimensionality Reduction**: Techniques like PCA use eigenvalues and eigenvectors to project high-dimensional data onto lower-dimensional subspaces, reducing

noise and computational complexity while retaining the most informative variance (Chaps. 8 and 10). Singular value decomposition (SVD) similarly decomposes matrices to identify latent structures, useful in applications like image compression or topic modeling in text data, preventing overfitting and improving model generalization (Chap. 11).

- **Optimization**: Gradient descent (GD), a fundamental algorithm for minimizing loss functions, relies on vector operations to compute gradients and update model parameters iteratively, often involving matrix-vector products in multivariable settings (Chap. 12). Linear algebra concepts like norms measure convergence, and Hessian matrices (second derivatives) enable advanced methods like Newton's method for faster optimization in convex problems.
- **Recommender Systems**: Matrix factorization techniques decompose user-item interaction matrices into lower-rank approximations, revealing latent factors such as user preferences or item attributes, which are then used to predict missing ratings (Chap. 11). This approach, central to systems like Netflix recommendations, handles sparse data efficiently and scales to millions of users through algorithms like alternating least squares or SVD-based methods.

These applications demonstrate how linear algebra not only provides the tools for computation but also offers geometric intuitions, such as distances in vector spaces for similarity measures or transformations for data augmentation, that deepen our understanding of ML models. As explored in subsequent chapters, mastering these concepts through both theory and Python implementations empowers practitioners to build and innovate in AI.

1.2 Overview of Key Concepts

This book covers linear algebra topics essential for ML, introduced here and explored in detail in later chapters:

- **Vectors (Chap.** 2): Represent data points or directions in $\mathbb{R}^n$. Operations like addition, scaling, dot products, and cross products are used in feature engineering, similarity measures, and optimization.
- **Matrices (Chap.** 3): Encode linear transformations, such as neural network weights or data transformations. Matrix operations, including addition, multiplication, and transposition, drive computations in deep learning.
- **Tensors (Chap.** 4): Generalize scalars, vectors, and matrices to represent multidimensional data, with operations like addition, contraction, unfolding, and decompositions enabling applications in ML, computer vision, signal processing, and medical imaging.
- **Linear Systems (Chap.** 5): Solve systems of equations, critical for linear regression (LR), optimization problems, and model identifiability.
- **Linear Transformations (Chap.** 6): Map vectors between spaces, modeling data transformations in ML pipelines, including rotations, projections, and shears.

- **Determinants (Chap.** 7): Measure volume scaling and matrix invertibility, used in optimization, stability analysis, and solving linear systems via Cramer's rule.
- **Eigenvalues and Eigenvectors (Chap.** 8): Capture data variance in PCA and stability in dynamical systems, with applications in diagonalization and spectral decomposition.
- **Vector Spaces and Subspaces (Chap.** 9): Provide the framework for feature spaces, with concepts like span, basis, and orthogonal projections used in data preprocessing.
- **Orthogonality (Chap.** 10): Describes perpendicular vectors, norms, dot products, orthogonal complements, and projections, fundamental in feature decorrelation, least squares approximations, and simplifying computations in ML.
- **Matrix Decompositions: Factorization and SVD (Chap.** 11): Enable dimensionality reduction, data compression, and robust solutions to linear systems through techniques like QR, LU, and SVD.
- **Optimization (Chap.** 12): Uses gradients, Hessians, and methods like GD to train ML models.
- **Advanced Topics (Chap.** 13): Cover generalized inverses, low-rank matrix approximations, PCA, randomized linear algebra, and matrix manifolds, with applications in high-dimensional data analysis and optimization.

1.3   Python Tools for Linear Algebra

Python is the preferred language for ML due to its rich ecosystem of libraries. In this book, Python is frequently used as a practical tool to illustrate mathematical concepts through visualization, enabling readers to directly connect abstract theory with graphical interpretation. Key Python libraries for linear algebra include:

- **NumPy**: Provides efficient array operations, matrix computations, and linear algebra functions (e.g., `np.linalg`).
- **SciPy**: Extends NumPy with advanced scientific computing, including sparse matrices and optimization.
- **SymPy**: Enables symbolic mathematics, including exact solutions for matrices, determinants, eigenvalues, and algebraic manipulations.
- **Matplotlib**: Visualizes data and transformations, crucial for understanding linear algebra concepts.
- **Scikit-learn**: Implements ML algorithms like PCA and LR, built on NumPy.
- **TensorFlow/PyTorch**: Support tensor operations for deep learning, leveraging linear algebra for neural networks.

All code accompanying the book is publicly available at [https://github.com/jalil-piran/Linear-Algebra-with-Applications-in-Machine-Learning].

As an example, the following Python code demonstrates the distinction between linear and nonlinear functions through plotting. Using NumPy, it generates a range of values from -3 to 3 and computes two functions: a linear function $y = 2x$ and a nonlinear function $y = x^2$. Matplotlib is then employed to plot these functions, with the linear function shown as a solid blue line and the nonlinear function as a dashed red curve. The code also adds horizontal and vertical reference lines at the x- and y-axes, includes a grid for better readability, and uses a legend and title for clarity. The resulting plot (Fig. 1.1) effectively contrasts the constant rate of change in the linear function with the varying curvature of the quadratic function, offering a visual comparison of their fundamental differences.

Python Example: Linear vs. Non-linear Functions

```python
import numpy as np
import matplotlib.pyplot as plt

x = np.linspace(-3, 3, 100)
y_linear = 2 * x
y_nonlinear = x**2

plt.plot(x, y_linear, label='y = 2x',
    color='blue')
plt.plot(x, y_nonlinear, label='y = x^2',
    color='red', linestyle='--')
plt.axhline(0, color='gray', linewidth=0.5)
plt.axvline(0, color='gray', linewidth=0.5)
plt.legend()
plt.title('Linear vs. Non-linear Functions')
plt.grid(True)
plt.show()
```

Output:

See Figure 1.1.

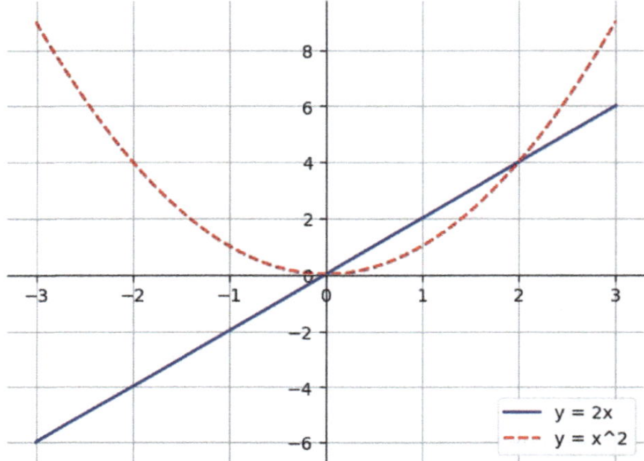

Fig. 1.1 Python visualization of the linear function $y = 2x$ (blue solid line) and the nonlinear function $y = x^2$ (red dashed curve)

1.4   How to Use This Book

This book is designed for students, researchers, and practitioners seeking to master linear algebra for ML. Each chapter builds on the previous, progressing from foundational concepts to advanced applications:

- **Structure**: Chaps. 2–10 cover core linear algebra topics (vectors, matrices, tensors, linear systems, transformations, determinants, eigenvalues, vector spaces, and orthogonality), Chap. 11 explores matrix decompositions including factorization and SVD, Chap. 12 focuses on optimization and gradients, and Chap. 13 delves into advanced topics.
- **Learning Tools**: Each chapter includes examples, visualizations, Python implementations with separate code and output, and exercises. All code accompanying the book is publicly available at [https://github.com/jalil-piran/Linear-Algebra-with-Applications-in-Machine-Learning].
- **Prerequisites**: Familiarity with high school algebra and basic Python programming is assumed.
- **Exercises**: End-of-chapter exercises reinforce concepts through theoretical problems, computations, and coding tasks.
- **Applications**: Real-world ML examples (e.g., PCA, covariance matrices, GD) connect theory to practice.

To maximize learning:

- Read each chapter sequentially, as concepts build progressively.
- Work through examples and verify computations manually or with Python.

- Use visualizations to develop geometric intuition.
- Complete exercises to test understanding and apply concepts.

Chapter Summary

This chapter provided a foundational introduction to linear algebra as a vital tool for ML, emphasizing its role in modeling relationships, processing high-dimensional data, and enabling efficient algorithms. It began with the etymology and core properties of linearity, such as additivity and homogeneity, as illustrated in Example 1.1 and Figure 1.1, and highlighted the distinction between linear and nonlinear functions, which underpinned the design of models from simple regressions to deep neural networks.

Section 1.1 explored linear algebra's centrality to ML through key applications. The section detailed how linear algebra facilitated:

- Representation of data using vectors, matrices, and tensors for scalable handling of high-dimensional datasets;
- Model training via matrix operations in forward and backward propagation of neural networks;
- Dimensionality reduction techniques like PCA and SVD to minimize noise and computational complexity;
- Optimization algorithms such as GD, relying on vector and matrix computations for parameter updates;
- Recommender systems through matrix factorization to uncover latent factors in user-item interactions.

These applications not only drove computational efficiency but also provided geometric insights, such as vector distances for similarity measures and transformations for data augmentation.

The overview of key concepts in Sect. 1.2 outlined the book's progression, covering the following topics, each with direct ties to ML challenges:

- Vectors for representing data points and directions in $\mathbb{R}^n$;
- Matrices for encoding linear transformations like neural network weights;
- Tensors for multidimensional data in computer vision and signal processing;
- Linear systems for solving equations in regression and optimization;
- Linear transformations for data preprocessing and augmentation;
- Determinants for measuring matrix invertibility and volume scaling;
- Eigenvalues and eigenvectors for capturing data variance in PCA;
- Vector spaces and subspaces for understanding feature spaces and projections;
- Orthogonality for feature decorrelation and least squares approximations;
- Matrix decompositions (factorization and SVD) for dimensionality reduction and data compression;

- Optimization using gradients and Hessians for model training;
- Advanced topics including kernel methods and numerical stability for high-dimensional analysis.

Section 1.3 introduced essential Python libraries for linear algebra in ML, demonstrating their practical use through a code example that replicated the linear vs. non-linear plot in Fig. 1.1. The libraries covered included:

- NumPy for efficient array and matrix operations;
- SciPy for advanced scientific computing and optimization;
- SymPy for symbolic mathematics and exact solutions;
- Matplotlib for visualizing transformations and data;
- Scikit-learn for implementing ML algorithms like PCA;
- TensorFlow and PyTorch for tensor operations in deep learning.

Finally, Sect. 1.4 guided readers on leveraging the book's structure, from sequential chapter reading and hands-on exercises to visualizations. By learning these elements, learners gained the tools to innovate in AI, transforming abstract mathematics into actionable ML solutions.

Exercises

1. Explain why linear algebra is essential for ML, providing two specific examples from Sect. 1.1.
2. Sketch the graph of an affine function $y = 3x - 1$ and a nonlinear function $y = x^3$ on the same plot. Explain their differences, and verify that $y = 3x - 1$ is not a linear map in the sense of Example 1.1 by checking whether additivity holds.
3. Identify three Python libraries mentioned in Sect. 1.3 and their roles in linear algebra for ML.
4. Define what a vector and a matrix represent in the context of ML. Give one real-world example of each.

Chapter 2
Vectors

Introduction

Let's start with something simple yet incredibly powerful: vectors. They're everywhere. Whether you're building a machine learning (ML) model, analyzing a massive dataset, or just trying to understand how artificial neural networks (ANNs) learn, you're working with vectors, even if you don't realize it. They form the language that much of modern science and engineering speaks.

In ML and data science, vectors aren't just arrows floating in space or abstract collections of numbers. They carry meaning. They represent everything from rows in a dataset and weights in a model to pixels in an image or word embeddings in natural language processing (NLP). Think of them as compact containers, lean, efficient, and packed with structure. Understanding vectors isn't optional; it's essential. You'll find them at the heart of nearly every algorithm, from the simplest Linear Regression (LR) to the deepest of ANNs.

So, what exactly is a vector? At its most basic, it's an ordered list of numbers. But that undersells its role. A vector tells you not just where something is, but where it's going. It has direction and magnitude. Originally rooted in physical intuition, forces, velocities, that kind of thing, vectors have evolved into a powerful abstraction that can describe just about anything in a structured, high-dimensional space. In $\mathbb{R}^n$, a vector is a point or movement in a space of n dimensions, each coordinate giving you part of the full picture.

This chapter is designed to walk you through the essentials of vector thinking with a special focus on how it connects to ML and data science. We'll start with the basics: definitions, notations, and the dual lens of algebra and geometry. You'll see how vectors are both mathematical objects and visual entities, and we'll back that up with plenty of illustrations and intuitive examples.

Then we'll move into operations, vector addition, scalar multiplication, and the all-important dot product. These are not just mechanical procedures; they're tools

© The Author(s), under exclusive license to Springer Nature Singapore Pte Ltd. 2026
Md. Jalil Piran, *Linear Algebra with Applications in Machine Learning*,
https://doi.org/10.1007/978-981-95-5167-5_2

that underpin deeper ideas like orthogonality, similarity, and projection, all of which are foundational in techniques like PCA or cosine similarity.

We'll also highlight some special vectors including the zero vector, unit vectors, basis vectors, and more. These may sound like technical footnotes, but in practice, they show up everywhere. For example, standard basis vectors are at the heart of one-hot encoding, and unit vectors steer the direction of optimization in gradient descent.

By the end of this chapter, you'll have more than just a working knowledge of vectors, you'll see how they function as a bridge between mathematical theory and real-world ML. To keep things practical, we'll use Python's NumPy and Matplotlib to illustrate each concept in code, so you can immediately apply what you're learning.

This is your first step toward mastering linear algebra for ML. Everything from here, matrices, transformations, eigenvalues, gradients, builds on the solid vector intuition we'll develop together.

Topics Covered

This chapter is organized as follows:

- **2.1 Vectors: Definition and Basics**: Defining vectors algebraically and geometrically, with notation and representations.
- **2.2 Vector Components and Geometry**: Understanding components as projections and visualizing vectors in 1D, 2D, and 3D.
- **2.3 Magnitude and Direction**: Computing Euclidean norms and unit vectors to capture length and orientation.
- **2.4 Vector Operations**: Exploring addition, subtraction, scalar multiplication, and the dot product, with geometric insights.
- **2.5 Special Vectors**: Examining zero vectors, unit vectors, standard basis vectors, and polynomials as vectors.
- **2.6 Vector Sets and $\mathbb{R}^n$**: Defining vector sets, spans, and subspaces in n-dimensional spaces.
- **2.7 Vectors in ML**: Connecting vectors to data representation, model parameters, and algorithms like gradient descent.
- **Summary and Exercises**: Recaps key concepts and provides practical and theoretical problems to reinforce concepts.

Let's dive into the world of vectors, where numbers meet geometry and computation. All code accompanying the chapter is publicly available at https://github.com/jalil-piran/Linear-Algebra-with-Applications-in-Machine-Learning/blob/main/Chapter_02_Vectors.ipynb.

Acronyms

ML Machine Learning
NLP Natural Language Processing

ANNs Artificial Neural Networks
 PCA Principal Component Analysis
  1D One-Dimensional
  2D Two-Dimensional
  3D Three-Dimensional
  GD Gradient Descent

2.1 Vectors: Definition and Basics

A vector is an element of a vector space, where addition and scalar multiplication satisfy specific axioms. In this chapter, we first focus on $\mathbb{R}^n$ before generalizing. Formally, a vector $\mathbf{v} \in \mathbb{R}^n$ is an array of n real numbers, representing a point or displacement in n-dimensional real space.

> **Definition**
> A **vector** is an ordered list of numbers that encodes both *magnitude* (length) and *direction*.

Geometrically, vectors are visualized as directed arrows from the origin to a point, where the arrow's length indicates magnitude and its orientation indicates direction. Algebraically, vectors are manipulated through operations like addition, scalar multiplication, and the dot product.

Vectors are versatile, representing:

- Positions or displacements in 2D or 3D space (e.g., coordinates of an object).
- Physical quantities with direction, such as velocity or force.
- Abstract data points in high-dimensional spaces (e.g., feature sets in datasets).

2.1.1 Vector Notation and Representation

In linear algebra texts, vectors are typically represented using consistent notational conventions to distinguish them from scalars. Throughout this book, we adopt the convention of using boldface lowercase letters for vectors, such as $\mathbf{v}$, $\mathbf{u}, \mathbf{x}$, and $\mathbf{w}$, which clearly differentiates them from scalar quantities like a, b, or c. This notation is particularly effective in ML contexts where vectors represent feature vectors, model parameters, or gradients. Alternatively, the arrow notation $\vec{v}$ is commonly used in physics and engineering texts to emphasize the directional nature of vectors, while the subscript notation $\mathbf{v} = (v_1, v_2, \ldots, v_n)$ provides explicit component-wise representation for computational clarity.

Vectors are represented in two primary forms including column vector and row vector.

- **Column Vector**:

$$\mathbf{v} = \begin{bmatrix} v_1 \\ v_2 \\ \vdots \\ v_n \end{bmatrix}, \quad v_i \in \mathbb{R}$$

This is standard in linear algebra, especially for matrix operations. In this book, unless otherwise specified, vectors are treated as column vectors.

- **Row Vector**:

$$\mathbf{v} = [v_1, v_2, \ldots, v_n]$$

Row vectors are common in computational libraries like NumPy for data processing.

Each v_i is a *component*, representing the vector's projection along the i-th axis. The number n is the *dimension* of the vector space.

Example 2.1 Consider $\mathbf{v} = \begin{bmatrix} 3 \\ 4 \end{bmatrix} \in \mathbb{R}^2$. As shown in Fig. 2.1, this vector represents a displacement of 3 units along the x-axis and 4 units along the y-axis, pointing to the point (3,4). This vector is originating from the origin $(0, 0)$ and terminating at the point $(3, 4)$ in $\mathbb{R}^2$. The vector is represented in blue, and its orthogonal projections onto the coordinate axes are shown using dashed gray lines.

The horizontal dashed line from the vector's tip to the y-axis represents the component $v_2 = 4$, while the vertical dashed line to the x-axis indicates the component $v_1 = 3$. These components represent the contributions of the vector in the x- and y-directions, respectively.

This diagram effectively illustrates how any vector in $\mathbb{R}^2$ can be decomposed into its orthogonal components. In Chap. 10, we will study the concept of orthogonality in detail. Understanding these projections is critical in applications such as:

- computing the *magnitude* of the vector: $\|\mathbf{v}\| = \sqrt{3^2 + 4^2} = 5$ (see Sect. 2.3),
- analyzing *direction*, such as the angle with the x-axis: $\theta = \tan^{-1}(4/3)$ (see Sect. 2.3.2),
- and solving problems involving forces, velocities, or gradients in two dimensions.

Example 2.2 A 3D vector $\mathbf{x} = \begin{bmatrix} 2 \\ 3 \\ 4 \end{bmatrix} \in \mathbb{R}^3$ represents a point (2,3,4) in three-dimensional space.

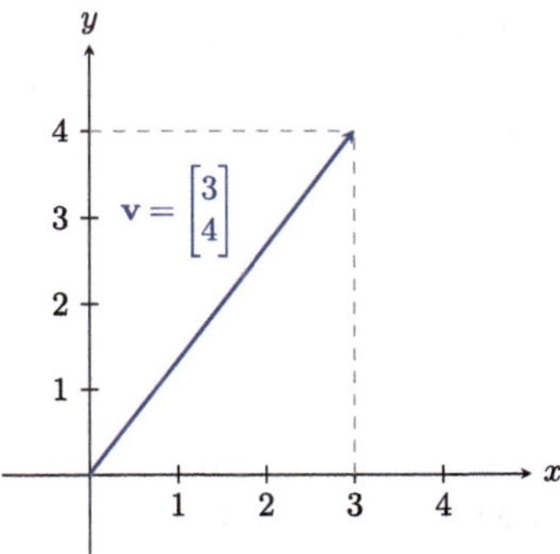

Fig. 2.1 Vector $\mathbf{v} = \begin{bmatrix} 3 \\ 4 \end{bmatrix}$ in $\mathbb{R}^2$, with projections onto the x- and y-axes

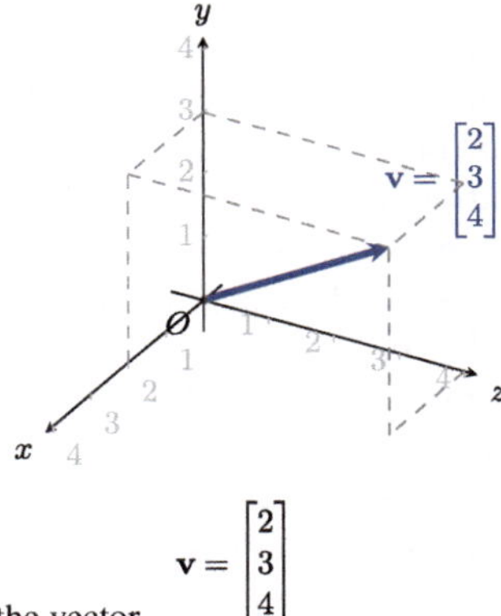

Fig. 2.2 3D representation of the vector $\mathbf{v} = \begin{bmatrix} 2 \\ 3 \\ 4 \end{bmatrix}$

Figure 2.2 illustrates this vector, which originates at the origin $(0, 0, 0)$ and extends to the point $(2, 3, 4)$ in $\mathbb{R}^3$. The vector is shown in blue, and a series of dashed gray lines depict its orthogonal projections onto the xy-, xz-, and yz-coordinate planes.

These projections help visualize how the vector decomposes into its component directions:

- The projection onto the xy-plane lies at $(2, 3, 0)$,
- The projection onto the xz-plane lies at $(2, 0, 4)$,
- The projection onto the yz-plane lies at $(0, 3, 4)$.

Geometrically, each projection shows how much of $\mathbf{v}$ lies along two axes while ignoring the third. This decomposition is fundamental in vector analysis, where a vector's components are analyzed independently in each dimension.

Understanding these projections is essential in applications like physics (e.g., force decomposition), computer graphics (e.g., lighting and shading), and multivariable calculus (e.g., surface integrals and directional derivatives).

This visualization highlights how components combine to form the vector's position. In higher dimensions, while visualization becomes challenging, the algebraic interpretation of components as coordinates remains consistent.

> **Definition**
>
> A high-dimensional vector, such as $\mathbf{z} = \begin{bmatrix} 1 \\ 0 \\ \vdots \\ 0 \end{bmatrix} \in \mathbb{R}^{100}$, is called a ***sparse vector***
>
> because most of its components are zero. Such vectors are common in applications requiring efficient storage, like text processing or signal analysis, where only a few components carry significant information.

Python Implementation

Note that geometric, algebraic, and data-centric interpretations of vectors are equivalent representations of the same mathematical object. The following Python example demonstrates the fundamental process of creating vectors using NumPy, the primary library for numerical computing in Python. The code begins by importing NumPy with the conventional alias "np," which provides efficient array operations essential for linear algebra computations. We then create a 2D vector $\mathbf{v} = [3, 4]$ and a 3D vector $\mathbf{x} = [0.1, -7.2, 5]$ using the "np.array()" constructor, which converts Python lists into NumPy arrays. These arrays support vectorized operations and are the foundation for all subsequent linear algebra computations in ML. The "print()" statements display the vectors, showing NumPy's default formatting where integers appear without decimal points and floating-point numbers show six decimal places. This simple creation process is the starting point for representing data points, feature vectors, and model parameters throughout the book.

Python Example: Creating Vectors

```python
import numpy as np

v = np.array([3, 4])
print("2D vector:", v)

x = np.array([0.1, -7.2, 5])
print("3D vector:", x)
```

Output:

```
2D vector: [3 4]
3D vector: [ 0.1 -7.2  5. ]
```

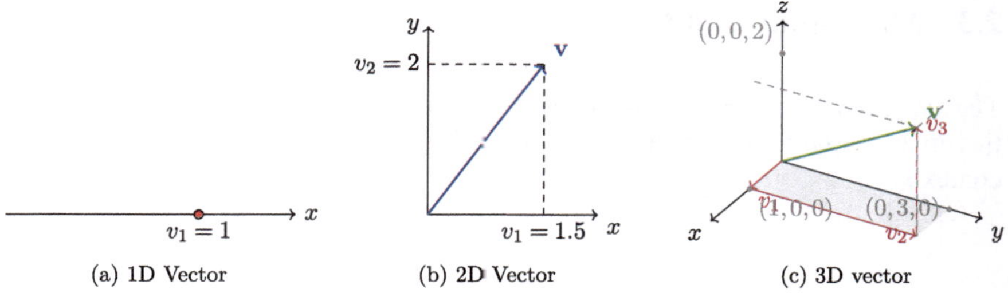

Fig. 2.3 Vector components in 1D, 2D, and 3D spaces, showing projections onto axes

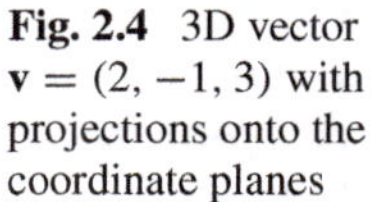

Fig. 2.4 3D vector $\mathbf{v} = (2, -1, 3)$ with projections onto the coordinate planes

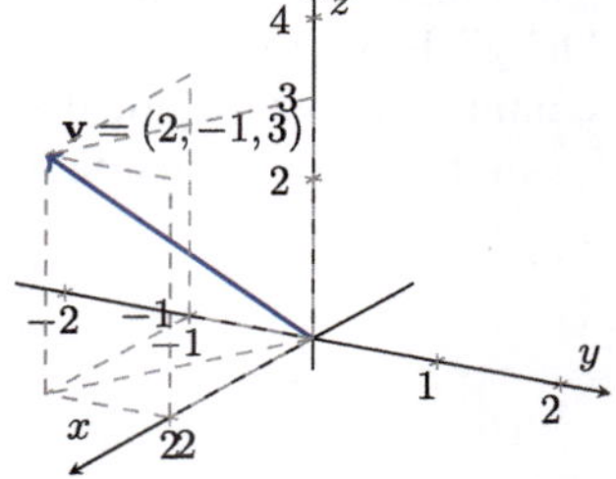

2.2 Vector Components and Geometry

The components of a vector $\mathbf{v} = [v_1, v_2 \ldots, v_n]$ represent its projections onto the coordinate axes. Geometrically, they describe how far the vector extends along each axis, combining to define its position or displacement. For a 3D vector $\mathbf{v} = [1, 2, 3]$, the components are:

- $v_1 = 1$: displacement along the x-axis.
- $v_2 = 2$: displacement along the y-axis.
- $v_3 = 3$: displacement along the z-axis.

Figure 2.3 illustrates how components build vectors across dimensions. In 1D, the vector is a point on a line. In 2D, components form a right triangle, and in 3D, they define a path through space, with projections onto coordinate planes clarifying their contributions.

Example 2.3 For $\mathbf{v} = [2, -1, 3]$, the components indicate a movement of 2 units along the x-axis, -1 units along the y-axis, and 3 units along the z-axis. This can be visualized as a sequence of steps in 3D space, as shown in Fig. 2.4.

2.3   Magnitude and Direction

The *magnitude* and *direction* of a vector define its size and orientation. These properties are essential for understanding vector behavior in geometric and computational contexts.

2.3.1   Magnitude (Euclidean Norm)

The **magnitude** or **length** of a vector $\mathbf{v} = [v_1, v_2, \ldots, v_n]$ in $\mathbb{R}^n$ is a measure of how "long" the vector is. The magnitude can be calculated by the **Euclidean norm**. Other common norms include the **Manhattan norm**, **Maximum norm**, and p**-norm**.

In the case of **Euclidean norm**, we calculate the magnitude as follows.

$$\|\mathbf{v}\| = \sqrt{v_1^2 + v_2^2 + \cdots + v_n^2}.$$

This formula is a direct generalization of the **Pythagorean theorem** in higher-dimensional space. In two dimensions, for instance, the length of a vector $\mathbf{v} = [v_1, v_2]$ is:

$$\|\mathbf{v}\| = \sqrt{v_1^2 + v_2^2},$$

which corresponds to the hypotenuse of a right triangle with legs v_1 and v_2.

Geometrically, the magnitude represents the **straight-line distance** from the origin $\mathbf{0} = [0, 0, \ldots, 0]$ to the point defined by $\mathbf{v}$ in space. It quantifies how far the tip of the vector is from the origin, regardless of the direction.

Example 2.4 Let $\mathbf{v} = \begin{bmatrix} 3 \\ 4 \end{bmatrix}$. Then its magnitude is:

$$\|\mathbf{v}\| = \sqrt{3^2 + 4^2} = \sqrt{9 + 16} = \sqrt{25} = 5.$$

This is the same as the length of the hypotenuse of a right triangle with sides of length 3 and 4.

In addition to the Euclidean norm, there are several other commonly used norms:

- **Manhattan norm** (or L^1-norm): This norm measures distance as the sum of the absolute values of the components. For example: For $\mathbf{v} = [3, -4]$,

$$\|\mathbf{v}\|_1 = |3| + |-4| = 7.$$

- **Maximum norm** (or L^∞-norm): This norm takes the largest absolute value among the components. For example: For $\mathbf{v} = [3, -4]$,

$$\|\mathbf{v}\|_\infty = \max(|3|, |-4|) = 4.$$

- *p*-**norm**: This generalizes norms by raising each component to the p-th power, summing them, and taking the p-th root. For example: For $\mathbf{v} = [3, -4]$ and $p = 3$,

$$\|\mathbf{v}\|_3 = \left(|3|^3 + |-4|^3\right)^{1/3} = (27 + 64)^{1/3} = 91^{1/3} \approx 4.497.$$

2.3.2 Direction (Unit Vector)

In addition to magnitude, every vector in Euclidean space has a **direction**, which tells us where the vector is pointing. To isolate direction from magnitude, we use the concept of a **unit vector**. A unit vector has a length (magnitude) of exactly one and points in the same direction as the original vector. It is computed by dividing the vector by its magnitude:

$$\hat{\mathbf{v}} = \frac{\mathbf{v}}{\|\mathbf{v}\|}.$$

This operation is known as **normalization**. It scales the vector without altering its direction, effectively "standardizing" it to lie on the unit circle in $\mathbb{R}^2$, or the unit sphere in higher dimensions.

Unit vectors are fundamental in many areas of ML and data science, including:

- Computing direction of gradients in optimization.
- Representing normalized features in vector space models.
- Projecting data onto unit-length axes in dimensionality reduction.

In addition to normalization, another important way of analyzing the direction of a vector is by computing the angle it makes with a coordinate axis. For a 2D vector

$$\mathbf{v} = \begin{bmatrix} v_1 \\ v_2 \end{bmatrix},$$

the angle θ between the vector and the x-axis can be obtained using the arctangent function:

$$\theta = \tan^{-1}\left(\frac{v_2}{v_1}\right).$$

This formula is correct when $v_1 > 0$. However, it fails to properly determine the angle when $v_1 \leq 0$, since the standard arctangent function cannot distinguish between different quadrants of the plane. For a complete and robust solution, one uses the two-argument arctangent function:

$$\theta = \text{atan2}(v_2, v_1),$$

which correctly accounts for the signs of both v_1 and v_2 and therefore determines the appropriate quadrant of the angle. This angle provides a direct geometric interpretation of the vector's direction in the plane, and it is frequently used in applications such as navigation, robotics, and physics. The unit vector $\hat{\mathbf{v}}$ then points in the same direction as the angle θ, reinforcing the close connection between angular and normalized representations of direction.

$$
\theta = \text{atan2}(v_2, v_1) = \begin{cases} \arctan\left(\dfrac{v_2}{v_1}\right), & v_1 > 0, \\[2mm] \arctan\left(\dfrac{v_2}{v_1}\right) + \pi, & v_1 < 0,\ v_2 \geq 0, \\[2mm] \arctan\left(\dfrac{v_2}{v_1}\right) - \pi, & v_1 < 0,\ v_2 < 0, \\[2mm] +\frac{\pi}{2}, & v_1 = 0,\ v_2 > 0, \\[2mm] -\frac{\pi}{2}, & v_1 = 0,\ v_2 < 0, \\[2mm] \text{undefined}, & v_1 = 0,\ v_2 = 0. \end{cases}
$$

The function $\text{atan2}(v_2, v_1)$ provides a robust way to compute the angle θ of a vector $\mathbf{v} = [v_1, v_2]^T$ with respect to the x-axis. Unlike the simple arctangent $\arctan(v_2/v_1)$, which only works correctly for $v_1 > 0$, the two-argument arctangent accounts for the signs of both v_1 and v_2, thereby determining the correct quadrant of the vector in the plane. As shown in the piecewise definition above, the function extends $\arctan(v_2/v_1)$ by adding or subtracting π when $v_1 < 0$, and assigns $\pm\frac{\pi}{2}$ when $v_1 = 0$. This ensures that θ spans the full range $(-\pi, \pi]$, providing an unambiguous and complete representation of the vector's direction.

Example 2.5 Consider the vector $\mathbf{v} = \begin{bmatrix} 3 \\ 4 \end{bmatrix}$. We compute its direction using the following steps:

- Step 1-Magnitude:

$$
\|\mathbf{v}\| = \sqrt{3^2 + 4^2} = \sqrt{9 + 16} = \sqrt{25} = 5.
$$

- Step 2-Compute Unit Vector:

$$
\hat{\mathbf{v}} = \frac{1}{5}\begin{bmatrix} 3 \\ 4 \end{bmatrix} = \begin{bmatrix} 0.6 \\ 0.8 \end{bmatrix}.
$$

- Step 3-Verification of Unit Length:

$$
\|\hat{\mathbf{v}}\| = \sqrt{(0.6)^2 + (0.8)^2} = \sqrt{0.36 + 0.64} = \sqrt{1} = 1.
$$

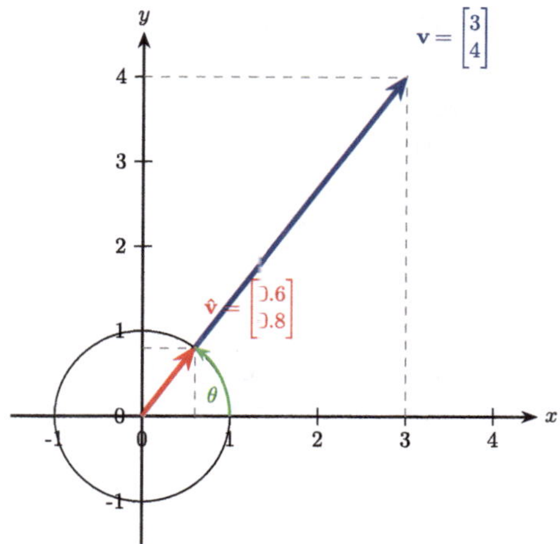

Fig. 2.5 Vector $\mathbf{v} = \begin{bmatrix} 3 \\ 4 \end{bmatrix}$, its unit vector, and the angle θ with respect to the x-axis

- Step 4-Compute Direction Angle:

$$\theta = \tan^{-1}\left(\frac{4}{3}\right) \approx 53.13°.$$

This confirms that the unit vector $\hat{\mathbf{v}}$ maintains the direction of $\mathbf{v}$ while having unit length, and its angle relative to the x-axis is approximately 53.13° (Fig. 2.5).

2.4 Vector Operations

Vector operations enable algebraic and geometric manipulation, forming the basis for many computations in linear algebra and ML. Through operations such as addition, scalar multiplication, and the dot product, vectors can represent transformations, directions, and relationships in high-dimensional spaces. These operations are not only essential for understanding spatial intuition, such as projecting one vector onto another or measuring angles between them, but also underpin key ML algorithms, including gradient descent, similarity measurement, and feature transformation. A solid understanding of vector operations is a critical step toward building and analyzing models that learn from data.

2.4.1   Vector Addition and Subtraction

Vector addition and subtraction are foundational operations in linear algebra, used extensively in ML to manipulate feature vectors, model updates, and gradient computations.

Addition combines two vectors $\mathbf{u}, \mathbf{v} \in \mathbb{R}^n$ element-wise:

$$\mathbf{u} + \mathbf{v} = [u_1 + v_1, u_2 + v_2, \ldots, u_n + v_n].$$

Each component of the resulting vector is the sum of the corresponding components of $\mathbf{u}$ and $\mathbf{v}$. This operation is commutative ($\mathbf{u} + \mathbf{v} = \mathbf{v} + \mathbf{u}$) and associative (($\mathbf{u} + \mathbf{v}) + \mathbf{w} = \mathbf{u} + (\mathbf{v} + \mathbf{w})$).

Subtraction follows a similar element-wise pattern:

$$\mathbf{u} - \mathbf{v} = [u_1 - v_1, u_2 - v_2, \ldots, u_n - v_n].$$

This computes a new vector that when vectors are positioned tail-to-tail then points from the tip of $\mathbf{v}$ to the tip of $\mathbf{u}$, effectively describing the relative displacement from $\mathbf{v}$ to $\mathbf{u}$.

Vector addition can be visualized using the tip-to-tail rule: place the tail of $\mathbf{v}$ at the tip of $\mathbf{u}$, and the vector from the tail of $\mathbf{u}$ to the tip of $\mathbf{v}$ represents $\mathbf{u} + \mathbf{v}$. For subtraction, the vector $\mathbf{u} - \mathbf{v}$ is drawn from the tip of $\mathbf{v}$ to the tip of $\mathbf{u}$, capturing the direction and magnitude needed to move from one vector to another.

Example 2.6 For $\mathbf{u} = \begin{bmatrix} 2 \\ 1 \end{bmatrix}, \mathbf{v} = \begin{bmatrix} 1 \\ 2 \end{bmatrix}$:

$$\mathbf{u} + \mathbf{v} = \begin{bmatrix} 2 \\ 1 \end{bmatrix} + \begin{bmatrix} 1 \\ 2 \end{bmatrix} = \begin{bmatrix} 3 \\ 3 \end{bmatrix},$$

$$\mathbf{u} - \mathbf{v} = \begin{bmatrix} 2 \\ 1 \end{bmatrix} - \begin{bmatrix} 1 \\ 2 \end{bmatrix} = \begin{bmatrix} 1 \\ -1 \end{bmatrix}.$$

Figure 2.6 illustrates the geometric process of vector addition in $\mathbb{R}^2$. The blue vector represents $\mathbf{u} = \begin{bmatrix} 2 \\ 1 \end{bmatrix}$ and the red vector represents $\mathbf{v} = \begin{bmatrix} 1 \\ 2 \end{bmatrix}$, both originating from the origin. The green vector represents the sum: $\begin{bmatrix} 3 \\ 3 \end{bmatrix}$.

Geometrically, this operation is visualized using the *parallelogram rule* or *tip-to-tail method*. The dashed lines complete the parallelogram formed by the vectors $\mathbf{u}$ and $\mathbf{v}$, and the vector $\mathbf{u} + \mathbf{v}$ corresponds to the diagonal of that parallelogram starting from the origin.

This interpretation is fundamental in linear algebra and physics, where vector addition models combined forces, velocities, and displacements.

Figure 2.6 shows how vector subtraction works geometrically in $\mathbb{R}^2$. The blue vector $\mathbf{u} = \begin{bmatrix} 2 \\ 1 \end{bmatrix}$ and the red vector $\mathbf{v} = \begin{bmatrix} 1 \\ 2 \end{bmatrix}$ are both anchored at the origin. The green vector represents the result of subtracting $\mathbf{v}$ from $\mathbf{u}$, computed as: $\begin{bmatrix} 1 \\ -1 \end{bmatrix}$.

Geometrically, vector subtraction $\mathbf{u} - \mathbf{v}$ corresponds to placing the tail of $\mathbf{v}$ at the head of the result, as shown by the dashed line from the tip of $\mathbf{u} - \mathbf{v}$ to the tip of $\mathbf{u}$. This construction shows how vector subtraction effectively "reverses" $\mathbf{v}$ and adds it to $\mathbf{u}$.

Python Example: Vector Addition and Subtraction

```python
import numpy as np
import matplotlib.pyplot as plt

u = np.array([2, 1])
v = np.array([1, 2])
sum_uv = u + v
diff_uv = u - v

print("Vector u:", u)
print("Vector v:", v)
print("Sum of u and v:", sum_uv)
print("Difference of u and v:", diff_uv)

plt.quiver(0, 0, u[0], u[1], color='blue',
    angles='xy', scale_units='xy', scale=1,
    label='u')
plt.quiver(0, 0, v[0], v[1], color='red',
    angles='xy', scale_units='xy', scale=1,
    label='v')
plt.quiver(0, 0, sum_uv[0], sum_uv[1],
    color='green', angles='xy', scale_units='xy',
    scale=1, label='u+v')
plt.quiver(0, 0, diff_uv[0], diff_uv[1],
    color='purple', angles='xy',
    scale_units='xy', scale=1, label='u-v')
plt.grid(True)
plt.legend()
plt.xlim(-1, 4)
plt.ylim(-1, 4)
plt.show()
```

Output:

```
Vector u: [2 1]
Vector v: [1 2]
```

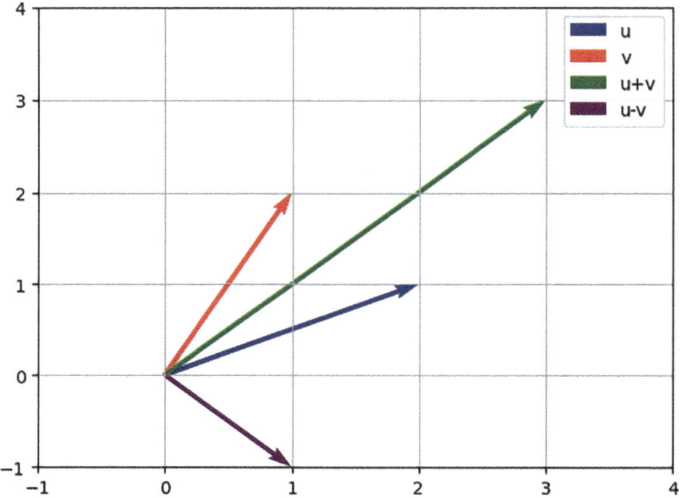

Fig. 2.6 Visualization of vector addition and subtraction using Matplotlib

```
Sum of u and v: [3 3]
Difference of u and v: [ 1 -1]
```

Figure 2.6 shows the vectors **u** and **v** along with their addition and subtraction drawn using Matplotlib library in Python.

2.4.2   Scalar Multiplication

Scalar multiplication scales a vector **v** by a scalar c:

$$c\mathbf{v} = [cv_1, cv_2, \ldots, cv_n].$$

- If $c > 1$, the vector stretches;
- If $0 < c < 1$, it shrinks;
- If $c < 0$, it reverses direction.

Example 2.7 For $\mathbf{v} = \begin{bmatrix} 1.5 \\ 1.5 \end{bmatrix}, c = 2$:

$$c \times \mathbf{v} = 2 \times \begin{bmatrix} 1.5 \\ 1.5 \end{bmatrix} = \begin{bmatrix} 3 \\ 3 \end{bmatrix}.$$

Fig. 2.7 Scalar
multiplication: 2**v** doubles
the magnitude of **v**

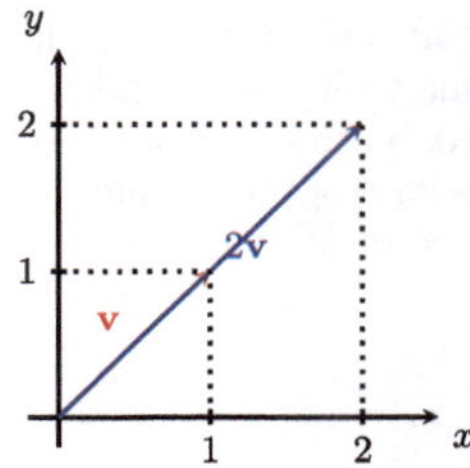

Geometrically, scalar multiplication with a positive scalar $c > 1$ stretches the vector, preserving its direction but increasing its magnitude. In this case, 2**v** lies along the same line as **v**, but it is exactly twice as long (Fig. 2.7).

The dotted lines help visualize the projections of both vectors onto the x- and y-axes, confirming the scaling effect. Such visualizations are crucial in understanding vector operations, linear transformations, and the behavior of linear systems under scaling.

Example 2.8 For **v** $= [1, 1]$, compare $c = 0.5$ and $c = -1$:

$$0.5 \times \mathbf{v} = 0.5 \times [1, 1] = [0.5, 0.5],$$

$$-1 \times \mathbf{v} = -1 \times [1, 1] = [-1, -1].$$

Figure 2.8 demonstrates how scalar multiplication affects a vector $\mathbf{v} \in \mathbb{R}^2$. The original vector $\mathbf{v} = \begin{bmatrix} 1 \\ 1 \end{bmatrix}$ is shown in blue, and two scaled versions of **v** are visualized:

- The red vector represents $0.5\mathbf{v} = \begin{bmatrix} 0.5 \\ 0.5 \end{bmatrix}$, which lies in the same direction as **v** but has half the magnitude. This illustrates how scalar multiplication with $0 < c < 1$ shrinks a vector.
- The green vector shows $-\mathbf{v} = \begin{bmatrix} -1 \\ -1 \end{bmatrix}$, which points in the exact opposite direction of **v**. This highlights the effect of multiplying a vector by a negative scalar: it reverses the direction while preserving the magnitude (up to the scalar factor).

Geometrically, scalar multiplication stretches, shrinks, or reverses a vector while maintaining its linear alignment. This concept is foundational in understanding vector spans, linear combinations, and linear transformations in vector spaces.

Figure 2.8 illustrates the impact of scalar multiplication on a vector. Scaling by a factor less than one, as in $0.5\mathbf{v}$, reduces its magnitude while preserving direction, whereas multiplication by a negative scalar, $-\mathbf{v}$, reverses the orientation of the vector. The dashed lines show the projections of each scaled vector onto the coordinate axes, highlighting their components in the x- and y-directions.

Fig. 2.8 Effects of scalar multiplication: shrinking (0.5**v**) and reversing (−**v**), with projections onto coordinate axes

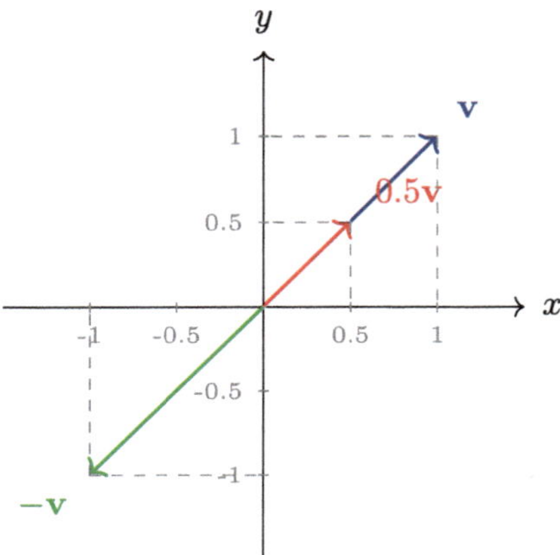

Python Example: Scalar Multiplication

```python
import numpy as np

v = np.array([1.5, 1.5])
c = 2
scaled_v = c * v
print("Scaled Vector:", scaled_v)
```

Output:

```
Scaled Vector: [3. 3.]
```

2.4.3   Vector Multiplication: Dot Product and Cross Product

Vector multiplication encompasses two fundamental operations: the **dot product** (scalar product) and the **cross product** (vector product). These operations bridge algebraic computations with geometric interpretations, enabling applications in physics, computer graphics, and ML. The dot product measures the degree of alignment between two vectors, quantifying how much one vector extends in the direction of another. It plays a critical role in projections, angle computations, and similarity measures, particularly in high-dimensional spaces such as those encountered in data science and neural networks.

In contrast, the cross product produces a new vector orthogonal to the plane defined by the original two vectors, capturing notions of orientation and rotational direction in three-dimensional space. This is especially valuable in 3D modeling, robotics, and physics simulations, where understanding torque, surface normals, and angular momentum is essential. Together, these operations not only enrich the mathematical toolkit for linear algebra but also form the foundation for practical algorithms in areas ranging from computer vision to mechanical engineering.

Dot Product (Inner Product)

The **dot product**, also known as the **inner product** in Euclidean space, is a fundamental operation that combines two vectors and returns a scalar quantity. For vectors $\mathbf{u}, \mathbf{v} \in \mathbb{R}^n$, the dot product is defined algebraically as:

$$\mathbf{u} \cdot \mathbf{v} = \sum_{i=1}^{n} u_i v_i = u_1 v_1 + u_2 v_2 + \cdots + u_n v_n.$$

This operation captures how much one vector "goes in the direction" of another. It forms the basis for numerous applications, including measuring similarity, computing projections, evaluating orthogonality, and optimizing functions in ML and numerical analysis. Importantly, the dot product is commutative and distributive, and it directly relates to the concept of vector length when a vector is dotted with itself.

Beyond its algebraic form, the dot product has a rich geometric meaning. It quantifies the degree of alignment between two vectors through the angle θ formed between them:

$$\mathbf{u} \cdot \mathbf{v} = \|\mathbf{u}\| \|\mathbf{v}\| \cos\theta,$$

where $\|\mathbf{u}\| = \sqrt{\mathbf{u} \cdot \mathbf{u}}$ and $\|\mathbf{v}\| = \sqrt{\mathbf{v} \cdot \mathbf{v}}$ denote the Euclidean norms (magnitudes) of the respective vectors.

This formulation connects the dot product with angular relationships:

- **Parallel vectors** ($\theta = 0°$): The vectors point in the same direction, and the dot product is positive and maximized.
- **Orthogonal vectors** ($\theta = 90°$): The vectors are perpendicular, and the dot product is zero, indicating no directional overlap.
- **Opposite vectors** ($\theta = 180°$): The vectors point in exactly opposite directions, and the dot product is negative and minimized.

This geometric view provides powerful insight into vector relationships, especially in high-dimensional spaces where visual intuition is limited. In ML, for example, the cosine of the angle derived from the dot product is used to assess the similarity between feature vectors or embeddings. In physics, it models work done by a force applied in the direction of motion. As such, the dot product serves as a critical bridge between algebraic manipulation and spatial understanding.

Example 2.9 Let $\mathbf{u} = \begin{bmatrix} 2 \\ -3 \end{bmatrix}$ and $\mathbf{v} = \begin{bmatrix} 5 \\ 1 \end{bmatrix}$:

$$\mathbf{u} \cdot \mathbf{v} = (2)(5) + (-3)(1) = 7.$$

The angle θ between them:

$$\cos\theta = \frac{\mathbf{u}.\mathbf{v}}{\|\mathbf{u}\|\,\|\mathbf{v}\|} = \frac{7}{\sqrt{13}\cdot\sqrt{26}} \approx 0.3806 \implies \theta \approx 67.6°.$$

Since $\mathbf{u}\cdot\mathbf{v} > 0$, the angle is **acute**, meaning the vectors point in generally the same direction and the angle between them is less than $90°$.

Example 2.10 For $\mathbf{u} = \begin{bmatrix} 4 \\ 2 \end{bmatrix}$ and $\mathbf{v} = \begin{bmatrix} -1 \\ 2 \end{bmatrix}$:

$$\mathbf{u}\cdot\mathbf{v} = 4(-1) + 2(2) = -4 + 4 = 0 \implies \text{Orthogonal (perpendicular).}$$

Two vectors are said to be **orthogonal** if their dot product equals zero. Geometrically, this means the angle between them is $90°$, and they meet at a right angle. This property is fundamental in linear algebra and ML, especially in contexts involving projection, orthonormal bases, and dimensionality reduction, where orthogonal vectors are desirable for independence and minimal redundancy. (In Chap. 10, we will study the concept of orthogonality in detail.)

Figure 2.9 displays two vectors in $\mathbb{R}^2$:

$$\mathbf{u} = \begin{bmatrix} 4 \\ 2 \end{bmatrix}, \quad \mathbf{v} = \begin{bmatrix} -1 \\ 2 \end{bmatrix}.$$

The vectors are shown as originating from the origin and extending in different directions. A right-angle marker is drawn at their point of intersection, visually confirming that they are orthogonal.

Two vectors are said to be **orthogonal** if the angle between them is $90°$, or equivalently, if their dot product is zero:

$$\mathbf{u}\cdot\mathbf{v} = 4\cdot(-1) + 2\cdot 2 = -4 + 4 = 0.$$

This calculation confirms that $\mathbf{u} \perp \mathbf{v}$, meaning the two vectors are perpendicular.

Orthogonality is a fundamental concept in linear algebra, with applications in projections, decompositions (such as the Gram–Schmidt process, Chap. 11), and in defining orthonormal bases. It also plays a key role in optimization and ML, where orthogonal directions often represent statistically uncorrelated or independent features.

Fig. 2.9 Vectors $\mathbf{u} = \begin{bmatrix} 4 \\ 2 \end{bmatrix}$ and $\mathbf{v} = \begin{bmatrix} -1 \\ 2 \end{bmatrix}$ form a right angle at the origin, with projections shown on coordinate axes

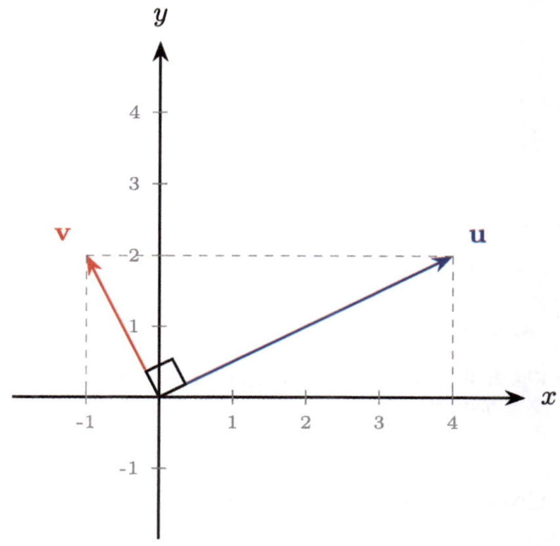

Python Example: Dot Product and Angle

```python
import numpy as np

# Define vectors
u = np.array([1, 0])
v = np.array([1, 1])

# Compute dot product
dot_product = np.dot(u, v)

# Compute angle in degrees
angle = np.arccos(dot_product /
    (np.linalg.norm(u) * np.linalg.norm(v))) *
    180 / np.pi

print(f"Dot Product: {dot_product}, Angle:
    {angle:.1f}°")
```

Output:

```
Dot Product: 1, Angle: 45.0°
```

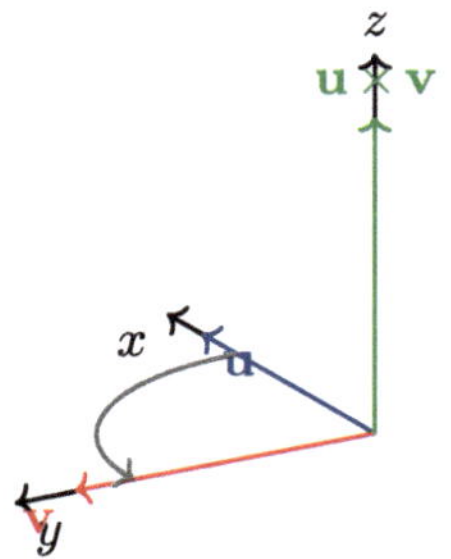
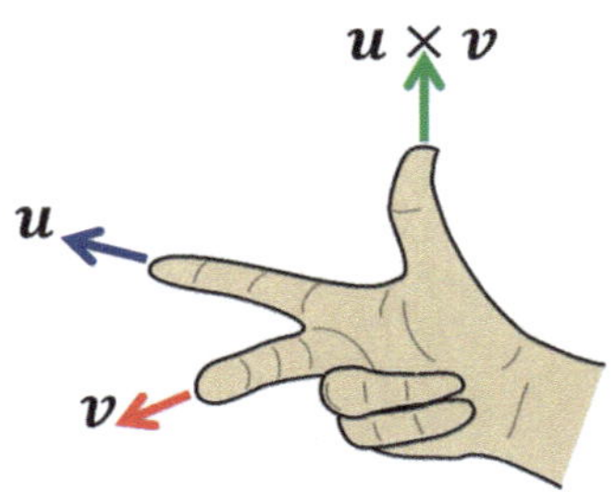

(a) Cross product direction using vectors.   (b) Right-hand rule illustration.

Fig. 2.10 Visualizing the direction of the cross product using **(a)** a 3D vector diagram and **(b)** a right-hand rule sketch

Cross Product (Vector Product)

The cross product applies only to $\mathbb{R}^3$ and yields a vector orthogonal to both inputs. For $\mathbf{u}, \mathbf{v} \in \mathbb{R}^3$:

$$
\mathbf{u} \times \mathbf{v} = \begin{bmatrix} u_2 v_3 - u_3 v_2 \\ u_3 v_1 - u_1 v_3 \\ u_1 v_2 - u_2 v_1 \end{bmatrix}.
$$

Geometric Properties

- **Magnitude**: $\|\mathbf{u} \times \mathbf{v}\| := \|\mathbf{u}\| \|\mathbf{v}\| \sin\theta$ (area of the parallelogram spanned by $\mathbf{u}$ and $\mathbf{v}$).
- **Direction**: Follows the **right-hand rule** (orthogonal to the plane of $\mathbf{u}$ and $\mathbf{v}$).

Note that the cross product is limited to three-dimensional space and is rarely used in machine learning. Figure 2.10 illustrates the concept of the cross product between two vectors $\mathbf{u}$ and $\mathbf{v}$ in three-dimensional space. The left subfigure depicts the vectors $\mathbf{u}$ (blue) and $\mathbf{v}$ (red) positioned along the x- and y-axes, respectively, with their cross product $\mathbf{u} \times \mathbf{v}$ (green) pointing along the z-axis, perpendicular to both. This orientation is confirmed visually by the right-hand arc, representing the rotation from $\mathbf{u}$ to $\mathbf{v}$. The right subfigure reinforces this geometric interpretation by showing the right-hand rule, a mnemonic that determines the direction of the cross product vector: curling the fingers of the right hand from $\mathbf{u}$ toward $\mathbf{v}$ causes the extended thumb to point in the direction of $\mathbf{u} \times \mathbf{v}$. Together, these visuals clarify the fundamental geometric nature of the cross product, which is essential in areas such as physics, computer graphics, and 3D ML models involving vector operations.

Example 2.11 Let

$$\mathbf{u} = \begin{bmatrix} 1 \\ 2 \\ 0 \end{bmatrix}, \quad \mathbf{v} = \begin{bmatrix} 0 \\ 1 \\ 1 \end{bmatrix}.$$

Then the cross product is:

$$\mathbf{u} \times \mathbf{v} = \begin{bmatrix} 2 \cdot 1 - 0 \cdot 1 \\ 0 \cdot 0 - 1 \cdot 1 \\ 1 \cdot 1 - 2 \cdot 0 \end{bmatrix} = \begin{bmatrix} 2 \\ -1 \\ 1 \end{bmatrix}.$$

Verification of Orthogonality:

$$(\mathbf{u} \times \mathbf{v}) \cdot \mathbf{u} = 2 \cdot 1 + (-1) \cdot 2 + 1 \cdot 0 = 2 - 2 + 0 = 0,$$

$$(\mathbf{u} \times \mathbf{v}) \cdot \mathbf{v} = 2 \cdot 0 + (-1) \cdot 1 + 1 \cdot 1 = 0 - 1 + 1 = 0.$$

Python Example: Cross Product

```python
import numpy as np
u = np.array([1, 2, 0])
v = np.array([0, 1, 1])
cross_product = np.cross(u, v)
print("Cross Product:", cross_product)
```

Output:

```
Cross Product: [ 2 -1  1]
```

Figure 2.11 illustrates the geometric interpretation of the cross product. The result is perpendicular to both input vectors and lies in the direction determined by the right-hand rule. The magnitude corresponds to the area of the parallelogram defined by $\mathbf{u}$ and $\mathbf{v}$.

Dot Product vs. Cross Product; Key Differences

Table 2.1 highlights key differences between the dot product and the cross product. The dot product yields a scalar and is defined in any $\mathbb{R}^n$ space, making it widely applicable for measuring angles, projections, and similarity in arbitrary dimensions. In contrast, the cross product is specific to $\mathbb{R}^3$ and results in a vector that is orthogonal to both input vectors, capturing notions of rotational direction and surface area. While the dot product evaluates alignment between vectors, returning zero when they are orthogonal, the cross product constructs a new vector perpendicular to the plane formed by the inputs. These differences make each operation suitable for distinct applications in physics, geometry, and ML.

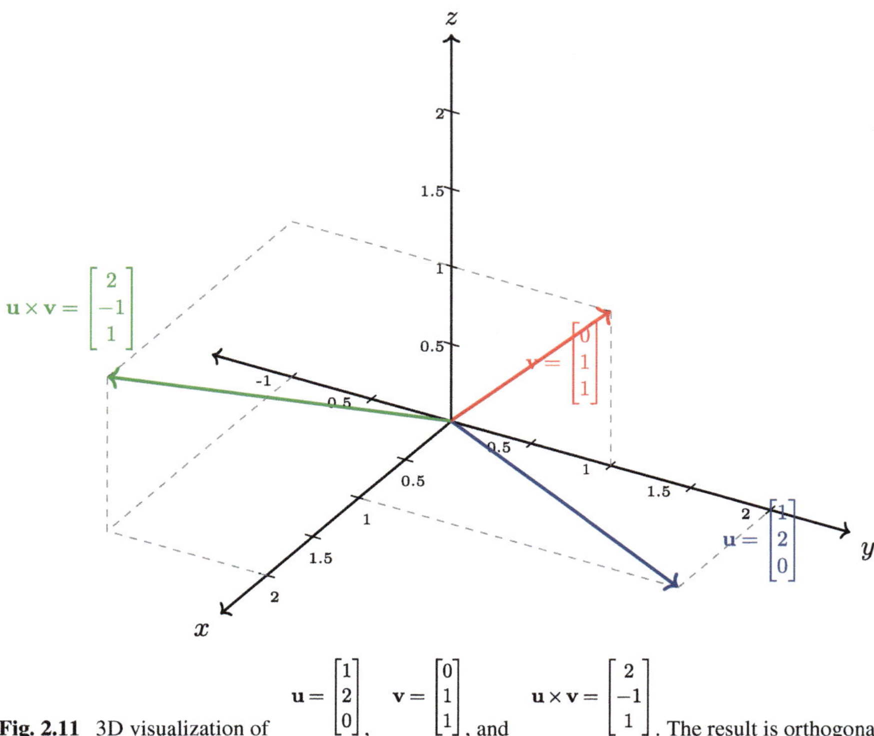

Fig. 2.11 3D visualization of $\mathbf{u} = \begin{bmatrix} 1 \\ 2 \\ 0 \end{bmatrix}$, $\mathbf{v} = \begin{bmatrix} 0 \\ 1 \\ 1 \end{bmatrix}$, and $\mathbf{u} \times \mathbf{v} = \begin{bmatrix} 2 \\ -1 \\ 1 \end{bmatrix}$. The result is orthogonal to both input vectors

Table 2.1 Comparison between dot product and cross product

Property	Dot Product	Cross Product
Output	Scalar	Vector
Dimension	$\mathbb{R}^n$	$\mathbb{R}^3$ only
Orthogonality	Zero if vectors are orthogonal	Output is orthogonal to inputs
Application	Projections, angles	Torque, area calculations

2.5 Special Vectors

Certain vectors have unique properties that simplify computations. For example, *unit vectors* have a length of one and are often used to indicate direction without scaling magnitude. *Zero vectors* contain all zero components and act as the additive identity in vector spaces. Additionally, *orthogonal vectors* are perpendicular to each other, which makes their dot product zero and plays a key role in decomposing vectors and simplifying projections. Understanding these special vectors helps streamline many linear algebra operations and is fundamental in applications such as coordinate transformations, ML feature engineering, and optimization.

2.5.1 Zero Vector

The zero vector $\mathbf{0} = [0, \ldots, 0]$ is the additive identity:

$$\mathbf{v} + \mathbf{0} = \mathbf{v}, \quad 0\mathbf{v} = \mathbf{0} \,(\text{scalar zero times vector})$$

Python Example: Zero Vector

```python
import numpy as np

v = np.array([3, -1, 2])
zero = np.zeros_like(v)
print("v+zero:", v + zero)
print("zero*v:", zero * v)
```

Output:

```
v+zero: [ 3 -1  2]
zero*v: [0 0 0]
```

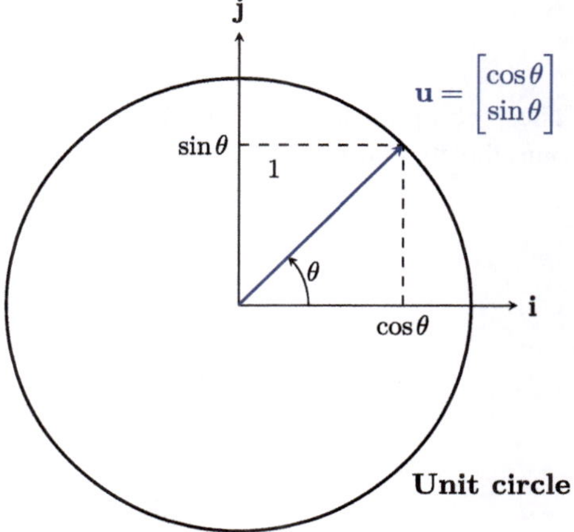

Fig. 2.12 Unit vector **u** on the unit circle in 2D, with components $\cos\theta$ and $\sin\theta$

2.5.2 Unit Vectors

In Sect. 2.3.2, we learned that a unit vector has magnitude one:

$$\hat{\mathbf{v}} = \frac{\mathbf{v}}{\|\mathbf{v}\|}.$$

Example 2.12 For $\mathbf{v} = \begin{bmatrix} 1 \\ 1 \\ 1 \\ 1 \end{bmatrix}$:

$$\|\mathbf{v}\| = \sqrt{4} = 2,$$

$$\hat{\mathbf{v}} = \frac{\mathbf{v}}{\|\mathbf{v}\|} = \frac{1}{2} \times \begin{bmatrix} 1 \\ 1 \\ 1 \\ 1 \end{bmatrix} = \begin{bmatrix} 0.5 \\ 0.5 \\ 0.5 \\ 0.5 \end{bmatrix}.$$

Figure 2.12 depicts a unit vector $\mathbf{u} \in \mathbb{R}^2$ positioned on the unit circle, which is the set of all vectors in $\mathbb{R}^2$ with length 1:

$$\|\mathbf{u}\| = 1.$$

The unit circle provides a geometric representation of direction vectors, where each point on the circle corresponds to a unit vector emanating from the origin.

In this case, the vector $\mathbf{u}$ makes an angle θ with the positive x-axis and has the following coordinate representation:

$$\mathbf{u} = \begin{bmatrix} \cos\theta \\ \sin\theta \end{bmatrix}.$$

These components represent the horizontal and vertical projections of the vector $\mathbf{u}$ onto the standard basis vectors $\mathbf{i}$ and $\mathbf{j}$, respectively.

Python Example: Unit Vector

```python
import numpy as np

v = np.array([1, 1, 1, 1])
unit_v = v / np.linalg.norm(v)
print(unit_v)
```

Output:

```
[0.5 0.5 0.5 0.5]
```

2.5.3 Standard Basis Vectors

The standard basis vectors form the fundamental coordinate system for Euclidean space $\mathbb{R}^n$, providing a natural and intuitive framework for representing any vector as a linear combination of unit vectors aligned with the coordinate axes. These vectors are particularly important in ML, where they serve as the default representation for feature vectors, enable efficient indexing operations, and form the basis for coordinate transformations and orthogonal projections. Their orthonormal properties make them ideal for numerical computations and theoretical analysis.

Standard basis vectors in $\mathbb{R}^n$ are:

$$\mathbf{e}_1 = \begin{bmatrix} 1 \\ 0 \\ \vdots \\ 0 \end{bmatrix}, \quad \mathbf{e}_2 = \begin{bmatrix} 0 \\ 1 \\ \vdots \\ 0 \end{bmatrix}, \quad \mathbf{e}_3 = \begin{bmatrix} 0 \\ 0 \\ 1 \\ \vdots \\ 0 \end{bmatrix}, \quad \ldots, \quad \mathbf{e}_n = \begin{bmatrix} 0 \\ \vdots \\ 0 \\ 1 \end{bmatrix}.$$

Any vector is a linear combination:

$$\mathbf{v} = v_1 \mathbf{e}_1 - \cdots + v_n \mathbf{e}_n.$$

Figure 2.13 illustrates the standard basis vectors in $\mathbb{R}^3$, denoted as

$$\mathbf{e}_1 = \begin{bmatrix} 1 \\ 0 \\ 0 \end{bmatrix}, \quad \mathbf{e}_2 = \begin{bmatrix} 0 \\ 1 \\ 0 \end{bmatrix}, \quad \mathbf{e}_3 = \begin{bmatrix} 0 \\ 0 \\ 1 \end{bmatrix}.$$

Fig. 2.13 Standard basis vectors $\mathbf{e}_1$, $\mathbf{e}_2$, $\mathbf{e}_3$ in $\mathbb{R}^3$

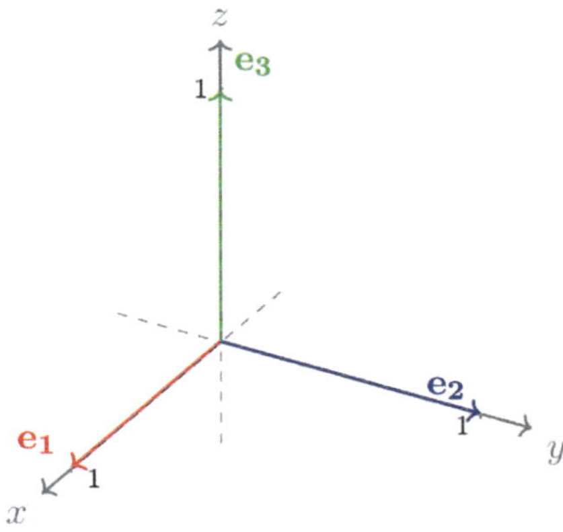

These vectors form an orthonormal basis for three-dimensional Euclidean space, meaning they are mutually perpendicular (orthogonal) and each has unit length. Any vector $\mathbf{v} \in \mathbb{R}^3$ can be expressed uniquely as a linear combination of these basis vectors:

$$\mathbf{v} = x\mathbf{e}_1 + y\mathbf{e}_2 + z\mathbf{e}_3,$$

where $x, y, z \in \mathbb{R}$ are the coordinates of $\mathbf{v}$ in this basis.

The red, blue, and green arrows in the figure represent $\mathbf{e}_1$, $\mathbf{e}_2$, $\mathbf{e}_3$, respectively, oriented along the x-, y-, and z-axes. This basis underlies much of linear algebra and vector calculus, serving as the default coordinate system in $\mathbb{R}^3$.

Python Example: Standard Basis Vectors

```python
import numpy as np

n = 3
basis = np.eye(n)
print("Basis vectors:")
print("e_1:", basis[0])
print("e_2:", basis[1])
print("e_3:", basis[2])
```

Output:

```
Basis vectors:
e_1: [1. 0. 0.]
e_2: [0. 1. 0.]
e_3: [0. 0. 1.]
```

2.5.4 *Polynomials as Vectors*

A polynomial $p(x) = a_0 + a_1 x + a_2 x^2 + \cdots + a_n x^n$ can be represented as a vector of its coefficients:

$$\mathbf{p} = \begin{bmatrix} a_0 \\ a_1 \\ \vdots \\ a_n \end{bmatrix} \in \mathbb{R}^{n+1}.$$

For example, $p(x) = 2 + 3x + x^2$ corresponds to $\mathbf{p} = \begin{bmatrix} 2 \\ 3 \\ 1 \end{bmatrix}$. Polynomials form a vector space under addition and scalar multiplication, enabling linear algebra techniques to be applied. Polynomials can be represented as vectors relative to a chosen basis (e.g., $\{1, x, x^2\}$).

Python Example: Polynomial as Vector

```python
import numpy as np
import matplotlib.pyplot as plt

# Coefficients of p(x) = -x^2 + 3x + 1
coeffs = np.array([-1, 3, 1])

x = np.linspace(0, 3, 100)
y = np.polyval(coeffs, x)

plt.plot(x, y, label='$p(x) = -x^2 + 3x + 1$',
    color='magenta')
plt.grid(True)
plt.legend()
plt.xlabel('$x$')
plt.ylabel('$p(x)$')
plt.show()
```

Output:

See Figure 2.14.

A polynomial can be expressed not only as an algebraic function but also as a vector of its coefficients. For instance, the quadratic polynomial

$$p(x) = -x^2 + 3x + 1,$$

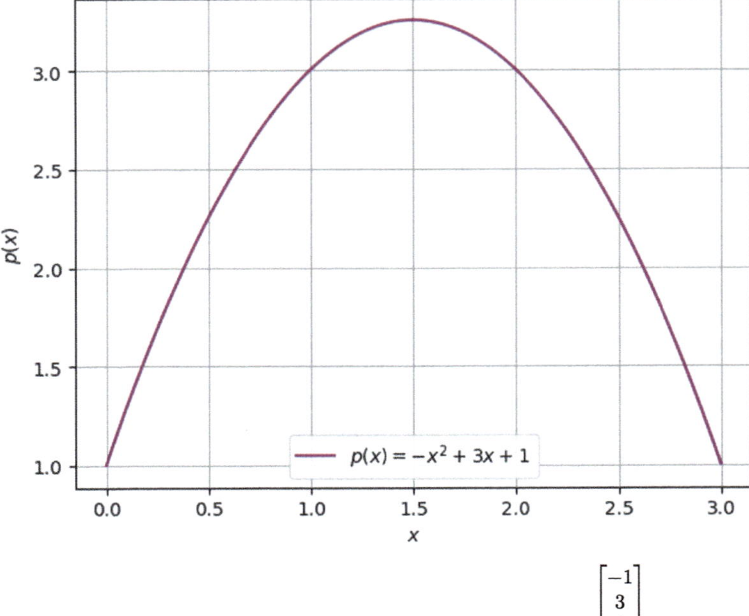

Fig. 2.14 Polynomial $p(x) = -x^2 + 3x + 1$ represented as the vector $\begin{bmatrix} -1 \\ 3 \\ 1 \end{bmatrix}$

is written in terms of the standard polynomial basis $\{1, x, x^2\}$, where each element of the basis corresponds to a power of x. Thus, the coefficient of x^2 is -1, the coefficient of x is 3, and the constant term is 1. By collecting these coefficients, we represent the polynomial as the vector $\begin{bmatrix} -1 \\ 3 \\ 1 \end{bmatrix}$.

In general, a degree-n polynomial

$$p(x) = a_n x^n + a_{n-1} x^{n-1} + \cdots + a_1 x + a_0,$$

can be represented as a vector in $\mathbb{R}^{n+1}$:

$$\mathbf{p} = \begin{bmatrix} a_n \\ a_{n-1} \\ \vdots \\ a_1 \\ a_0 \end{bmatrix},$$

with respect to the basis $\{x^n, x^{n-1}, \ldots, x, 1\}$. This perspective is particularly useful in linear algebra, since operations on polynomials (such as addition, scalar multiplication, and differentiation) can be studied through vector and matrix operations.

Example 2.13 Consider the set $\mathcal{P}_2$, consisting of all real polynomials of degree at most 2:

$$\mathcal{P}_2 = \left\{ a_0 + a_1 x + a_2 x^2 \mid a_0, a_1, a_2 \in \mathbb{R} \right\}.$$

This set forms a **three-dimensional vector space** over $\mathbb{R}$, with a natural basis:

$$\left\{ 1, x, x^2 \right\}.$$

Let's try the addition:

$$p(x) = 1 + 2x + 3x^2,$$

$$q(x) = -2 + x - x^2.$$

Then their sum is:

$$p(x) + q(x) = (1 - 2) + (2 + 1)x + (3 - 1)x^2 = -1 + 3x + 2x^2.$$

Since the result is again a degree ≤ 2 polynomial, this confirms **closure under addition**. That is, when we add two polynomials of degree at most 2, the sum is also a polynomial whose degree does not exceed 2. This means the set of all polynomials of degree less than or equal to 2 is closed under the operation of addition, no matter which two elements we choose from the set, their sum remains within the set. Closure under addition is a fundamental property required for a set to be considered a vector space.

Example 2.14 Let $\alpha = 4$. Then:

$$\alpha \cdot p(x) = 4(1 + 2x + 3x^2) = 4 + 8x + 12x^2.$$

Again, the result lies in $\mathcal{P}_2$, confirming **closure under scalar multiplication**. This means that when any polynomial of degree at most 2 is multiplied by a scalar (a real number), the resulting polynomial still has degree less than or equal to 2. Therefore, the set $\mathcal{P}_2$, which consists of all polynomials of degree at most 2, is closed under scalar multiplication. This property is essential for satisfying the axioms of a vector space, ensuring that scalar scaling does not produce elements outside the set.

The vector space structure allows us to apply **linear algebra techniques** to polynomials, treating them as algebraic objects that can be manipulated systematically:

- **Representing polynomials as vectors**: Each polynomial in $\mathcal{P}_2$ can be uniquely expressed as a linear combination of the basis elements $\{1, x, x^2\}$. This enables us to map polynomials to vectors in $\mathbb{R}^3$, where the coefficients become vector components. For example:

$$p(x) = 1 + 2x + 3x^2 \quad \longleftrightarrow \quad \begin{bmatrix} 1 \\ 2 \\ 3 \end{bmatrix}.$$

This representation facilitates computation and visualization using familiar matrix and vector operations.

- **Performing matrix operations on systems of polynomials**: Once polynomials are represented as vectors, we can apply matrix multiplication, compute linear combinations, and construct systems of equations involving polynomial expressions. This opens the door to efficient algorithmic processing and symbolic manipulation.
- **Solving polynomial approximation problems using least squares methods**: Many real-world problems involve fitting polynomial models to data. Viewing polynomials as vectors allows us to use least squares approximation, a linear algebra technique, to find the best-fitting polynomial by minimizing the error between the model and the observed data.
- **Applying linear transformations such as differentiation**: Differentiation is a linear transformation on $\mathcal{P}_2$, meaning it preserves vector space structure. For example:

$$D(p(x)) = \frac{d}{dx}(1 + 2x + 3x^2) = 2 + 6x \in \mathcal{P}_2.$$

The output is still a polynomial of degree at most 2, demonstrating that differentiation acts linearly on this space and respects closure.

Therefore, the set of polynomials $\mathcal{P}_2$ not only satisfies the axioms of a vector space, such as closure under addition and scalar multiplication, but also enables the application of powerful tools from linear algebra. This perspective unifies polynomial algebra with vector space theory, allowing techniques from one domain to inform and solve problems in the other.

2.6 Vector Sets and $\mathbb{R}^n$

$\mathbb{R}^n$ is a vector space under standard addition and scalar multiplication. The space $\mathbb{R}^n$ includes all n-dimensional vectors, where each vector consists of n real-valued components. This space forms the foundation of many concepts in linear algebra and geometry. Within $\mathbb{R}^n$, a vector set can be either finite or infinite, depending on the context and the number of vectors under consideration. Such sets are central to defining more structured constructs, including spans, which represent all possible linear combinations of a given set of vectors, and subspaces, which are subsets of $\mathbb{R}^n$ that themselves satisfy all the properties of a vector space. These structures are crucial for analyzing the dimensionality, basis, and independence of vector systems, and they underpin many practical applications such as solving linear systems, projecting data, and understanding geometric transformations.

Example 2.15 Finite set:

$$S = \left\{ \begin{bmatrix} 1 \\ 2 \\ 3 \end{bmatrix}, \begin{bmatrix} 4 \\ 5 \\ 6 \end{bmatrix}, \begin{bmatrix} 6 \\ 8 \\ 9 \end{bmatrix} \right\}.$$

This is an example of a **finite set of vectors** in $\mathbb{R}^3$, meaning it contains a countable number of distinct vectors, in this case, three. Such finite sets are often used to generate subspaces through linear combinations, forming what is known as the *span* of the set. By analyzing properties like linear independence among the vectors in S, we can determine the dimension of the subspace they span and identify whether they form a basis for that subspace. Finite sets like this play a key role in practical computations and theoretical analysis in linear algebra.

Python Example: Finite Vector Set

```python
import numpy as np

S = [np.array([1, 2, 3]),
np.array([4, 5, 6]),
np.array([6, 8, 9])]
print("v_1", S[0])
print("v_2", S[1])
print("v_3", S[2])
```

Output:

```
v_1 [1 2 3]
v_2 [4 5 6]
v_3 [6 8 9]
```

Example 2.16 Infinite set:

$$\mathcal{L} = \left\{ \begin{bmatrix} x_1 \\ x_2 \end{bmatrix} : x_1 + x_2 = 1 \right\}.$$

This is an example of an **infinite set of vectors** in $\mathbb{R}^2$, defined by a linear condition on its components. The set includes all vectors whose components satisfy the equation $x_1 + x_2 = 1$, which describes a line in the plane. Since there are infinitely many pairs (x_1, x_2) that satisfy this equation, the set $\mathcal{L}$ contains infinitely many vectors. Such sets often represent geometric objects like lines or planes and are used to model subspaces or affine spaces, depending on whether they pass through the origin. In this case, $\mathcal{L}$ does not pass through the origin and thus forms an affine subset rather than a subspace of $\mathbb{R}^2$.

Fig. 2.15 Set $\mathcal{L}$ forming a
line in $\mathbb{R}^2$

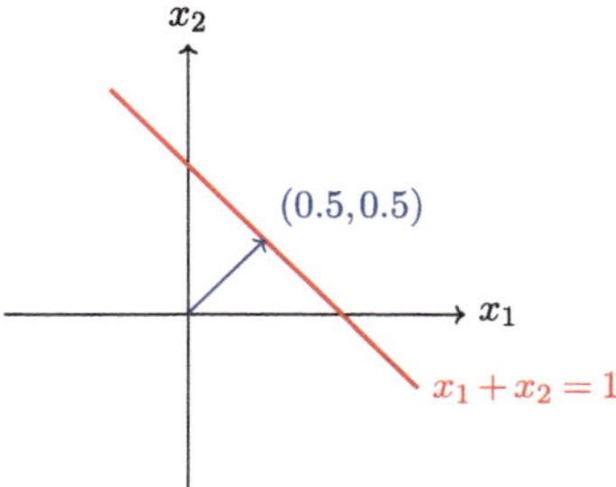

Figure 2.15 illustrates the set

$$\mathcal{L} = \left\{ \mathbf{x} \in \mathbb{R}^2 \mid x_1 + x_2 = 1 \right\},$$

which forms a straight line in the plane. Each point on this line satisfies the linear equation $x_1 + x_2 = 1$, meaning that the set $\mathcal{L}$ represents an infinite number of solutions to a single linear constraint.

The red line represents the entire solution set, while the blue arrow indicates a specific solution vector, $\mathbf{x} = [0.5, 0.5]^T$, which lies on the line. This demonstrates that any point along the line is a valid solution, and the line itself is an affine subspace (not passing through the origin).

2.7   Vectors in ML

Vectors are the backbone of ML, representing data, parameters, and computations.

2.7.1   Data Representation

Each data sample is expressed as a vector in $\mathbb{R}^n$, where n corresponds to the number of features describing the sample. For instance, in image recognition, each pixel intensity can be a feature component. In NLP, vectors encode words, sentences, or documents as numerical representations that enable models to capture semantic relationships and similarities, and in text analysis these vectors are often sparse because only a small subset of possible words appears in a document, as in the bag-of-words model. This vector representation allows algorithms to efficiently process, compare, and learn from data points in a high-dimensional space.

2.7.2 Model Parameters

Model parameters, such as weights in linear regression or neural networks, are organized as vectors. These parameter vectors interact with input data vectors through operations like the dot product $\mathbf{w} \cdot \mathbf{x}$, which produces scalar predictions or intermediate activations. Adjusting these parameter vectors during training enables the model to fit data patterns and improve performance.

2.7.3 Vector Operations

- **Addition and Subtraction**: Vector addition combines feature values or aggregates changes, while subtraction is used to compute differences such as prediction errors or residuals during model evaluation.
- **Scalar Multiplication**: Multiplying a vector by a scalar scales all its components uniformly. This operation is fundamental in optimization algorithms, where gradient vectors are scaled by a learning rate to update model parameters.
- **Dot Product**: The dot product quantifies the similarity between vectors and is widely used to compute predictions in linear models or to measure cosine similarity between feature vectors in clustering and classification tasks.

2.7.4 Normalization

Normalization transforms vectors into unit vectors, which have length one, preserving direction but standardizing magnitude. This is crucial in algorithms like k-nearest neighbors or when computing cosine similarity, where the angle between vectors rather than their length determines similarity:

$$\text{sim}(\mathbf{u}, \mathbf{v}) = \frac{\mathbf{u} \cdot \mathbf{v}}{\|\mathbf{u}\| \|\mathbf{v}\|},$$

where $\text{sim}(\mathbf{u}, \mathbf{v})$ is the cosine similarity. Normalized vectors ensure fair comparisons by eliminating bias caused by varying scales.

2.7.5 Gradient Descent

In optimization, gradients are vectors that point in the direction of the steepest increase of a loss function. Gradient descent uses these vectors to iteratively update model parameters:

$$\mathbf{w} \leftarrow \mathbf{w} - \eta \nabla L,$$

where η is the learning rate and ∇L is the gradient vector. This process guides the model toward parameter values that minimize errors.

2.7.6   Standard Basis Vectors

Standard basis vectors, which have a single component equal to one and the rest zero, serve as building blocks in vector spaces. In ML, they are used in one-hot encoding to represent categorical variables, allowing models to handle discrete data by converting categories into vectors suitable for computation.

2.7.7   Polynomial Features

Polynomial regression models nonlinear relationships by expanding input features into polynomial terms. Each polynomial coefficient corresponds to a vector component, allowing the model to learn complex patterns by combining these weighted terms. This vector-based formulation extends linear models to capture curvature in data.

Python Example: Vectorized Prediction

```python
import numpy as np

x = np.array([1.5, 2.0, 3.0])
w = np.array([0.2, -0.4, 0.6])
y_hat = np.dot(x, w)
print("Predicted value:", y_hat)
```

Output:

```
Predicted value: 1.3
```

Python Example: Cosine Similarity

```python
import numpy as np

u = np.array([1, 2, 3])
v = np.array([2, 4, 6])
```

```
5 cos_sim = np.dot(u, v) / (np.linalg.norm(u) *
     np.linalg.norm(v))
6 print(f"Cosine Similarity: {cos_sim:.2f}")
```

Output:

```
Cosine Similarity: 1.00
```

Chapter Summary

This chapter introduced the foundational concepts of vectors, establishing them as essential building blocks in linear algebra and ML. It began with a comprehensive exploration of vector definitions, notations, and geometric interpretations, emphasizing their role as ordered arrays of numbers that encode both magnitude and direction in $\mathbb{R}^n$. Through examples and visualizations, such as the 2D vector $\mathbf{v} = \begin{bmatrix} 3 \\ 4 \end{bmatrix}$ in Fig. 2.1 and the 3D vector in Fig. 2.2, the chapter illustrated how vector components represent projections onto coordinate axes, providing a bridge between algebraic manipulation and spatial intuition.

The discussion extended to magnitude and direction, where the Euclidean norm was computed as $\|\mathbf{v}\| = \sqrt{\sum v_i^2}$ and unit vectors were derived via normalization $\hat{\mathbf{v}} = \mathbf{v}/\|\mathbf{v}\|$, as demonstrated in Figure 2.5. Vector operations were thoroughly examined, including addition and subtraction via element-wise rules and tip-to-tail geometry, scalar multiplication for scaling and reversal (Fig. 2.8), and the dot product for measuring alignment $\mathbf{u} \cdot \mathbf{v} = \|\mathbf{u}\|\|\mathbf{v}\| \cos\theta$ (Fig. 2.9). The cross product, unique to $\mathbb{R}^3$, was introduced for orthogonal vector generation, with its direction governed by the right-hand rule (Fig. 2.10).

Special vectors were highlighted for their simplifying properties, including the zero vector as the additive identity, unit vectors for direction isolation (Fig. 2.12), standard basis vectors as orthonormal building blocks (Fig. 2.13), and polynomials represented as coefficient vectors, enabling linear algebra techniques on functions (Fig. 2.14). The chapter also covered vector sets in $\mathbb{R}^n$, distinguishing finite collections and infinite spans like lines defined by linear equations (Fig. 2.15).

Finally, the applications of vectors in ML were connected through practical examples, such as data representation, model parameters, normalization for cosine similarity, and gradient descent updates, with Python implementations using NumPy for computations like dot products and visualizations via Matplotlib. Throughout, Python code snippets reinforced theoretical concepts, allowing readers to implement and verify operations hands-on.

Takeaways
Key takeaways from the chapter included:

- Vectors as versatile representations of data points, directions, and features in high-dimensional spaces, with components enabling decomposition and analysis.
- Core operations, addition, scalar multiplication, dot product, and cross product, that underpin geometric insights and computational efficiency in algorithms.
- Special vectors like unit and basis vectors that facilitate normalization, encoding, and basis expansions essential for ML pipelines.
- The structure of $\mathbb{R}^n$ and vector sets, laying groundwork for spans, subspaces, and linear independence explored in later chapters.
- Practical integration with Python tools, from creating arrays to plotting projections, bridging abstract theory with implementable code for model training and optimization.

Matrices (discussed in the next chapter) can be viewed as collections of vectors, either as columns or rows, forming the foundation for linear transformations.

Exercises

1. Write a vector in $\mathbb{R}^5$ with integer components and describe its geometric interpretation as a point in five-dimensional space.
2. Represent a point with coordinates (height, weight, age) as a vector in $\mathbb{R}^3$.
3. Convert "2 units along x-axis, 0 along y-axis, -1 along z-axis" to a vector in $\mathbb{R}^3$.
4. For $\mathbf{a} = [2, 3]$, $\mathbf{b} = [-1, 4]$, compute $\mathbf{a} + \mathbf{b}$, $\mathbf{a} - \mathbf{b}$, and $\mathbf{a} \cdot \mathbf{b}$.
5. For $\mathbf{v} = [5, -2]$, compute $3\mathbf{v}$ and $\|\mathbf{v}\|$.
6. Write Python code to compute the dot product of $\mathbf{u} = [1, 2, 3]$, $\mathbf{v} = [4, -1, 0]$.
7. Plot the 2D vector $\mathbf{v} = [4, -1]$ using Matplotlib, labeling components.
8. Visualize the 3D vector $\mathbf{w} = [1, 2, 3]$ in Python, showing its projection onto the xy-plane by dropping the third component.
9. Express $\mathbf{v} = [2, -1, 4]$ as a linear combination of standard basis vectors.
10. Verify in Python that $\mathbf{0} + \mathbf{v} = \mathbf{v}$ for $\mathbf{v} = [1, 2, 3]$.
11. Confirm that $\mathbf{u} = \left[\frac{1}{\sqrt{2}}, \frac{1}{\sqrt{2}}\right]$ is a unit vector using Python.
12. Generate standard basis vectors in $\mathbb{R}^4$ using NumPy.
13. Define three linearly independent vectors in $\mathbb{R}^3$.
14. Create a set of 5 random vectors in $\mathbb{R}^2$ using NumPy and plot them.
15. Plot the set $\{[x_1, x_2] : x_1 + x_2 = 1\}$ in $\mathbb{R}^2$ using Matplotlib.

16. Compute the angle between $\mathbf{u} = [2, 1]$ and $\mathbf{v} = [1, 2]$ in degrees using the dot product.
17. For $\mathbf{v} = [1, -2, 3]$, compute its unit vector and verify its magnitude in Python.
18. Visualize the effect of scaling $\mathbf{v} = [1, 1]$ by $c = -2, 0.5, 2$ in a single Matplotlib plot.
19. Represent the polynomial $p(x) = 2 - x + 4x^2$ as a vector and compute $3p(x)$ as a vector operation.
20. Write Python code to check if two vectors $\mathbf{u} = [3, 0]$ and $\mathbf{v} = [0, 2]$ are orthogonal.
21. Plot a set of vectors in $\mathbb{R}^2$ that lie on the circle $x_1^2 + x_2^2 = 1$ using Matplotlib.

Chapter 3
Matrices

Introduction

When you first meet a matrix, it might look like nothing more than a neatly arranged grid of numbers. But don't be fooled; this grid is one of the most powerful mathematical tools we have. In linear algebra, matrices take the role of both structure and operator. They store data, define systems, perform transformations, and enable optimization. In the world of ML, where we juggle massive datasets, train multi-layered models, and refine complex cost functions, matrices are absolutely everywhere.

At their essence, matrices extend the concept of vectors. A vector may represent a single data point or a direction in space, whereas a matrix can capture an entire dataset or describe a transformation applied to many points simultaneously. For instance, a dataset with rows corresponding to individuals and columns representing features such as height, weight, and age naturally forms a matrix. Similarly, a transformation that rotates every point in a 3D space can be expressed as a matrix.

In this chapter, we set out to explore matrices from both conceptual and practical angles. We'll start with their definition, what it means for a matrix to have dimensions $m \times n$, how entries are organized, and how to interpret those entries depending on context. We'll approach matrices not just as static data containers, but also as dynamic transformations that map inputs to outputs, reshape geometry, or manipulate structured data.

We'll also dive into viewing matrices as collections of vectors, columns as features, and rows as observations. This duality is more than an abstraction; it's fundamental to ML. Algorithms like LR, PCA, or neural networks treat matrices both as representations of data and as the means of transforming it.

Operations on matrices, like addition, subtraction, scalar multiplication, transposition, and both matrix-vector and matrix-matrix multiplication, will be introduced not as dry algebraic procedures, but as intuitive, visualizable tools. You'll see how these operations support everything from combining models to computing predictions and

© The Author(s), under exclusive license to Springer Nature Singapore Pte Ltd. 2026

Md. Jalil Piran, *Linear Algebra with Applications in Machine Learning*,

https://doi.org/10.1007/978-981-95-5167-5_3

updating parameters during optimization. Even the humble Hadamard product, often overlooked in basic courses, plays a big role in element-wise computations within neural networks and attention mechanisms.

Along the way, we'll identify special matrix types such as square, diagonal, identity, symmetric, triangular matrices, and show where they appear in real-world scenarios. These are not just formal categories; they help us compute more efficiently, understand structure more deeply, and apply the right tools in the right situations.

As always in this book, theory is paired with practice. Every concept is supported by Python code examples to bridge the gap between abstract math and ML implementation.

By the end of this chapter, you'll be able to do more than just define or manipulate matrices, you'll understand how they operate behind the scenes in your ML models. You'll see how data flows through them, how parameters are structured, and how transformations are performed. And you'll be ready for what comes next: matrix decompositions, linear systems, and the geometry of learning.

This isn't just an introduction to matrices. It's your first real step into the machinery of modern AI.

This chapter is structured to provide a comprehensive understanding of matrices, blending theoretical rigor with practical applications.

Topics Covered

This chapter is organized as follows:

- **Section** 3.1 **Matrices; Definition and Basics:** What matrices are, their dimensions, and how to interpret them as collections of vectors or grids.
- **Section** 3.2 **Matrix Operations:** Addition, subtraction, scalar multiplication, transposition, matrix-vector and matrix-matrix multiplication, and the Hadamard product.
- **Section** 3.3 **Types of Matrices:** Square, diagonal, identity, symmetric, and triangular matrices, and their significance.
- **Section** 3.4 **Augmentation and Partition of Matrices:** Combining and dividing matrices for efficient computation.
- **Section** 3.5 **Block Matrix Multiplication:** Treating submatrices as elements for large-scale operations.
- **Section** 3.6 **Matrices in ML:** How matrices store data, represent transformations, and drive algorithms like LR and neural networks.
- **Summary and Exercises**: Recaps key concepts and provides practical and theoretical problems to reinforce concepts.

All code accompanying the chapter is publicly available at https://github.com/jalil-piran/Linear-Algebra-with-Applications-in-Machine-Learning/blob/main/Chapter_03_Matrices.ipynb.

Through detailed explanations, geometric visualizations, and Python code, we aim to make matrices intuitive and actionable. Each section builds on the previous one, ensuring a smooth progression from foundational concepts to advanced applications. By the end, you'll see matrices not just as mathematical objects, but as the machinery powering modern computation.

Acronyms

ML	Machine Learning
LR	Linear Regression
PCA	Principal Component Analysis
SVMs	Support Vector Machines
CNNs	Convolutional Neural Networks
LU	Lower-Upper (decomposition)
RREF	Reduced Row Echelon Form
2D/3D	Two-Dimensional / Three-Dimensional

3.1　Matrices: Definition and Basics

Matrices act as powerful tools in mathematics and computer science, allowing us to model complex systems, perform transformations, and efficiently solve large-scale problems.

> **Definition**
> A **matrix** is a rectangular array of numbers arranged in m rows and n columns, denoted as $M_{m \times n}$ or $\mathbb{R}^{m \times n}$.

Each element a_{ij} represents the entry at the intersection of the i-th row and j-th column. Matrices generalize vectors, enabling the representation of complex datasets or transformations.

$$A = \begin{bmatrix} a_{11} & a_{12} & \cdots & a_{1n} \\ a_{21} & a_{22} & \cdots & a_{2n} \\ \vdots & \vdots & \ddots & \vdots \\ a_{m1} & a_{m2} & \cdots & a_{mn} \end{bmatrix}.$$

From linear transformation point of view, a matrix $A \in \mathbb{R}^{m \times n}$ represents a linear transformation $\mathbb{R}^m$.

Example 3.1 **A** 3×3 **Matrix**

Consider this matrix:

$$A = \begin{bmatrix} 2 & 3 & 5 \\ 3 & 1 & -1 \\ -2 & 1 & 1 \end{bmatrix},$$

where $m = 3$ is the number of rows, and $n = 3$ is the number of the columns, and specific entries are:

- $a_{12} = 3$ (row 1, column 2),
- $a_{31} = -2$ (row 3, column 1),
- $a_{33} = 1$ (row 3, column 3).

The row-first, column-second indexing is intuitive: think of a_{ij} as the value at the intersection of the i-th row and j-th column, like finding a cell in a grid.

Geometrically, matrices can represent data (e.g., a table of measurements) or transformations (e.g., rotating a 3D object). Geometrically, a matrix $A \in M_{m \times n}$ maps vectors from $\mathbb{R}^n$ to $\mathbb{R}^m$. Algebraically, it defines systems of linear equations, such as $A\mathbf{x} = \mathbf{b}$, where, A is the coefficient matrix, $\mathbf{x} \in \mathbb{R}^n$ is the vector of unknowns, and $\mathbf{b} \in \mathbb{R}^m$ is the vector of outputs or constants.

3.1.1 Matrices as Collections of Vectors

A matrix can be viewed as a collection of column or row vectors:

$$
A = \begin{bmatrix} \mathbf{a}_1 & \mathbf{a}_2 & \cdots & \mathbf{a}_n \end{bmatrix}, \quad \text{or} \quad A = \begin{bmatrix} \mathbf{r}_1 \\ \mathbf{r}_2 \\ \vdots \\ \mathbf{r}_m \end{bmatrix},
$$

where $\mathbf{a}_j \in \mathbb{R}^m$ are columns and $\mathbf{r}_i \in \mathbb{R}^n$ are rows.

Example 3.2 For column vectors:

$$
\mathbf{a}_1 = \begin{bmatrix} 2 \\ 3 \\ -2 \end{bmatrix}, \quad \mathbf{a}_2 = \begin{bmatrix} 3 \\ 1 \\ 1 \end{bmatrix}, \quad \mathbf{a}_3 = \begin{bmatrix} 5 \\ -1 \\ 1 \end{bmatrix},
$$

the matrix is:

$$
A = [\mathbf{a}_1, \mathbf{a}_2, \mathbf{a}_3] = \begin{bmatrix} 2 & 3 & 5 \\ 3 & 1 & -1 \\ -2 & 1 & 1 \end{bmatrix}.
$$

Each column is a vector in $\mathbb{R}^3$, and each row is a vector in $\mathbb{R}^3$.

This perspective is crucial for understanding matrices as datasets (rows as samples, columns as features) or transformations (columns as basis vectors).

The following Python example demonstrates how to construct a matrix from individual vectors using NumPy. Three vectors, a_1, a_2, and a_3, are defined as one-dimensional arrays. The function `np.column_stack` stacks these vectors as columns to form the matrix A. As a result, the output shows that A is a 3×3 matrix whose columns are exactly the given vectors.

Python Example: Matrix from Vectors

```python
import numpy as np

a1 = np.array([2, 3, -2])
a2 = np.array([3, 1, 1])
a3 = np.array([5, -1, 1])
A = np.column_stack((a1, a2, a3))

print("Matrix A:\n", A)
```

Output:

```
[[ 2  3  5]
 [ 3  1 -1]
 [-2  1  1]]
```

3.1.2 Matrix Dimensions

The **dimensions** of a matrix are denoted $m \times n$, where m is the number of rows and n is the number of columns. This determines the matrix's shape and compatibility for operations like multiplication.

Example 3.3 **Matrix Dimensions:**

- **A 2×3 matrix (2 rows, 3 columns)**: This matrix belongs to the space $M_{2\times3}$, meaning it has 2 rows and 3 columns. It's commonly used to represent two data samples, each with three features.

$$A = \begin{bmatrix} 1 & 2 & 3 \\ 4 & 5 & 6 \end{bmatrix} \in M_{2\times3}.$$

- **A 3×2 matrix (3 rows, 2 columns)**: This type is useful for stacking three vectors of 2 dimensions each. It's often encountered when reshaping data or performing linear transformations.

$$B = \begin{bmatrix} 1 & 4 \\ 2 & 5 \\ 3 & 6 \end{bmatrix} \in M_{3\times2}.$$

- **Square Matrix** ($m = n$): When the number of rows and columns are equal, we get a square matrix, essential for many core linear algebra concepts, such as computing the determinant, trace, or eigenvalues.

$$C = \begin{bmatrix} 1 & 2 & 3 \\ 4 & 5 & 6 \\ 7 & 8 & 9 \end{bmatrix} \in M_{3\times3}.$$

- **Row Vector** ($1 \times n$): A matrix with a single row and multiple columns. It can represent a feature vector in ML or a single data point laid out horizontally.

$$\mathbf{r} = \begin{bmatrix} 1 & 2 & 3 \end{bmatrix} \in M_{1\times3}.$$

- **Column Vector** ($m \times 1$): A matrix with one column and multiple rows. Commonly used for parameter vectors, gradients, or a single data point written vertically.

$$\mathbf{c} = \begin{bmatrix} 1 \\ 2 \\ 3 \end{bmatrix} \in M_{3\times1}.$$

- **Zero Matrix**: All elements are zero. This is useful in initialization or in theoretical derivations as the additive identity.

$$\mathbf{0}_{2\times3} = \begin{bmatrix} 0 & 0 & 0 \\ 0 & 0 & 0 \end{bmatrix}.$$

- **Identity Matrix** ($n \times n$): A square matrix with ones on the diagonal and zeros elsewhere. Acts as the multiplicative identity in matrix algebra.

$$I_3 = \begin{bmatrix} 1 & 0 & 0 \\ 0 & 1 & 0 \\ 0 & 0 & 1 \end{bmatrix} \in M_{3\times3}.$$

- **Diagonal Matrix**: A square matrix with nonzero entries only on its diagonal. Efficient for storage and multiplication.

$$D = \begin{bmatrix} 2 & 0 & 0 \\ 0 & 5 & 0 \\ 0 & 0 & 7 \end{bmatrix} \in M_{3\times3}.$$

Fig. 3.1 3×3 matrix visualized as a grid, corresponding to
$$A = \begin{bmatrix} 2 & 3 & 5 \\ 3 & 1 & -1 \\ -2 & 1 & 1 \end{bmatrix}$$

	Col 1	Col 2	Col 3
Row 1	2	3	5
Row 2	3	1	-1
Row 3	-2	1	1

3.1.3 Visualization of Matrices

Matrices can be visualized as grids, as shown in Fig. 3.1, which illustrates the structure of the matrix from Example 3.1. Each cell represents an element a_{ij}, with rows and columns labeled for clarity.

3.2 Matrix Operations

Matrix operations are the algebraic tools that manipulate matrices, enabling data combination, transformation, and analysis. This section covers addition, subtraction, scalar multiplication, transposition, vector operations like dot and cross products, matrix-vector and matrix-matrix multiplication, and the Hadamard product.

3.2.1 Addition and Subtraction

For matrices $A, B \in M_{m \times n}$, addition and subtraction are element-wise:

$$(A + B)_{ij} = a_{ij} + b_{ij},$$

$$(A - B)_{ij} = a_{ij} - b_{ij}.$$

This means that two matrices can only be added or subtracted if they have the same dimensions, that is, the same number of rows and columns. Each corresponding entry from the two matrices is combined through standard arithmetic addition or subtraction. These operations preserve the structure of the matrix and maintain compatibility with the vector space properties of matrix sets. Matrix addition and subtraction are fundamental in linear algebra and are used extensively in applications such as solving systems of equations, performing updates in iterative algorithms, and modeling linear combinations of data or transformations.

Example 3.4 For the following matrices:

$$A = \begin{bmatrix} 1 & 4 \\ 2 & 5 \\ 3 & 6 \end{bmatrix}, \quad B = \begin{bmatrix} 6 & 9 \\ 8 & 0 \\ 9 & 2 \end{bmatrix},$$

we compute $A + B$ and $A - B$ as follows:

$$A + B = \begin{bmatrix} 1 & 4 \\ 2 & 5 \\ 3 & 6 \end{bmatrix} + \begin{bmatrix} 6 & 9 \\ 8 & 0 \\ 9 & 2 \end{bmatrix} = \begin{bmatrix} 1+6 & 4+9 \\ 2+8 & 5+0 \\ 3+9 & 6+2 \end{bmatrix} = \begin{bmatrix} 7 & 13 \\ 10 & 5 \\ 12 & 8 \end{bmatrix},$$

$$A - B = \begin{bmatrix} 1 & 4 \\ 2 & 5 \\ 3 & 6 \end{bmatrix} - \begin{bmatrix} 6 & 9 \\ 8 & 0 \\ 9 & 2 \end{bmatrix} = \begin{bmatrix} 1-6 & 4-9 \\ 2-8 & 5-0 \\ 3-9 & 6-2 \end{bmatrix} = \begin{bmatrix} -5 & -5 \\ -6 & 5 \\ -6 & 4 \end{bmatrix}.$$

The following Python example illustrates matrix addition and subtraction using NumPy. Two matrices, A and B, of the same dimension are defined. The operations A + B and A - B are performed element-wise, meaning each entry in the resulting matrices is obtained by adding or subtracting the corresponding entries of A and B. The output confirms that matrix addition and subtraction follow straightforward element-wise rules.

Python Example: Matrix Addition and Subtraction

```python
import numpy as np

A = np.array([[1, 4], [2, 5], [3, 6]])
B = np.array([[6, 9], [8, 0], [9, 2]])
add_result = A + B
sub_result = A - B

print("A + B:\n", add_result)
print("A - B:\n", sub_result)
```

Output:

```
A + B:
[[ 7 13]
 [10  5]
 [12  8]]
A - B:
[[-5 -5]
 [-6  5]
 [-6  4]]
```

3.2.2 *Scalar Multiplication*

Scalar multiplication scales every element by a scalar s:

$$(sA)_{ij} = s \cdot a_{ij}.$$

In this operation, each entry of the matrix A is multiplied by the same scalar value s, resulting in a new matrix of the same size. This uniformly stretches or compresses the matrix values, depending on the magnitude and sign of the scalar. Scalar multiplication preserves the overall structure of the matrix while modifying its scale, and it is a key operation that supports the definition of a matrix as a vector space element. This operation is essential in numerous linear algebra applications, including forming linear combinations, adjusting weights in optimization algorithms, and applying gain or attenuation in signal processing and systems modeling.

Example 3.5 For a matrix like $B = \begin{bmatrix} 6 & 9 \\ 8 & 0 \\ 9 & 2 \end{bmatrix}$, and a scalar like $s = 9$:

$$9B = 9 \times \begin{bmatrix} 6 & 9 \\ 8 & 0 \\ 9 & 2 \end{bmatrix} = \begin{bmatrix} 6 \times 9 & 9 \times 9 \\ 8 \times 9 & 0 \times 9 \\ 9 \times 9 & 2 \times 9 \end{bmatrix} = \begin{bmatrix} 54 & 81 \\ 72 & 0 \\ 81 & 18 \end{bmatrix}.$$

The following code demonstrates scalar multiplication of a matrix in NumPy. Each element of matrix B is multiplied by the scalar value 9, producing a new matrix where all entries are scaled accordingly.

Python Example: Scalar Multiplication

```python
import numpy as np

B = np.array([[6, 9], [8, 0], [9, 2]])
scalar = 9
result = scalar * B

print(result)
```

Output:

```
[[54 81]
 [72  0]
 [81 18]]
```

3.2.3   *Transpose*

The transpose A^T of an $m \times n$ matrix A is an $n \times m$ matrix where $(A^T)_{ij} = a_{ji}$. This means that the rows of the original matrix become the columns of the transposed matrix, and vice versa. The transpose converts column vectors to row vectors and reverses the direction of linear mappings. Transposition reflects the matrix across its main diagonal and is a fundamental operation in linear algebra. It is widely used in expressing symmetric matrices, defining inner products, and simplifying matrix expressions. Transposes are also essential in many computational procedures, such as solving systems of equations, working with covariance matrices in statistics, and implementing linear transformations. The operation preserves key structural properties and plays a central role in defining concepts like orthogonality, rank, and adjoint matrices.

Example 3.6 For:

$$A = \begin{bmatrix} 6 & 9 \\ 8 & 0 \\ 9 & 2 \end{bmatrix},$$

$$A^T = \begin{bmatrix} 6 & 8 & 9 \\ 9 & 0 & 2 \end{bmatrix}.$$

The original matrix A is a 3×2 matrix, meaning it has 3 rows and 2 columns. Its transpose A^T is a 2×3 matrix, obtained by flipping the rows and columns of A. Specifically, each row of A becomes a column in A^T, and each column of A becomes a row in A^T. This operation is useful for reorienting data, computing dot products in matrix form, and constructing symmetric matrices when needed. Transposition also plays a vital role in defining and verifying properties such as symmetry and orthogonality in matrix algebra.

The following Python code shows how to compute the transpose of a matrix in NumPy. The attribute $A.T$ flips the rows and columns of A, producing A^T.

Python Example: Transpose

```python
import numpy as np

A = np.array([[6, 9], [8, 0], [9, 2]])
AT = A.T

print("A^T:\n", AT)
```

Output:

```
A^T:
```

```
[[6 8 9]
 [9 0 2]]
```

Properties of Matrix Transpose

The transpose operation satisfies several important properties that simplify algebraic manipulations and are essential for understanding matrix operations in ML. These properties hold for all matrices of compatible dimensions and are particularly useful when deriving algorithms for gradient computations, covariance analysis, and data transformations.

- **Double Transpose**: $(A^T)^T = A$, Transposing twice returns the original matrix.
- **Scalar Multiplication**: $(sA)^T = sA^T$, The scalar can be factored out before or after transposition.
- **Addition**: $(A + B)^T = A^T + B^T$, The transpose of a sum is the sum of the transposes.

Next, we will verify each property using examples.

Example 3.7 Consider the matrix:

$$A = \begin{bmatrix} 1 & 2 \\ 3 & 4 \end{bmatrix}.$$

First we compute the transpose:

$$A^T = \begin{bmatrix} 1 & 3 \\ 2 & 4 \end{bmatrix}.$$

Now we transpose again:

$$(A^T)^T = \begin{bmatrix} 1 & 2 \\ 3 & 4 \end{bmatrix} = A.$$

The result matches the original matrix, confirming that $(A^T)^T = A$.

Example 3.8 Using the same matrix

$$A = \begin{bmatrix} 1 & 2 \\ 3 & 4 \end{bmatrix}, \quad s = 2,$$

we check the property $(sA)^T = s(A^T)$.
First we compute sA:

$$sA = 2 \begin{bmatrix} 1 & 2 \\ 3 & 4 \end{bmatrix} = \begin{bmatrix} 2 & 4 \\ 6 & 8 \end{bmatrix}.$$

Now we transpose:

$$(sA)^T = \begin{bmatrix} 2 & 6 \\ 4 & 8 \end{bmatrix}.$$

Alternatively, we transpose first then multiply:

$$s(A^T) = 2 \begin{bmatrix} 1 & 3 \\ 2 & 4 \end{bmatrix} = \begin{bmatrix} 2 & 6 \\ 4 & 8 \end{bmatrix}.$$

Both results are identical, verifying that

$$(sA)^T = s(A^T).$$

Example 3.9 For:

$$A = \begin{bmatrix} 1 & 2 \\ 3 & 4 \end{bmatrix}, \quad B = \begin{bmatrix} 5 & 6 \\ 7 & 8 \end{bmatrix},$$

we compute $(A + B)^T$ and $A^T + B^T$:

$$A + B = \begin{bmatrix} 6 & 8 \\ 10 & 12 \end{bmatrix}, \quad (A + B)^T = \begin{bmatrix} 6 & 10 \\ 8 & 12 \end{bmatrix},$$

$$A^T = \begin{bmatrix} 1 & 3 \\ 2 & 4 \end{bmatrix}, \quad B^T = \begin{bmatrix} 5 & 7 \\ 6 & 8 \end{bmatrix}, \quad A^T + B^T = \begin{bmatrix} 6 & 10 \\ 8 & 12 \end{bmatrix}.$$

Both results match, confirming the property $(A + B)^T = A^T + B^T$.

Python Example: Verifying Transpose Properties

```python
import numpy as np

A = np.array([[1, 2], [3, 4]])
B = np.array([[5, 6], [7, 8]])
s = 2

# Double transpose
double_transpose = (A.T).T
print("Double transpose equals original:",
    np.allclose(double_transpose, A))

# Scalar multiplication property
scalar_left = (s * A).T
scalar_right = s * A.T
print("Scalar multiplication property:",
    np.allclose(scalar_left, scalar_right))

# Addition property
```

```
17 add_left  =  (A + B).T
18 add_right =  A.T + B.T
19 print("Addition property:",
         np.allclose(add_left, add_right))
```

These properties are crucial in ML for operations like computing gradients of matrix expressions and manipulating covariance matrices, where transpose operations frequently appear in analytical derivations and numerical implementations.

3.2.4 Matrix-Vector Product

The matrix-vector product is a core operation in linear algebra, where an $m \times n$ matrix A multiplies a vector $\mathbf{x} \in \mathbb{R}^n$, producing a new vector in $\mathbb{R}^m$. This operation defines a linear transformation, mapping input vectors from an n-dimensional space to an m-dimensional space. It is foundational to solving linear systems, performing projections, and modeling transformations in various applications.

There are two equally valid ways to interpret the matrix-vector product. The first is the **column perspective**, in which the product is viewed as a linear combination of the columns of A. Each entry of $\mathbf{x}$ acts as a scalar coefficient, scaling the corresponding column of A. The result is a weighted sum of the columns, illustrating how the matrix encodes a transformation via its column space.

The second interpretation is the **row perspective**, where the product is computed as a series of dot products between the rows of A and the vector $\mathbf{x}$. Each entry of the resulting output vector corresponds to the interaction between one row of the matrix and the input vector. This view is especially useful in solving systems of linear equations, where each equation corresponds to a row and defines a constraint on the solution.

Formally, for an $m \times n$ matrix A with entries a_{ij} and a vector $\mathbf{x} = [x_1, x_2, \ldots, x_n]^T \in \mathbb{R}^n$, the matrix-vector product $A\mathbf{x}$ can be expressed in two equivalent ways that provide complementary geometric and computational insights.

The first perspective treats the product as a *linear combination of columns*:

$$A\mathbf{x} = \sum_{j=1}^{n} x_j \mathbf{a}_j = x_1 \begin{bmatrix} a_{11} \\ a_{21} \\ \vdots \\ a_{m1} \end{bmatrix} + x_2 \begin{bmatrix} a_{12} \\ a_{22} \\ \vdots \\ a_{m2} \end{bmatrix} + \cdots + x_n \begin{bmatrix} a_{1n} \\ a_{2n} \\ \vdots \\ a_{mn} \end{bmatrix},$$

where $\mathbf{a}_j$ represents the j-th column of A. The i-th entry of the result is then:

$$(A\mathbf{x})_i = \sum_{j=1}^{n} a_{ij} x_j.$$

The second perspective views the product as a *collection of row-vector dot products*:

$$A\mathbf{x} = \begin{bmatrix} \mathbf{r}_1 \cdot \mathbf{x} \\ \mathbf{r}_2 \cdot \mathbf{x} \\ \vdots \\ \mathbf{r}_m \cdot \mathbf{x} \end{bmatrix},$$

where $\mathbf{r}_i = [a_{i1}, a_{i2}, \ldots, a_{in}]$ is the i-th row of A. Each entry becomes:

$$\mathbf{r}_i \cdot \mathbf{x} = \sum_{j=1}^{n} a_{ij} x_j.$$

Both formulations yield identical results, highlighting the dual nature of matrix multiplication. The column interpretation is particularly intuitive in ML, where $A\mathbf{x}$ represents a weighted combination of feature transformations, while the row interpretation emphasizes each output dimension as a linear combination of input features.

Example 3.10 For:

$$A = \begin{bmatrix} 2 & -1 & 3 \\ 4 & 0 & 1 \end{bmatrix}, \quad \mathbf{x} = \begin{bmatrix} 1 \\ 2 \\ -1 \end{bmatrix},$$

$$A\mathbf{x} = \begin{bmatrix} 2 & -1 & 3 \\ 4 & 0 & 1 \end{bmatrix} \times \begin{bmatrix} 1 \\ 2 \\ -1 \end{bmatrix} = \begin{bmatrix} (2)(1) + (-1)(2) + (3)(-1) \\ (4)(1) + (0)(2) + (1)(-1) \end{bmatrix} = \begin{bmatrix} -3 \\ 3 \end{bmatrix}.$$

Python Example: Matrix-Vector Product

```python
import numpy as np

A = np.array([[2, -1, 3], [4, 0, 1]])
x = np.array([1, 2, -1])
result = A @ x

print("Ax:", result)
```

Output:

```
Ax: [-3  3]
```

3.2.5 *Matrix-Matrix Product*

For matrices $A \in M_{m \times n}$ and $B \in M_{n \times p}$, their product AB is defined as:

$$(AB)_{ij} = \sum_{k=1}^{n} a_{ik} b_{kj}.$$

This definition means that each entry in the resulting $m \times p$ matrix AB is computed by taking the dot product of the i-th row of A with the j-th column of B. Conceptually, this operation captures how the linear transformation represented by B is followed by the transformation represented by A, effectively composing the two transformations into a single operation.

Matrix multiplication is not performed element-wise; instead, it synthesizes information across entire rows and columns, making it a more complex and powerful operation. Each column of B can be viewed as a vector being transformed by A, and the entire matrix product can be interpreted as applying A to all columns of B simultaneously.

Moreover, matrix multiplication underlies many core concepts in linear algebra, including change of basis, diagonalization, and eigenvalue decomposition. Its structure also reflects deep algebraic properties, such as associativity and distributivity, although it is notably not commutative.

Example 3.11 For:

$$A = \begin{bmatrix} 1 & 2 \\ 1 & 1 \end{bmatrix}, \quad B = \begin{bmatrix} 1 & 1 \\ 1 & -1 \end{bmatrix},$$

we compute AB and BA:

$$AB = \begin{bmatrix} (1)(1) + (2)(1) & (1)(1) + (2)(-1) \\ (1)(1) + (1)(1) & (1)(1) + (1)(-1) \end{bmatrix} = \begin{bmatrix} 3 & -1 \\ 2 & 0 \end{bmatrix},$$

$$BA = \begin{bmatrix} (1)(1) + (1)(1) & (1)(2) + (1)(1) \\ (1)(1) + (-1)(1) & (1)(2) + (-1)(1) \end{bmatrix} = \begin{bmatrix} 2 & 3 \\ 0 & 1 \end{bmatrix}.$$

Since $AB \neq BA$, matrix multiplication is not commutative.

Python Example: Matrix Multiplication

```python
import numpy as np

A = np.array([[1, 2], [1, 1]])
B = np.array([[1, 1], [1, -1]])
AB = A @ B
BA = B @ A
```

```
7
8  print ( "AB:\n" ,  AB )
9  print ( "BA:\n" ,  BA )
```

Output:

```
AB:
[[ 3 -1]
 [ 2  0]]
BA:
[[2 3]
 [0 1]]
```

Linear Combination Perspectives

Matrix multiplication can be interpreted through two complementary lenses: as linear combinations of columns or as linear combinations of rows. These perspectives provide valuable geometric and algebraic insights into how matrix products operate and are particularly helpful for understanding transformations and designing algorithms.

(a) Column-Wise View: In this interpretation, each column of the product matrix $C = AB$ is expressed as a linear combination of the columns of matrix A. The weights for this combination come from the corresponding column of matrix B. That is, to compute the j-th column of C, we take the j-th column of B, use its entries as scalar coefficients, and apply them to the columns of A. Formally:

$$C_j = b_{1j}\mathbf{a}_1 + b_{2j}\mathbf{a}_2 + \cdots + b_{nj}\mathbf{a}_n.$$

This view emphasizes how matrix B specifies how to mix the columns of A to produce each column of the resulting matrix. It aligns naturally with interpreting A as a transformation and B as a set of input vectors being transformed.

(b) Row-Wise View: Alternatively, matrix multiplication can be seen from a row perspective. Here, each row of the product matrix C is a linear combination of the rows of matrix B, with weights provided by the entries of the corresponding row of matrix A. This means that to compute the i-th row of C, we scale each row of B by the respective entry in the i-th row of A, and then sum the results:

$$C_i = a_{i1}\mathbf{b}_1 + a_{i2}\mathbf{b}_2 + \cdots + a_{in}\mathbf{b}_n.$$

This perspective is particularly useful in computational implementations, where operations are often optimized for row-wise traversal and memory access. It also highlights how the action of A distributes across the structure of B.

Together, these dual perspectives underscore the flexibility and richness of matrix multiplication. Whether viewed column-wise or row-wise, the operation encodes the idea of combining vector spaces under structured linear transformations, forming the algebraic backbone of linear systems, regression models, and neural network layers.

Example 3.12 Given:

$$
A = \begin{bmatrix} 1 & 2 \\ 3 & 4 \\ 5 & 6 \end{bmatrix}, \quad B = \begin{bmatrix} -1 & 1 \\ 3 & 2 \end{bmatrix},
$$

we compute the matrix product AB using both the **column view** and the **row view**.

(a) Column View Interpretation: Each column of AB is a linear combination of the columns of A, weighted by the corresponding column of B.

- First column of B is $\begin{bmatrix} -1 \\ 3 \end{bmatrix}$: We multiply the first column of A by -1 and the second column by 3, then add:

$$
-1 \begin{bmatrix} 1 \\ 3 \\ 5 \end{bmatrix} + 3 \begin{bmatrix} 2 \\ 4 \\ 6 \end{bmatrix} = \begin{bmatrix} 5 \\ 9 \\ 13 \end{bmatrix}.
$$

- Second column of B is $\begin{bmatrix} 1 \\ 2 \end{bmatrix}$: We multiply the first column of A by 1 and the second column by 2, then we add:

$$
1 \begin{bmatrix} 1 \\ 3 \\ 5 \end{bmatrix} + 2 \begin{bmatrix} 2 \\ 4 \\ 6 \end{bmatrix} = \begin{bmatrix} 5 \\ 11 \\ 17 \end{bmatrix}.
$$

Thus, the matrix product is:

$$
AB = \begin{bmatrix} 5 & 5 \\ 9 & 11 \\ 13 & 17 \end{bmatrix}.
$$

(b) Row View Interpretation: Each row of AB is a linear combination of the rows of B, weighted by the corresponding row of A.

- First row of A is $[1, 2]$: Multiply the first row of B by 1 and the second row by 2, then add:

$$
1 \begin{bmatrix} -1 & 1 \end{bmatrix} + 2 \begin{bmatrix} 3 & 2 \end{bmatrix} = \begin{bmatrix} 5 & 5 \end{bmatrix}.
$$

- Second row of A is $[3, 4]$: Multiply the first row of B by 3 and the second row by 4, then add:

$$3\begin{bmatrix} -1 & 1 \end{bmatrix} + 4\begin{bmatrix} 3 & 2 \end{bmatrix} = \begin{bmatrix} 9 & 11 \end{bmatrix}.$$

- Third row of A is $[5, 6]$: Multiply the first row of B by 5 and the second row by 6, then add:

$$5\begin{bmatrix} -1 & 1 \end{bmatrix} + 6\begin{bmatrix} 3 & 2 \end{bmatrix} = \begin{bmatrix} 13 & 17 \end{bmatrix}.$$

Again, we obtain:

$$AB = \begin{bmatrix} 5 & 5 \\ 9 & 11 \\ 13 & 17 \end{bmatrix}.$$

This example demonstrates how both column-wise and row-wise interpretations of matrix multiplication arrive at the same result, reinforcing the consistency and flexibility of matrix algebra.

Function Composition Interpretation

Matrix multiplication is conceptually analogous to function composition in mathematics. Just as one function can be applied to the result of another, one matrix can be applied to the output of another matrix-vector transformation. Suppose we have a function $f : \mathbb{R}^p \rightarrow \mathbb{R}^n$, represented by matrix B, and a function $g : \mathbb{R}^n \rightarrow \mathbb{R}^m$, represented by matrix A. Then the composition $g \circ f$, which maps $\mathbb{R}^p \rightarrow \mathbb{R}^m$, corresponds to the matrix product AB.

In this framework, applying the composed transformation to a vector $\mathbf{x} \in \mathbb{R}^p$ means first applying B to obtain $B\mathbf{x} \in \mathbb{R}^n$, and then applying A to the result:

$$(AB)\mathbf{x} = A(B\mathbf{x}).$$

This interpretation reveals the order of operations in matrix multiplication: although AB is written with A on the left and B on the right, the actual computation proceeds from right to left, just as with function composition. This view is especially important in applications involving layered transformations, such as in computer graphics pipelines, neural networks, and systems of differential equations. Each matrix corresponds to a transformation, and multiplying them corresponds to chaining those transformations in sequence.

Example 3.13 Given:

$$A = \begin{bmatrix} 2 & 0 \\ 0 & 1 \end{bmatrix}, \quad B = \begin{bmatrix} 1 & 0 \\ 0 & 1 \end{bmatrix},$$

we compute the matrix product AB.

Step 1-Matrix Multiplication:

Multiply A by B. Since B is the identity matrix, the product AB leaves matrix A unchanged:

$$AB = \begin{bmatrix} 2 & 0 \\ 0 & 1 \end{bmatrix}.$$

Step 2-Apply to a Vector:

Let $\mathbf{x} = \begin{bmatrix} x_1 \\ x_2 \end{bmatrix}$ be an arbitrary vector. Applying the transformation AB to $\mathbf{x}$ gives:

$$AB\mathbf{x} = \begin{bmatrix} 2x_1 \\ x_2 \end{bmatrix}.$$

This transformation scales the first component x_1 by a factor of 2, while leaving the second component x_2 unchanged. Geometrically, this represents an **anisotropic scaling transformation**, a stretch along the x-axis and no change along the y-axis. Such transformations are common in graphics, modeling, and coordinate manipulations.

Multiplication of Multiple Matrices

Matrix multiplication is an **associative operation**, meaning that the order in which the multiplications are grouped does not affect the final result:

$$(AB)C = A(BC).$$

This property is essential when dealing with sequences of linear transformations. If A, B, and C are matrices of compatible dimensions, the product of all three can be computed by first multiplying A and B, then multiplying the result with C, or by first multiplying B and C, then applying A to the result. In either case, the outcome is the same.

Associativity allows us to simplify and reorder complex matrix expressions without changing their meaning, which is particularly useful in both theoretical analysis and computational implementations. It also reflects the associativity of function composition: composing three functions in any associative grouping produces the same overall transformation. This makes matrix associativity a powerful tool in applications ranging from numerical linear algebra and optimization to computer graphics and ML pipelines.

Example 3.14 For:

$$A = \begin{bmatrix} 2 & 3 & 1 \\ 4 & 1 & 2 \\ 5 & 0 & 3 \end{bmatrix}, \quad B = \begin{bmatrix} 1 & 5 & 2 \\ 3 & 2 & 1 \\ 0 & 2 & 1 \end{bmatrix}, \quad C = \begin{bmatrix} 3 & 2 & 1 \\ 0 & 1 & 4 \\ 2 & 3 & 1 \end{bmatrix},$$

we compute $(AB)C$ as follows:

$$AB = \begin{bmatrix} 2 & 3 & 1 \\ 4 & 1 & 2 \\ 5 & 0 & 3 \end{bmatrix} \times \begin{bmatrix} 1 & 5 & 2 \\ 3 & 2 & 1 \\ 0 & 2 & 1 \end{bmatrix} = \begin{bmatrix} 11 & 18 & 8 \\ 7 & 26 & 11 \\ 5 & 31 & 13 \end{bmatrix},$$

$$(AB)C = \begin{bmatrix} 11 & 18 & 8 \\ 7 & 26 & 11 \\ 5 & 31 & 13 \end{bmatrix} \times \begin{bmatrix} 3 & 2 & 1 \\ 0 & 1 & 4 \\ 2 & 3 & 1 \end{bmatrix} = \begin{bmatrix} 49 & 64 & 91 \\ 43 & 73 & 122 \\ 41 & 80 & 142 \end{bmatrix}.$$

Python Example: Multiple Matrix Multiplication

```python
import numpy as np

A = np.array([[2, 3, 1], [4, 1, 2], [5, 0, 3]])
B = np.array([[1, 5, 2], [3, 2, 1], [0, 2, 1]])
C = np.array([[3, 2, 1], [0, 1, 4], [2, 3, 1]])
ABC = (A @ B) @ C

print("ABC:\n", ABC)
```

Output:

```
ABC:
[[ 49  64  91]
 [ 43  73 122]
 [ 41  80 142]]
```

3.2.6 Hadamard Product

Unlike matrix multiplication, the Hadamard product operates element-wise and requires identical dimensions. The **Hadamard product** $A \odot B$ is defined as the element-wise product of two matrices of the same dimensions. For matrices $A, B \in M_{m \times n}$, the Hadamard product is given by:

$$(A \odot B)_{ij} = a_{ij} \cdot b_{ij}.$$

Unlike standard matrix multiplication, which involves dot products of rows and columns, the Hadamard product performs multiplication independently at each corresponding entry. This operation preserves the shape of the original matrices and is only defined when both matrices have the same size.

The Hadamard product is especially useful in contexts where interactions between individual components of matrices are needed without introducing the global structure imposed by standard matrix multiplication. It frequently appears in element-wise operations in ML, neural networks, signal processing, and numerical computing. For example, it is used in masking, feature-wise scaling, and attention mechanisms, where computations must be carried out independently for each component of a matrix.

Mathematically, the Hadamard product retains many algebraic properties, such as commutativity and associativity under certain conditions, and it supports efficient parallel computation due to its entry-wise nature.

Example 3.15 For:

$$A = \begin{bmatrix} 0 & 2 \\ 1 & 4 \end{bmatrix}, \quad B = \begin{bmatrix} 1 & 3 \\ 2 & 1 \end{bmatrix},$$

$$A \odot B = \begin{bmatrix} 0 & 2 \\ 1 & 4 \end{bmatrix} \odot \begin{bmatrix} 1 & 3 \\ 2 & 1 \end{bmatrix} = \begin{bmatrix} 0 \times 1 & 2 \times 3 \\ 1 \times 2 & 4 \times 1 \end{bmatrix} = \begin{bmatrix} 0 & 6 \\ 2 & 4 \end{bmatrix}.$$

Python Example: Hadamard Product

```python
import numpy as np

A = np.array([[0, 2], [1, 4]])
B = np.array([[1, 3], [2, 1]])
hadamard = A * B

print(hadamard)
```

Output:

```
[[0 6]
 [2 4]]
```

3.2.7 Properties of Matrix Operations

Matrix operations follow algebraic rules that simplify computations and ensure consistency across linear algebra applications. These rules not only make computation easier but also underpin core concepts in ML and data science.

- **Addition (Commutative):** $A + B = B + A$
  The order in which matrices are added does not affect the outcome. This holds as addition is performed entry-wise.

Example 3.16 Consider the following matrices:

$$A = \begin{bmatrix} 2 & 3 \\ 1 & 4 \end{bmatrix}, \quad B = \begin{bmatrix} 5 & -1 \\ 0 & 2 \end{bmatrix}.$$

We want to check that $A + B$ is equal to $B + A$,

$$A + B = \begin{bmatrix} 2+5 & 3+(-1) \\ 1+0 & 4+2 \end{bmatrix} = \begin{bmatrix} 7 & 2 \\ 1 & 6 \end{bmatrix}.$$

$$B + A = \begin{bmatrix} 5+2 & (-1)+3 \\ 0+1 & 2+4 \end{bmatrix} = \begin{bmatrix} 7 & 2 \\ 1 & 6 \end{bmatrix}.$$

- **Addition (Associative):** $(A + B) + C = A + (B + C)$

  Grouping does not change the sum when adding three matrices.

Example 3.17 Let

$$A = \begin{bmatrix} 1 & 0 \\ 2 & 1 \end{bmatrix}, B = \begin{bmatrix} 3 & 1 \\ 0 & 2 \end{bmatrix}, C = \begin{bmatrix} -2 & 4 \\ 1 & 3 \end{bmatrix}.$$

To verify matrix addition is associative, we compute both $(A + B) + C$ and $A + (B + C)$:

First, compute $A + B$:

$$A + B = \begin{bmatrix} 1 & 0 \\ 2 & 1 \end{bmatrix} + \begin{bmatrix} 3 & 1 \\ 0 & 2 \end{bmatrix} = \begin{bmatrix} 1+3 & 0+1 \\ 2+0 & 1+2 \end{bmatrix} = \begin{bmatrix} 4 & 1 \\ 2 & 3 \end{bmatrix}.$$

Now we add C:

$$(A + B) + C = \begin{bmatrix} 4 & 1 \\ 2 & 3 \end{bmatrix} + \begin{bmatrix} -2 & 4 \\ 1 & 3 \end{bmatrix} = \begin{bmatrix} 4+(-2) & 1+4 \\ 2+1 & 3+3 \end{bmatrix} = \begin{bmatrix} 2 & 5 \\ 3 & 6 \end{bmatrix}.$$

Next, we compute $B + C$:

$$B + C = \begin{bmatrix} 3 & 1 \\ 0 & 2 \end{bmatrix} + \begin{bmatrix} -2 & 4 \\ 1 & 3 \end{bmatrix} = \begin{bmatrix} 3+(-2) & 1+4 \\ 0+1 & 2+3 \end{bmatrix} = \begin{bmatrix} 1 & 5 \\ 1 & 5 \end{bmatrix}.$$

Now we add A:

$$A + (B + C) = \begin{bmatrix} 1 & 0 \\ 2 & 1 \end{bmatrix} + \begin{bmatrix} 1 & 5 \\ 1 & 5 \end{bmatrix} = \begin{bmatrix} 1+1 & 0+5 \\ 2+1 & 1+5 \end{bmatrix} = \begin{bmatrix} 2 & 5 \\ 3 & 6 \end{bmatrix}.$$

Both $(A + B) + C$ and $A + (B + C)$ yield the same result:

$$\begin{bmatrix} 2 & 5 \\ 3 & 6 \end{bmatrix},$$

confirming that matrix addition is associative.

- **Scalar Multiplication**:

$$(st)A = s(tA),$$

$$s(A + B) = sA + sB,$$

$$(s + t)A = sA + tA.$$

These properties ensure scalar multiplication distributes over addition and is associative.

Example 3.18 Let $s = 2$, $t = 3$, and

$$A = \begin{bmatrix} 1 & 2 \\ 3 & 4 \end{bmatrix}, \quad B = \begin{bmatrix} 5 & 0 \\ -1 & 2 \end{bmatrix}.$$

$$(st)A = s(tA) = \begin{bmatrix} 6 & 12 \\ 18 & 24 \end{bmatrix}$$

$$s(A + B) = sA + sB = \begin{bmatrix} 12 & 4 \\ 4 & 12 \end{bmatrix}$$

$$(s + t)A = sA + tA = \begin{bmatrix} 5 & 10 \\ 15 & 20 \end{bmatrix}$$

verification left to the reader.

- **Matrix-Vector Product**:

$$A(\mathbf{u} + \mathbf{v}) = A\mathbf{u} + A\mathbf{v},$$

$$A(c\mathbf{u}) = c(A\mathbf{u}).$$

This expresses the linearity of matrix-vector multiplication.

Example 3.19 Let

$$A = \begin{bmatrix} 2 & 1 \\ 0 & 3 \end{bmatrix}, \mathbf{u} = \begin{bmatrix} 1 \\ 2 \end{bmatrix}, \mathbf{v} = \begin{bmatrix} 3 \\ -1 \end{bmatrix}, c = 4.$$

Then:

$$A(\mathbf{u} + \mathbf{v}) = A\mathbf{u} + A\mathbf{v} = \begin{bmatrix} 9 \\ 3 \end{bmatrix},$$

$$A(c\mathbf{u}) = c(A\mathbf{u}) = \begin{bmatrix} 16 \\ 24 \end{bmatrix}.$$

- **Matrix Distributivity**: $(A + B)\mathbf{u} = A\mathbf{u} + B\mathbf{u}$
  Matrix-vector product distributes over matrix addition.

Example 3.20 Let

$$A = \begin{bmatrix} 1 & 2 \\ 0 & 1 \end{bmatrix}, B = \begin{bmatrix} 3 & -1 \\ 2 & 0 \end{bmatrix}, \mathbf{u} = \begin{bmatrix} 1 \\ 2 \end{bmatrix}.$$

To verify the distributive property of matrix multiplication over addition, compute both $(A + B)\mathbf{u}$ and $A\mathbf{u} + B\mathbf{u}$:
First, we compute $A + B$:

$$A + B = \begin{bmatrix} 1 & 2 \\ 0 & 1 \end{bmatrix} + \begin{bmatrix} 3 & -1 \\ 2 & 0 \end{bmatrix} = \begin{bmatrix} 1+3 & 2+(-1) \\ 0+2 & 1+0 \end{bmatrix} = \begin{bmatrix} 4 & 1 \\ 2 & 1 \end{bmatrix}.$$

Now we multiply by $\mathbf{u}$:

$$(A + B)\mathbf{u} = \begin{bmatrix} 4 & 1 \\ 2 & 1 \end{bmatrix}\begin{bmatrix} 1 \\ 2 \end{bmatrix} = \begin{bmatrix} 4 \cdot 1 + 1 \cdot 2 \\ 2 \cdot 1 + 1 \cdot 2 \end{bmatrix} = \begin{bmatrix} 4+2 \\ 2+2 \end{bmatrix} = \begin{bmatrix} 6 \\ 4 \end{bmatrix}.$$

Next, we compute $A\mathbf{u}$:

$$A\mathbf{u} = \begin{bmatrix} 1 & 2 \\ 0 & 1 \end{bmatrix}\begin{bmatrix} 1 \\ 2 \end{bmatrix} = \begin{bmatrix} 1 \cdot 1 + 2 \cdot 2 \\ 0 \cdot 1 + 1 \cdot 2 \end{bmatrix} = \begin{bmatrix} 1+4 \\ 0+2 \end{bmatrix} = \begin{bmatrix} 5 \\ 2 \end{bmatrix}.$$

Now we compute $B\mathbf{u}$:

$$B\mathbf{u} = \begin{bmatrix} 3 & -1 \\ 2 & 0 \end{bmatrix}\begin{bmatrix} 1 \\ 2 \end{bmatrix} = \begin{bmatrix} 3 \cdot 1 + (-1) \cdot 2 \\ 2 \cdot 1 + 0 \cdot 2 \end{bmatrix} = \begin{bmatrix} 3-2 \\ 2+0 \end{bmatrix} = \begin{bmatrix} 1 \\ 2 \end{bmatrix}.$$

Finally, we add the results:

$$A\mathbf{u} + B\mathbf{u} = \begin{bmatrix} 5 \\ 2 \end{bmatrix} + \begin{bmatrix} 1 \\ 2 \end{bmatrix} = \begin{bmatrix} 5+1 \\ 2+2 \end{bmatrix} = \begin{bmatrix} 6 \\ 4 \end{bmatrix}.$$

Both $(A + B)\mathbf{u}$ and $A\mathbf{u} + B\mathbf{u}$ yield the same result: $\begin{bmatrix} 6 \\ 4 \end{bmatrix}$, confirming that matrix multiplication distributes over matrix addition.

- **Identity Matrix**: $I_n\mathbf{v} = \mathbf{v}$
  The identity matrix leaves any vector unchanged.

Example 3.21

$$I_3 \cdot \begin{bmatrix} 4 \\ -2 \\ 7 \end{bmatrix} = \begin{bmatrix} 1 & 0 & 0 \\ 0 & 1 & 0 \\ 0 & 0 & 1 \end{bmatrix} \cdot \begin{bmatrix} 4 \\ -2 \\ 7 \end{bmatrix} = \begin{bmatrix} 4 \\ -2 \\ 7 \end{bmatrix}.$$

- **Matrix Multiplication Is Not Commutative**: $AB \neq BA$
  In general, swapping matrix order changes the result or makes multiplication undefined.

Example 3.22 Consider the matrices:

$$A = \begin{bmatrix} 1 & 2 \\ 0 & 1 \end{bmatrix}, \quad B = \begin{bmatrix} 3 & 0 \\ 4 & 1 \end{bmatrix}.$$

First we compute AB:

$$AB = \begin{bmatrix} 1 & 2 \\ 0 & 1 \end{bmatrix} \begin{bmatrix} 3 & 0 \\ 4 & 1 \end{bmatrix} = \begin{bmatrix} 1 \cdot 3 + 2 \cdot 4 & 1 \cdot 0 + 2 \cdot 1 \\ 0 \cdot 3 + 1 \cdot 4 & 0 \cdot 0 + 1 \cdot 1 \end{bmatrix} = \begin{bmatrix} 3+8 & 0+2 \\ 0+4 & 0+1 \end{bmatrix} = \begin{bmatrix} 11 & 2 \\ 4 & 1 \end{bmatrix}.$$

Now we compute BA:

$$BA = \begin{bmatrix} 3 & 0 \\ 4 & 1 \end{bmatrix} \begin{bmatrix} 1 & 2 \\ 0 & 1 \end{bmatrix} = \begin{bmatrix} 3 \cdot 1 + 0 \cdot 0 & 3 \cdot 2 + 0 \cdot 1 \\ 4 \cdot 1 + 1 \cdot 0 & 4 \cdot 2 + 1 \cdot 1 \end{bmatrix} = \begin{bmatrix} 3+0 & 6+0 \\ 4+0 & 8+1 \end{bmatrix} = \begin{bmatrix} 3 & 6 \\ 4 & 9 \end{bmatrix}.$$

We observe that:

$$AB = \begin{bmatrix} 11 & 2 \\ 4 & 1 \end{bmatrix} \neq \begin{bmatrix} 3 & 6 \\ 4 & 9 \end{bmatrix} = BA.$$

This demonstrates that matrix multiplication is **not commutative,** the order of multiplication matters. In ML, this property is crucial when designing neural network architectures, where layer operations must be applied in specific sequences to achieve the desired transformations.

- **Distributive Law**: $(A + B)C = AC + BC, \quad C(P + Q) = CP + CQ$
  Matrix multiplication distributes over addition, allowing sums to be multiplied term-wise.

Example 3.23 With suitable A, B, and C, verify $(A + B)C = AC + BC$.
  Let

$$A = \begin{bmatrix} 1 & 2 \\ 0 & 1 \end{bmatrix}, \quad B = \begin{bmatrix} 3 & 0 \\ 1 & 2 \end{bmatrix}, \quad C = \begin{bmatrix} 1 & 0 \\ 0 & 1 \end{bmatrix}.$$

Then,

$$A + B = \begin{bmatrix} 4 & 2 \\ 1 & 3 \end{bmatrix}, \quad (A + B)C = \begin{bmatrix} 4 & 2 \\ 1 & 3 \end{bmatrix} \begin{bmatrix} 1 & 0 \\ 0 & 1 \end{bmatrix} = \begin{bmatrix} 4 & 2 \\ 1 & 3 \end{bmatrix},$$

$$AC = \begin{bmatrix} 1 & 2 \\ 0 & 1 \end{bmatrix} \begin{bmatrix} 1 & 0 \\ 0 & 1 \end{bmatrix} = \begin{bmatrix} 1 & 2 \\ 0 & 1 \end{bmatrix}, \quad BC = \begin{bmatrix} 3 & 0 \\ 1 & 2 \end{bmatrix} \begin{bmatrix} 1 & 0 \\ 0 & 1 \end{bmatrix} = \begin{bmatrix} 3 & 0 \\ 1 & 2 \end{bmatrix},$$

$$AC + BC = \begin{bmatrix} 1 & 2 \\ 0 & 1 \end{bmatrix} + \begin{bmatrix} 3 & 0 \\ 1 & 2 \end{bmatrix} = \begin{bmatrix} 4 & 2 \\ 1 & 3 \end{bmatrix}.$$

Thus, $(A + B)C = AC + BC$ is verified.

- **Scalar Multiplication and Product:** $s(AB) = (sA)B = A(sB)$.
  A scalar can be factored anywhere in a product.

Example 3.24 Let $s = 2$, and compute $2(AB)$ and $(2A)B$ to confirm their equality. Let:

$$A = \begin{bmatrix} 1 & 2 \\ 0 & 1 \end{bmatrix}, \quad B = \begin{bmatrix} 3 & 0 \\ 1 & 2 \end{bmatrix}.$$

First, we compute AB:

$$AB = \begin{bmatrix} 1 & 2 \\ 0 & 1 \end{bmatrix} \begin{bmatrix} 3 & 0 \\ 1 & 2 \end{bmatrix} = \begin{bmatrix} (1)(3) + (2)(1) & (1)(0) + (2)(2) \\ (0)(3) + (1)(1) & (0)(0) + (1)(2) \end{bmatrix} = \begin{bmatrix} 5 & 4 \\ 1 & 2 \end{bmatrix}.$$

Then, we compute $2(AB)$:

$$2(AB) = 2 \cdot \begin{bmatrix} 5 & 4 \\ 1 & 2 \end{bmatrix} = \begin{bmatrix} 10 & 8 \\ 2 & 4 \end{bmatrix}.$$

Now we compute $2A$:

$$2A = 2 \cdot \begin{bmatrix} 1 & 2 \\ 0 & 1 \end{bmatrix} = \begin{bmatrix} 2 & 4 \\ 0 & 2 \end{bmatrix}.$$

Then we compute $(2A)B$:

$$(2A)B = \begin{bmatrix} 2 & 4 \\ 0 & 2 \end{bmatrix} \begin{bmatrix} 3 & 0 \\ 1 & 2 \end{bmatrix} = \begin{bmatrix} (2)(3) + (4)(1) & (2)(0) + (4)(2) \\ (0)(3) + (2)(1) & (0)(0) + (2)(2) \end{bmatrix} = \begin{bmatrix} 10 & 8 \\ 2 & 4 \end{bmatrix}.$$

Since both $2(AB)$ and $(2A)B$ equal $\begin{bmatrix} 10 & 8 \\ 2 & 4 \end{bmatrix}$, the scalar-matrix multiplication rule $s(AB) = (sA)B$, is verified for $s = 2$, A, and B.

- **Matrix Identity:** $I_m A = A = A I_n$
  Multiplying by an identity matrix has no effect.

Example 3.25

$$I_2 = \begin{bmatrix} 1 & 0 \\ 0 & 1 \end{bmatrix},$$

$$I_2 \cdot \begin{bmatrix} 1 & 2 \\ 3 & 4 \end{bmatrix} = \begin{bmatrix} 1 & 2 \\ 3 & 4 \end{bmatrix}.$$

- **Zero Matrix**: $AO = OA = O$
  Any matrix multiplied by a zero matrix yields a zero matrix.

Example 3.26

$$A \cdot \begin{bmatrix} 0 & 0 \\ 0 & 0 \end{bmatrix} = \begin{bmatrix} 0 & 0 \\ 0 & 0 \end{bmatrix}.$$

- **Powers of a Matrix**: For square A,

$$A^k = A \cdot A \cdots A,$$

$$A^0 = I.$$

Powers represent repeated application of a transformation.

Example 3.27 If

$$A = \begin{bmatrix} 0 & 1 \\ 1 & 0 \end{bmatrix},$$

then

$$A^2 = I.$$

- **Transpose Rule**: $(AB)^T = B^T A^T$.
  Transpose of a product reverses the order and transposes each factor.

Example 3.28 For:

$$A = \begin{bmatrix} 1 & 2 \\ 3 & 4 \end{bmatrix}, \quad B = \begin{bmatrix} 5 & 6 \\ 7 & 8 \end{bmatrix},$$

we want to verify that $(AB)^T = B^T A^T$.

Step 1-We compute AB:

$$AB = \begin{bmatrix} 1 \cdot 5 + 2 \cdot 7 & 1 \cdot 6 + 2 \cdot 8 \\ 3 \cdot 5 + 4 \cdot 7 & 3 \cdot 6 + 4 \cdot 8 \end{bmatrix} = \begin{bmatrix} 19 & 22 \\ 43 & 50 \end{bmatrix}.$$

$$(AB)^T = \begin{bmatrix} 19 & 43 \\ 22 & 50 \end{bmatrix}.$$

Step 2-Then, we compute $B^T A^T$:

$$B^T = \begin{bmatrix} 5 & 7 \\ 6 & 8 \end{bmatrix}, \quad A^T = \begin{bmatrix} 1 & 3 \\ 2 & 4 \end{bmatrix},$$

$$B^T A^T = \begin{bmatrix} 5 \cdot 1 + 7 \cdot 2 & 5 \cdot 3 + 7 \cdot 4 \\ 6 \cdot 1 + 8 \cdot 2 & 6 \cdot 3 + 8 \cdot 4 \end{bmatrix} = \begin{bmatrix} 19 & 43 \\ 22 & 50 \end{bmatrix}.$$

Since $(AB)^T = B^T A^T$, the property holds.

3.3   Various Types of Matrices

Different matrix structures simplify computations and model specific relationships. This section covers key matrix types with examples and their significance.

- **Regular Matrix**: Any $m \times n$ matrix without special structure.

$$A = \begin{bmatrix} 1 & 2 & 3 \\ 4 & 5 & 6 \end{bmatrix}.$$

  A regular matrix represents general data or linear transformations, such as a dataset with 2 samples and 3 features. It supports basic operations like addition, scalar multiplication, and matrix multiplication.
- **Square Matrix**: A matrix with the same number of rows and columns ($m = n$).

$$A = \begin{bmatrix} 1 & 2 & 3 \\ 4 & 5 & 6 \\ 7 & 8 & 9 \end{bmatrix}.$$

  A square matrix enables special operations such as computing determinants, eigenvalues, and matrix inverses. Square matrices are central in solving systems of linear equations and analyzing linear transformations within the same vector space.
- **Diagonal Matrix**: A diagonal matrix has nonzero entries only on the diagonal. The main diagonal is the set of elements from the top-left to the bottom-right of the matrix.

$$A = \begin{bmatrix} 1 & 0 & 0 \\ 0 & 2 & 0 \\ 0 & 0 & 3 \end{bmatrix}.$$

Multiplication of diagonal matrices is:

$$(AB)_{ii} = a_{ii}b_{ii},$$

producing another diagonal matrix.

Example 3.29 For:

$$A = \begin{bmatrix} 1 & 0 & 0 \\ 0 & 2 & 0 \\ 0 & 0 & 3 \end{bmatrix}, \quad B = \begin{bmatrix} 3 & 0 & 0 \\ 0 & -1 & 0 \\ 0 & 0 & 2 \end{bmatrix},$$

$$AB = \begin{bmatrix} 1 \times 3 & 0 & 0 \\ 0 & 2 \times (-1) & 0 \\ 0 & 0 & 3 \times 2 \end{bmatrix} = \begin{bmatrix} 3 & 0 & 0 \\ 0 & -2 & 0 \\ 0 & 0 & 6 \end{bmatrix}.$$

Python Example: Diagonal Matrix Multiplication

```python
import numpy as np

A = np.diag([1, 2, 3])
B = np.diag([3, -1, 2])
AB = A @ B

print("AB:\n", AB)
```

Output:

```
AB:
 [[ 3  0  0]
 [ 0 -2  0]
 [ 0  0  6]]
```

A diagonal matrix efficiently represents scaling transformations, where each axis is scaled independently. Diagonal matrices are computationally simple and often arise in spectral decomposition.

- **Upper Triangular Matrix**: A square matrix with all entries below the main diagonal equal to zero.

$$U = \begin{bmatrix} 1 & 2 & 3 \\ 0 & 4 & 5 \\ 0 & 0 & 6 \end{bmatrix}.$$

Upper triangular matrix commonly appears during Gaussian elimination and LU. It allows for fast substitution when solving upper triangular systems.

- **Lower Triangular Matrix**: A square matrix with all entries above the main diagonal equal to zero.

$$L = \begin{bmatrix} 1 & 0 & 0 \\ 2 & 3 & 0 \\ 4 & 5 & 6 \end{bmatrix}.$$

Lower triangular matrices are used in forward substitution when solving linear systems and in decomposing matrices (e.g., LU, which will be discussed in Chap. 11).

- **Symmetric Matrix**: A symmetric matrix is a square matrix that is equal to its transpose. In other words, if you were to reflect a symmetric matrix across its main diagonal, the original matrix and its reflection would be identical. A matrix is **symmetric** if $A = A^T$. For:

$$A = \begin{bmatrix} 1 & 2 & 4 \\ 2 & 3 & -1 \\ 4 & -1 & 5 \end{bmatrix},$$

to verify that a matrix is symmetric we check whether $a_{ij} = a_{ji}$, e.g., $a_{12} = 2 = a_{21}$.

Products like $A^T A$ and $A A^T$ are symmetric:

$$(A^T A)^T = A^T (A^T)^T = A^T A,$$

$$(A A^T)^T = (A^T)^T A^T = A A^T.$$

Example 3.30 For:

$$A = \begin{bmatrix} 1 & 2 \\ 3 & 4 \end{bmatrix}, \quad A^T = \begin{bmatrix} 1 & 3 \\ 2 & 4 \end{bmatrix},$$

we compute:

$$A^T A = \begin{bmatrix} 1 \cdot 1 + 3 \cdot 3 & 1 \cdot 2 + 3 \cdot 4 \\ 2 \cdot 1 + 4 \cdot 3 & 2 \cdot 2 + 4 \cdot 4 \end{bmatrix} = \begin{bmatrix} 10 & 14 \\ 14 & 20 \end{bmatrix},$$

$$A A^T = \begin{bmatrix} 1 \cdot 1 + 2 \cdot 2 & 1 \cdot 3 + 2 \cdot 4 \\ 3 \cdot 1 + 4 \cdot 2 & 3 \cdot 3 + 4 \cdot 4 \end{bmatrix} = \begin{bmatrix} 5 & 11 \\ 11 & 25 \end{bmatrix}.$$

Both are symmetric.

Python Example: Symmetric Matrix Check

```python
import numpy as np

A = np.array([[1, 2, 4], [2, 3, -1], [4, -1, 5]])
is_symmetric = np.array_equal(A, A.T)

print("Is A symmetric?", is_symmetric)
```

Output:

```
Is A symmetric? True
```

- **Identity Matrix**: As already defined in Sect. 3.1.2, a diagonal matrix with all diagonal entries equal to 1.

$$I_3 = \begin{bmatrix} 1 & 0 & 0 \\ 0 & 1 & 0 \\ 0 & 0 & 1 \end{bmatrix}.$$

Identity matrices serve as the multiplicative identity for matrices: $I_m A = A I_n = A$, preserving matrix structure under multiplication.

- **Zero Matrix**: A matrix in which all entries are zero.

$$O = \begin{bmatrix} 0 & 0 & 0 \\ 0 & 0 & 0 \end{bmatrix}.$$

Zero matrices act as the additive identity in matrix addition: $A + O = A$. It is also used to represent the null output or the absence of transformation.

Python Example: Matrix Types

```python
import numpy as np

diagonal = np.diag([1, 5, 9])
identity = np.eye(3)
zero = np.zeros((2, 3))

print("Diagonal:\n", diagonal)
print("Identity:\n", identity)
print("Zero:\n", zero)
```

Output:

```
Diagonal:
[[1 0 0]
 [0 5 0]
 [0 0 9]]
Identity:
[[1. 0. 0.]
 [0. 1. 0.]
 [0. 0. 1.]]
Zero:
[[0. 0. 0.]
 [0. 0. 0.]]
```

3.4 Augmentation and Partition of Matrices

3.4.1 Augmentation

An augmented matrix is formed by horizontally concatenating two matrices, typically used in the context of solving systems of linear equations. Most commonly, it involves combining a coefficient matrix $A \in M_{m \times n}$ with a column matrix $B \in M_{m \times p}$, which contains the constants or right-hand side of the system. The resulting matrix captures both the structure of the equations and their outcomes in a single representation.

$$[A \mid B] = \begin{bmatrix} a_{11} & \cdots & a_{1n} & b_{11} & \cdots \\ \vdots & \ddots & \vdots & \vdots & \ddots \\ a_{m1} & \cdots & a_{mn} & b_{m1} & \cdots \end{bmatrix}.$$

This format is particularly useful for applying row operations during Gaussian elimination or Gauss–Jordan elimination. Instead of tracking the original system and its solution vector separately, the augmented matrix allows all manipulations to be performed in a unified framework. Once reduced to row echelon form (or RREF, Sect. 5.4), the augmented matrix reveals whether the system has a unique solution, infinitely many solutions, or no solution. It is also used when computing matrix inverses using the method of row reduction.

Example 3.31 For:

$$A = \begin{bmatrix} 1 & 2 \\ 3 & 4 \end{bmatrix}, \quad B = \begin{bmatrix} 5 \\ 6 \end{bmatrix},$$

we can construct an augmented as follows.

$$[A \mid B] = \begin{bmatrix} 1 & 2 & 5 \\ 3 & 4 & 6 \end{bmatrix}.$$

3.4.2 Partitioning

Partitioning divides a matrix into submatrices (blocks), which allows complex matrix operations to be broken down into simpler, more manageable computations. This technique is especially useful in theoretical proofs, algorithm design, and parallel computing. By treating blocks as single matrix elements, we can perform matrix addition, multiplication, and inversion more systematically, particularly when working with large-scale systems. Partitioning also facilitates structured methods such as block LU decomposition (Chap. 11), Schur complements, and efficient manipulation of sparse or hierarchical matrices. It is a powerful tool for both analysis and implementation in numerical linear algebra and scientific computing. Augmented matrices encode systems of equations, while partitioning enables efficient block computations in large-scale ML systems.

Example 3.32 Consider the following 3×4 matrix:

$$A = \begin{bmatrix} 1\ 2\ 3\ 4 \\ 5\ 6\ 7\ 8 \\ 4\ 3\ 2\ 1 \end{bmatrix} = \begin{bmatrix} A_{11} & A_{12} \\ A_{21} & A_{22} \end{bmatrix},$$

where the matrix is partitioned into four submatrices:

$$A_{11} = \begin{bmatrix} 1\ 2 \end{bmatrix}, \quad A_{12} = \begin{bmatrix} 3\ 4 \end{bmatrix}, \quad A_{21} = \begin{bmatrix} 5\ 6 \\ 4\ 3 \end{bmatrix}, \quad A_{22} = \begin{bmatrix} 7\ 8 \\ 2\ 1 \end{bmatrix}.$$

This block structure allows us to represent A in a more modular form and apply operations like matrix multiplication, inversion, or decomposition more efficiently by working with its submatrices.

Example 3.33 In this partitioning scheme, we divide the matrix into two vertical blocks, splitting the matrix by columns:

$$A = \left[\begin{array}{cc|cc} 1 & 2 & 3 & 4 \\ 5 & 6 & 7 & 8 \\ 4 & 3 & 2 & 1 \end{array} \right].$$

This format is useful in block matrix-vector multiplication, where each partition interacts independently with segments of the input vector.

Example 3.34 Here, we separate the matrix horizontally, dividing it by rows:

$$A = \left[\begin{array}{cccc} 1 & 2 & 3 & 4 \\ 5 & 6 & 7 & 8 \\ \hline 4 & 3 & 2 & 1 \end{array} \right].$$

Row partitions are helpful in iterative algorithms and hierarchical processing, allowing row-wise manipulation or storage.

Example 3.35 The matrix can also be partitioned into three vertical segments:

$$A = \left[\begin{array}{cc|cc|} 1 & 2 & 3 & 4 \\ 5 & 6 & 7 & 8 \\ 4 & 3 & 2 & 1 \end{array} \right].$$

This form enables fine-grained control in block algorithms, such as column-wise decomposition or pipelined computations.

Applications of Matrix Partitioning:

- **Block matrix multiplication**: Enables efficient implementation of large matrix products by reducing them to smaller block-level operations.
- **LU and QR (Chap.** 11): Partitioning allows recursive and block-based versions of these algorithms for better numerical stability and parallelization.
- **Sparse matrix storage**: Helps organize and compress large matrices with structured sparsity, especially in scientific computing and optimization.

Partitioning is not just a notational convenience, it is a powerful tool for structuring computations and optimizing performance in both symbolic and numerical linear algebra.

3.5 Block Matrix Multiplication

Block matrix multiplication enables scalable computation by dividing matrices into smaller submatrices processed independently. A block matrix is a matrix that is partitioned into smaller submatrices, or *blocks*, which are treated as individual matrix elements. This structure is particularly useful when performing large-scale computations or when matrices exhibit a repeating or hierarchical structure. Operations on block matrices follow the same rules as regular matrix operations, with each block acting as a single entity, as long as the blocks are conformable in size for the given operation.

For example, if:

$$A = \begin{bmatrix} A_{11} & A_{12} \\ A_{21} & A_{22} \end{bmatrix}, \quad B = \begin{bmatrix} B_{11} & B_{12} \\ B_{21} & B_{22} \end{bmatrix},$$

where each A_{ij} and B_{ij} are submatrices of appropriate sizes, then the product AB is computed as:

$$AB = \begin{bmatrix} A_{11}B_{11} + A_{12}B_{21} & A_{11}B_{12} + A_{12}B_{22} \\ A_{21}B_{11} + A_{22}B_{21} & A_{21}B_{12} + A_{22}B_{22} \end{bmatrix}.$$

This formulation mirrors the standard rule for matrix multiplication but at the level of submatrices. Each entry in the resulting block matrix is obtained by summing the products of corresponding row and column blocks.

Example 3.36 Consider the block matrix:

$$A = \begin{bmatrix} I_2 & 0 \\ B & 5I_2 \end{bmatrix}, \quad \text{where } B = \begin{bmatrix} 6 & 8 \\ -7 & 9 \end{bmatrix}.$$

Step 1- Let's interpret the structure:

A is a 4×4 matrix composed of four 2×2 blocks:

- I_2 is the 2×2 identity matrix.
- 0 is a 2×2 zero matrix.

- B is a general 2×2 matrix.
- $5I_2$ is a diagonal block scaled by 5.

Step 2- We compute $A^2 = A \cdot A$ as follows:
   Using block matrix multiplication rules:

$$A^2 = \begin{bmatrix} I_2 & 0 \\ B & 5I_2 \end{bmatrix} \begin{bmatrix} I_2 & 0 \\ B & 5I_2 \end{bmatrix} = \begin{bmatrix} I_2 \cdot I_2 + 0 \cdot B & I_2 \cdot 0 + 0 \cdot 5I_2 \\ B \cdot I_2 + 5I_2 \cdot B & B \cdot 0 + 5I_2 \cdot 5I_2 \end{bmatrix}.$$

   Simplifying each block:

- Top-left: I_2
- Top-right: 0
- Bottom-left: $B + 5B = 6B$
- Bottom-right: $25I_2$

$$A^2 = \begin{bmatrix} I_2 & 0 \\ 6B & 25I_2 \end{bmatrix}, \quad 6B = \begin{bmatrix} 36 & 48 \\ -42 & 54 \end{bmatrix}.$$

Step 3- Final expanded form:

$$A^2 = \begin{bmatrix} 1 & 0 & 0 & 0 \\ 0 & 1 & 0 & 0 \\ 36 & 48 & 25 & 0 \\ -42 & 54 & 0 & 25 \end{bmatrix}.$$

This example shows how block matrix multiplication simplifies the computation by operating on structured submatrices. Rather than multiplying the full 4×4 matrices directly, we leverage the known properties of identity, zero, and scalar multiplication within the blocks to compute A^2 efficiently.

The following code example constructs a block matrix using the 2×2 identity matrix I_2, a zero block, and a 2×2 matrix B, forming a structured 4×4 block matrix A. It then computes A^2 using the matrix multiplication operator @ in NumPy. This demonstrates how block matrices can be created and multiplied directly.

Python Example: Block Matrix Multiplication

```python
import numpy as np

I2 = np.eye(2)
B = np.array([[6, 8], [-7, 9]])
A = np.block([[I2, np.zeros((2, 2))], [B, 5 *
    I2]])
A2 = A @ A
```

```
8 print ("A^2:\n", A2)
```

Output:

```
A^2:
[[  1.    0.    0.    0.]
 [  0.    1.    0.    0.]
 [ 36.   48.   25.    0.]
 [-42.   54.    0.   25.]]
```

The following code example shows the explicit procedure of block matrix multiplication. Here, two 4×4 matrices A and B are partitioned into four 2×2 sub-blocks each. The multiplication is then carried out following the block multiplication rule:

$$C = \begin{bmatrix} A_{11}B_{11} + A_{12}B_{21} & A_{11}B_{12} + A_{12}B_{22} \\ A_{21}B_{11} + A_{22}B_{21} & A_{21}B_{12} + A_{22}B_{22} \end{bmatrix}.$$

Finally, the sub-results are combined with np.block to reconstruct the full product matrix C. This approach illustrates the theoretical block multiplication process in a practical coding example.

Python Example: Block Matrix Multiplication

```python
import numpy as np

A = np.array([[1,  2,  3,  0],
     [0,  1,  1,  2],
     [2, -1,  0,  1],
     [1,  0, -1,  3]])

B = np.array([[2,  1,  0,  -1],
     [-1,  0,  2,  1],
     [1,  3,  1,  0],
     [0, -2,  3,  2]])

# Partition A into blocks
A11, A12 = A[:2, :2], A[:2, 2:]
A21, A22 = A[2:, :2], A[2:, 2:]

# Partition B into blocks
B11, B12 = B[:2, :2], B[:2, 2:]
B21, B22 = B[2:, :2], B[2:, 2:]

# Perform block matrix multiplication
C11 = A11 @ B11 + A12 @ B21
C12 = A11 @ B12 + A12 @ B22
```

```
24  C21 = A21 @ B11 + A22 @ B21
25  C22 = A21 @ B12 + A22 @ B22
26
27  # Combine blocks into the final matrix
28  C = np.block([[C11, C12],
29      [C21, C22]])
30
31  print("Block Matrix Multiplication Result:\n", C)
```

Output:

```
Block Matrix Multiplication Result:
[[ 1  7  8  2]
 [ 1 -3  9  5]
 [ 3  2 -1  1]
 [ 1  7  8  5]]
```

SymPy for Matrices

So far, we used **Numpy** for working with matrices in Python. Another option is **SymPy**. SymPy is a Python library for symbolic mathematics. Unlike NumPy, which is numerical, SymPy allows exact computations with symbolic expressions. This is particularly useful in linear algebra for deriving formulas, computing determinants, inverses, eigenvalues, and performing symbolic simplifications. One key benefit of SymPy over NumPy is that it provides clean, making it easier to read, interpret, and present exact results compared to the purely numeric display of NumPy.

A matrix in SymPy can be defined using the Matrix class. The following examples illustrate how SymPy handles matrix definition, addition, and multiplication (Table 3.1).

Python Example: Matrix Definition in SymPy

```
1  from sympy import Matrix
2
3  A = Matrix([[1, 2],
4      [3, 4]])
5  A
```

Output:

$$\begin{bmatrix} 1 & 2 \\ 3 & 4 \end{bmatrix}$$

Python Example: Matrix Addition in SymPy

```
from sympy import Matrix

A = Matrix([[1, 2],
    [3, 4]])
B = Matrix([[5, 6],
    [7, 8]])

C = A + B
C
```

Output:

$$\begin{bmatrix} 1+5 & 2+6 \\ 3+7 & 4+8 \end{bmatrix} = \begin{bmatrix} 6 & 8 \\ 10 & 12 \end{bmatrix}$$

Python Example: Matrix Multiplication in SymPy

```
from sympy import Matrix

A = Matrix([[1, 2],
    [3, 4]])
B = Matrix([[5, 6],
    [7, 8]])

D = A * B
D
```

Output:

$$\begin{bmatrix} 1\cdot 5+2\cdot 7 & 1\cdot 6+2\cdot 8 \\ 3\cdot 5+4\cdot 7 & 3\cdot 6+4\cdot 8 \end{bmatrix} = \begin{bmatrix} 19 & 22 \\ 43 & 50 \end{bmatrix}$$

NumPy and SymPy both provide tools for working with matrices, but they serve different purposes. NumPy is optimized for fast numerical computation and is widely used in scientific computing, ML, and data analysis. Its matrix operations use the @ operator or `np.dot()` for multiplication, making it ideal for large-scale numerical

Table 3.1 Comparison of NumPy and SymPy for matrix operations

Feature	NumPy	SymPy

Matrix definition	`import numpy as np`
	`A = np.array([[1,`
	`2], [3, 4]])`
		`from sympy import`
		`Matrix`
		`A = Matrix([[1, 2],`
		`[3, 4]])`

Matrix multiplication	`A @ B` or `np.dot(A,B)`	`A * B`
Symbolic computation	Not supported (numerical only)	Fully supported (symbolic algebra, exact forms)
Output style	Compact numerical arrays (e.g., `[[1 2] [3 4]]`)	Pretty-printed, mathematical style (e.g., $\begin{bmatrix} 1 & 2 \\ 3 & 4 \end{bmatrix}$)
Use case	Efficient numerical linear algebra and large-scale computations	Symbolic mathematics, exact algebraic manipulation, clean representation

tasks. On the other hand, SymPy is a symbolic mathematics library designed for algebraic manipulation, exact computation, and human-readable output. Defining a matrix in SymPy automatically produces a clean, textbook-style representation, and matrix multiplication can be performed directly with the $*$ operator. Thus, NumPy is best for numerical efficiency, while SymPy excels in symbolic clarity and exactness.

3.6 Matrices in ML

Matrices are the foundation of ML, serving multiple roles: they act as data containers (storing datasets), transformation operators (applying mathematical operations), and computational tools (for implementing algorithms efficiently). This section explores these applications with concrete examples from real-world scenarios.

3.6.1 Data Representation

In ML, datasets are usually stored in matrix form. Each row represents one sample (e.g., a patient, a house, or an image), and each column represents a feature or variable measured (e.g., age, size, brightness). For instance:

$$A = \begin{bmatrix} 2 & 3 & 5 \\ 3 & 1 & -1 \\ -2 & 1 & 1 \end{bmatrix}.$$

Here, matrix A holds data for three samples, each with three features. Such representation is essential for feeding data into ML models.

3.6.2   *Feature Transformations*

Matrices are central to transforming features in models. For example, in LR:

$$\hat{y} = \mathbf{w}^T\mathbf{x},$$

the dot product of weight vector $\mathbf{w}$ and feature vector $\mathbf{x}$ gives a prediction $\hat{y}$. In neural networks:

$$\mathbf{y} = W\mathbf{x} + \mathbf{b},$$

a weight matrix W transforms input vector $\mathbf{x}$, and a bias vector $\mathbf{b}$ is added.

Python Example: Linear Model Prediction

```python
import numpy as np

X = np.array([[1, 2], [3, 4], [5, 6]])
w = np.array([0.5, 1.5])
predictions = X @ w

print("Predictions:", predictions)
```

Output:

```
Predictions: [ 3.5  7.5 11.5]
```

3.6.3 Matrix Operations in ML

Matrix operations are fundamental in ML workflows:

- **Addition and Subtraction**: Combine datasets or compute prediction errors (e.g., $\hat{y} - y$ in regression).
- **Scalar Multiplication**: Adjust magnitude of weights or scale learning rates.
- **Transpose**: Required for operations like computing covariance or least squares solutions (e.g., $X^T X$).
- **Matrix-Vector Product**: Used for computing predictions from input features.
- **Matrix-Matrix Product**: Facilitates multi-layer transformations in deep learning models.
- **Hadamard Product (Element-Wise)**: Common in attention mechanisms, element-wise weighting, or normalization.
- **Dot Product**: Used in similarity computations (e.g., cosine similarity) or LR models.
- **Cross Product**: Appears in computer vision or robotics, particularly for 3D vector calculations like torque or normals.

3.6.4 Special Matrices in ML

Special matrix structures simplify computation and ensure numerical stability:

- **Diagonal Matrices**: Have nonzero values only on the diagonal; used in regularization (e.g., λI for Ridge Regression).
- **Symmetric Matrices**: Arise in covariance and kernel matrices (principle component analysis (PCA), support vectors machines (SVMs)), with $A = A^T$.
- **Identity Matrices**: Neutral element in matrix multiplication; helps preserve dimensions during transformations (Fig. 3.2).

3.6.5 Grayscale Image Representation

In image processing, a grayscale image is represented as a 2D matrix of pixel intensities (values between 0 and 255). Each value corresponds to the brightness of a pixel.

100	150	200
50	75	100
25	50	75

Fig. 3.2 3×3 grayscale image matrix with pixel intensities

Fig. 3.3 Grayscale visualization of a 3×3 image matrix, where lighter values (e.g., 200) are brighter and darker values (e.g., 25) are dimmer. This illustrates how matrices can encode visual data

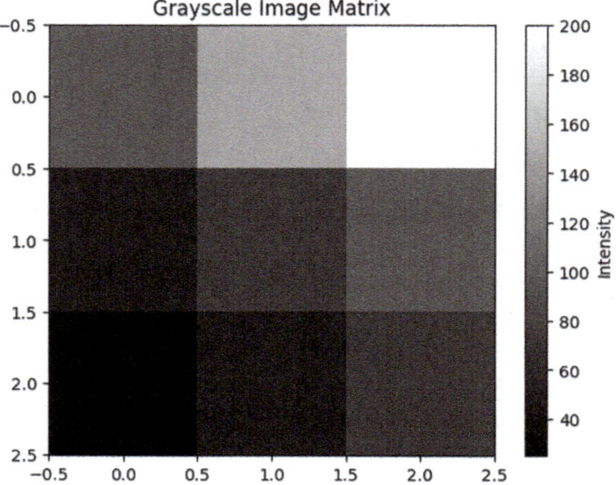

Python Example: Grayscale Image Matrix

```python
import numpy as np
import matplotlib.pyplot as plt

image = np.array([[100, 150, 200], [50, 75,
    100], [25, 50, 75]])
plt.imshow(image, cmap='gray')
plt.title('Grayscale Image Matrix')
plt.colorbar(label='Intensity')
plt.show()
```

Output: Visualizes the grayscale matrix from Figure 3.3, showing how pixel values translate to brightness.

Additionally, images can be represented as 3D matrices (tensors), where separate 2D matrices correspond to red, green, and blue color channels. These matrices are used in Convolutional Neural Networks (CNNs) to extract spatial features.

3.6.6 Covariance Matrix

The covariance matrix summarizes how features vary with respect to each other. For a centered data matrix X, the covariance matrix is computed as:

$$S = \frac{1}{n-1} X^T X.$$

This matrix is essential in PCA and other dimensionality reduction techniques.

Python Example: Covariance Matrix

```python
import numpy as np

X = np.array([[2, 3, 5], [3, 1, -1], [-2, 1, 1]])
X_centered = X - np.mean(X, axis=0)
S = (X_centered.T @ X_centered) / (X.shape[0] -
    1)

print("Covariance Matrix:\n", S)
```

Output:

```
[[ 7.      1.     -3.33]
 [1.      1.33    3.33]
 [-3.33   3.33    9.33]]
```

This code shows the variance along each feature (diagonal) and the covariance between features (off-diagonal).

Chapter Summary

This chapter provided a comprehensive introduction to matrices as essential structures in linear algebra and ML, extending the vector concepts from the previous chapter. It began with the definition of a matrix as a rectangular array of numbers in $M_{m \times n}$, illustrated through Example 3.1 and visualized as a grid in Fig. 3.1. Matrices were presented both algebraically, with entries a_{ij}, and geometrically, as transformations mapping vectors from $\mathbb{R}^n$ to $\mathbb{R}^m$. The dual perspective of matrices as collections of column or row vectors was emphasized, with Python examples using NumPy's *np.column_stack* to construct matrices from vectors.

Section 3.2 delved into matrix operations, starting with element-wise addition and subtraction for compatible dimensions, scalar multiplication for uniform scaling, and transposition A^T that swaps rows and columns, with properties like $(A^T)^T = A$ verified through examples. The matrix-vector product $A\mathbf{x}$ was explored via column and row perspectives, as in the computation yielding $\begin{bmatrix} -3 \\ 3 \end{bmatrix}$, and matrix-matrix multiplication AB highlighted non-commutativity with $AB \neq BA$. Linear combination views, column-wise as weighted sums of A's columns and row-wise as weighted sums of B's rows, were demonstrated, alongside associativity $(AB)C = A(BC)$ and the Hadamard product $A \odot B$ for element-wise multiplication. Properties such as distributivity and the transpose rule $(AB)^T = B^T A^T$ were covered extensively, supported by Python implementations.

Special matrix types were examined in Sect. 3.3, including square matrices for eigenvalue computations, diagonal matrices for efficient scaling (with multiplication simplifying to diagonal products), upper and lower triangular matrices for substitution methods, symmetric matrices $A = A^T$ arising in covariance, and identity and zero matrices as identities for multiplication and addition. Python examples using `np.diag`, `np.eye`, and symmetry checks reinforced these structures.

Section 3.4 introduced augmented matrices $[A \mid B]$ for linear systems and partitioning into submatrices for modular computations, with examples of row, column, and multi-partition schemes. Block matrix multiplication in Sect. 3.5 treated submatrices as elements, as in the computation of A^2 for a block form involving identity and scaled blocks, leveraging NumPy's `np.block`.

SymPy was introduced as a symbolic alternative to NumPy for exact computations, with examples of matrix definition, addition, and multiplication yielding symbolic outputs.

Finally, Sect. 3.6 connected matrices to ML, covering data representation (rows as samples, columns as features), feature transformations in linear models and neural networks, operations like transposes in covariance $X^T X$, and special matrices in regularization. Applications included grayscale image matrices visualized in Figs. 3.3 and 3.4.

Takeaways

Key concepts from the chapter are summarized as follows:

- Matrices as versatile tools for data storage and linear transformations, with dimensions $m \times n$ dictating operation compatibility and geometric mappings.
- Fundamental operations, addition, scalar multiplication, transposition, products (matrix-vector, matrix-matrix, Hadamard), enabling efficient computations, with properties like associativity and distributivity facilitating algebraic manipulations.
- Special types such as diagonal, symmetric, triangular, identity, and zero matrices that exploit structure for computational efficiency and model interpretability in ML.

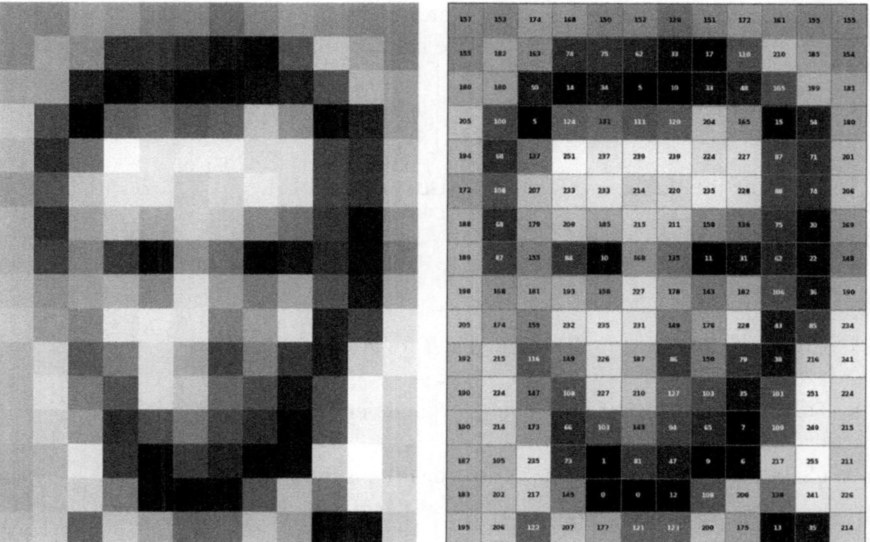

Fig. 3.4 Grayscale image represented as a 2D matrix of pixel intensities. Each square corresponds to a pixel with a value from 0 (black) to 255 (white)

- Augmentation and partitioning techniques for solving systems and modularizing large matrices, extending to block multiplication for hierarchical operations.
- Practical implementations in NumPy for numerical work and SymPy for symbolic derivations, bridging theory to code.
- Core applications in ML, from dataset representation and predictions via $W\mathbf{x} + \mathbf{b}$ to covariance analysis and image processing, underscoring matrices' role in scalable AI algorithms.

Through these elements, readers acquired the foundation to manipulate matrices intuitively, preparing for advanced topics like decompositions and optimization in subsequent chapters. Matrices provide a concrete representation of linear transformations, which will be studied formally in the next chapter.

Exercises

1. Compute $A + B$ for $A = \begin{bmatrix} 5 & 5 \\ 6 & 6 \end{bmatrix}$, $B = \begin{bmatrix} 7 & 7 \\ 8 & 8 \end{bmatrix}$.

2. Compute $2A$ for the matrix A in Exercise 1.

3. Verify $(A + B)^T = A^T + B^T$ for the matrices in Exercise 1.

4. Compute $A\mathbf{x}$ for $A = \begin{bmatrix} 1 & 2 & 3 \\ 4 & 5 & 6 \end{bmatrix}$, $\mathbf{x} = \begin{bmatrix} 1 \\ 1 \\ 1 \end{bmatrix}$.

5. Compute the Hadamard (element-wise) product of $A = \begin{bmatrix} 1 & 2 \\ 3 & 4 \end{bmatrix}$, $B = \begin{bmatrix} 2 & 1 \\ 0 & 3 \end{bmatrix}$.

6. Check if $\begin{bmatrix} 1 & 2 & 3 \\ 2 & 1 & 6 \\ 3 & 6 & 1 \end{bmatrix}$ is symmetric.

7. Compute $A + B$, $A - B$, $3A$, and AB for $A = \begin{bmatrix} 1 & 2 \\ 3 & 4 \end{bmatrix}$, $B = \begin{bmatrix} 5 & 6 \\ 7 & 8 \end{bmatrix}$.

8. Determine the size of $C = \begin{bmatrix} 1 & 0 & 2 \\ -1 & 3 & 4 \end{bmatrix}$ and verify it is in $C \in \mathbb{R}^{2\times 3}$

9. Classify each matrix (e.g., zero, identity, symmetric, etc.) (a) $\begin{bmatrix} 0 & 0 & 0 \\ 0 & 0 & 0 \end{bmatrix}$, (b) $\begin{bmatrix} 1 & 0 & 0 \\ 0 & 1 & 0 \\ 0 & 0 & 1 \end{bmatrix}$, (c) $\begin{bmatrix} 2 & -1 & 3 \\ -1 & 5 & 4 \\ 3 & 4 & 6 \end{bmatrix}$.

10. Compute $A\mathbf{x}$ for $A = \begin{bmatrix} 2 & 4 \\ 1 & 3 \end{bmatrix}$, $\mathbf{x} = \begin{bmatrix} 1 \\ -1 \end{bmatrix}$.

11. Verify $G + H = H + G$ for $G = \begin{bmatrix} 1 & 2 \\ 0 & 3 \end{bmatrix}$, $H = \begin{bmatrix} 4 & 1 \\ -2 & 5 \end{bmatrix}$.

12. Classify the matrix (e.g., diagonal, scalar, identity, etc.) $A = \begin{bmatrix} 7 & 0 & 0 \\ 0 & 5 & 0 \\ 0 & 0 & 3 \end{bmatrix}$.

13. Verify $A(\mathbf{u} + \mathbf{v}) = A\mathbf{u} + A\mathbf{v}$ for $A = \begin{bmatrix} 2 & -1 \\ 0 & 3 \end{bmatrix}$, $\mathbf{u} = \begin{bmatrix} 1 \\ 2 \end{bmatrix}$, $\mathbf{v} = \begin{bmatrix} 0 \\ -1 \end{bmatrix}$.

14. Check if $\begin{bmatrix} 1 & 3 & 5 \\ 3 & 2 & 4 \\ 5 & 4 & 6 \end{bmatrix}$ is symmetric.

15. Compute the dot product of $\mathbf{v} = \begin{bmatrix} 2 \\ -1 \end{bmatrix}$, $\mathbf{w} = \begin{bmatrix} 3 \\ 4 \end{bmatrix}$.

16. Compute the cross product of n $\mathbb{R}^3 = \begin{bmatrix} 1 \\ 2 \\ 3 \end{bmatrix}$, $\mathbf{c} = \begin{bmatrix} 0 \\ 1 \\ 0 \end{bmatrix}$.

17. For diagonal matrices $A = \begin{bmatrix} 1 & 0 & 0 \\ 0 & 2 & 0 \\ 0 & 0 & 3 \end{bmatrix}$, $B = \begin{bmatrix} 3 & 0 & 0 \\ 0 & -1 & 0 \\ 0 & 0 & 2 \end{bmatrix}$, compute A^2 and B^2.

18. Compute $(AB)^T$ and $B^T A^T$ for $A = \begin{bmatrix} 1 & 2 \\ 3 & 4 \end{bmatrix}$, $B = \begin{bmatrix} 5 & 6 \\ 7 & 8 \end{bmatrix}$.

19. For a block matrix $A = \begin{bmatrix} I_2 & 0_{2\times 2} \\ B & 2I_2 \end{bmatrix}$, $B = \begin{bmatrix} 1 & 2 \\ 3 & 4 \end{bmatrix}$, compute A^2.

20. Verify that the product is well-defined and compute ABC for $A = \begin{bmatrix} 1 & 2 \end{bmatrix}$, $B = \begin{bmatrix} 3 & 4 \\ 5 & 6 \end{bmatrix}$, $C = \begin{bmatrix} 7 \\ 8 \end{bmatrix}$.

21. Construct nontrivial 3×3 upper and lower triangular matrices, multiply each by I_3, and observe the result.

22. Compute AB for $A = \begin{bmatrix} 1 & 0 \\ 2 & 1 \end{bmatrix}$, $B = \begin{bmatrix} 0 & 1 \\ -1 & 0 \end{bmatrix}$ and interpret the result as a linear transformation (e.g., rotation, reflection, or shear).

Chapter 4
Tensors

Introduction

Tensors are versatile mathematical objects that generalize scalars, vectors, and matrices to higher dimensions, serving as containers for multidimensional data. Scalars (zeroth-order tensors) are single numbers, vectors (first-order tensors) are one-dimensional arrays, and matrices (second-order tensors) are two-dimensional arrays. Higher-order tensors, such as third-order or fourth-order tensors, organize data in three or more dimensions, resembling cubes or hypercubes. Tensors are foundational in computer science (e.g., ML, computer vision), data science (e.g., recommendation systems), and graph theory (e.g., network analysis). Their ability to model complex, multidimensional relationships makes them essential for analyzing systems with multiple interacting variables.

This chapter offers a thorough introduction to tensors, covering their definitions, dimensions, operations, fundamental properties, and various decomposition techniques, all supported by practical Python implementations using NumPy. Designed for undergraduate and early graduate students, it presents concepts with intuitive explanations, step-by-step computations, geometric interpretations, and real-world applications in ML. The chapter also explores advanced topics such as tensor transformations, mode-n products, tensor unfolding, tensor norms, rank estimation, and TT decomposition. Applications span across diverse fields, including ML, computer vision, signal processing, medical imaging, and graph theory. Rich examples, visual illustrations, and exercises are included to reinforce understanding and encourage hands-on learning.

© The Author(s), under exclusive license to Springer Nature Singapore Pte Ltd. 2026   95
Md. Jalil Piran, *Linear Algebra with Applications in Machine Learning*,
https://doi.org/10.1007/978-981-95-5167-5_4

Topics Covered

This chapter is organized as follows:

- **Section** 4.1 **Tensors; Definition and Basics**: Formal definition of tensors by order and shape, examples of scalars, vectors, matrices, and higher-order tensors, visualization as 3D grids or nested grids, and geometric interpretation of tensor transformations.
- **Section** 4.2 **Tensor Operations**: Tensor addition and subtraction, scalar multiplication, tensor product (outer product), tensor contraction, mode-n product, tensor unfolding (matricization), tensor norms (Frobenius and spectral), and tensor transformations under coordinate changes.
- **Section** 4.3 **Tensor Decompositions**: Canonical Polyadic (CP) Decomposition, Tucker decomposition, Higher-Order SVD (HOSVD), TT decomposition, and tensor rank estimation using numerical methods like alternating least squares (ALS).
- **Section** 4.4 **Properties of Tensors**: Symmetric tensors (invariance under index permutation), orthogonal tensors (generalizing orthogonal matrices), and tensor rank (minimum number of rank-one tensors in CP decomposition).
- **Section** 4.5 **Tensors in ML**: Applications in deep learning, computer vision, NLP, and recommendation systems.
- **Summary and Exercises:** Recaps key concepts and provides practical and theoretical problems to reinforce concepts.

All code accompanying the chapter is publicly available at https://github.com/jalil-piran/Linear-Algebra-with-Applications-in-Machine-Learning/blob/main/Chapter_04_Tensors.ipynb.

Acronyms

ML	Machine Learning
NLP	Natural Language Processing
CP	Canonical Polyadic (Decomposition)
CANDECOMP	Canonical Decomposition
PARAFAC	Parallel Factors
HOSVD	Higher-Order Singular Value Decomposition
SVD	Singular Value Decomposition
TT	Tensor Train
ALS	Alternating Least Squares
CNN(s)	Convolutional Neural Network(s)
RGB	Red, Green, Blue
EEG	Electroencephalography
MRI	Magnetic Resonance Imaging
CT	Computed Tomography
3D	Three-Dimensional

4.1 Tensors: Definition and Basics

Tensors extend the concept of matrices to higher dimensions, serving as versatile structures for representing and analyzing multidimensional data across fields like ML, physics, and computer vision. In this book, tensors are treated as multidimensional arrays (coordinate representations), although formally they are multilinear mappings.

Definition
A **tensor** is a mathematical object that generalizes scalars, vectors, and matrices. It is characterized by its order, also called the number of dimensions or, modes. The order of a tensor corresponds to the number of indices required to specify one of its elements. The shape of a tensor is represented by the tuple $(d_1, d_2, \ldots, d_n)$, where d_i denotes the size of the i-th mode (dimension).

For example:

- **Scalar** (zeroth-order tensor): A single value with no direction, e.g., $a = 5$, shape ().
- **Vector** (first-order tensor): An ordered array of numbers, represented by a single index, e.g., $\mathbf{v} = [v_1, v_2, \ldots, v_n]^T$, shape (n).
- **Matrix** (second-order tensor): A rectangular array representing a linear transformation between two vector spaces, e.g., $\mathbf{A}_{ij}$, shape (m, n).
- **Third-order tensor**: A multidimensional array connecting three sets of directions or indices, e.g., $\mathbf{T}_{ijk}$, shape (m, n, p).
- **Fourth-order tensor**: A higher-order structure encoding relationships along four dimensions, e.g., $\mathbf{S}_{ijkl}$, shape (m, n, p, q).
- **Fifth-order tensor**: An extension to five indices capturing interactions across five dimensions, e.g., $\mathbf{U}_{ijklm}$, shape (m, n, p, q, r).

Figure 4.1 visually represents these structures, with the scalar as a single block, the vector as a row of blocks, the matrix as a grid of blocks, and the tensor as a 3D stack of blocks.

Example 4.1 Consider a third-order tensor $\mathbf{T} \in \mathbb{R}^{2 \times 2 \times 2}$, which can be visualized as a collection of 2×2 matrices stacked along the third dimension:

$$\mathbf{T} = \left[\begin{bmatrix} 1 & 2 \\ 3 & 4 \end{bmatrix}, \begin{bmatrix} 5 & 6 \\ 7 & 8 \end{bmatrix} \right].$$

In tensor notation, each element is specified by three indices: $\mathbf{T}_{ijk}$, where:

$$\mathbf{T}(:,:,1) = \begin{bmatrix} 1 & 2 \\ 3 & 4 \end{bmatrix}, \quad \mathbf{T}(:,:,2) = \begin{bmatrix} 5 & 6 \\ 7 & 8 \end{bmatrix},$$

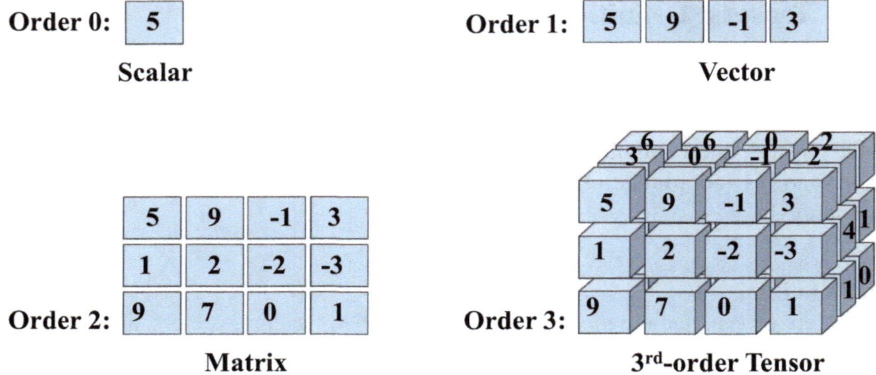

Fig. 4.1 Tensors with different orders

- **First index** (i): Layer or depth (1st dimension), selects which matrix layer.
- **Second index** (j): Row number (2nd dimension), selects the row within the matrix.
- **Third index** (k): Column number (3rd dimension), selects the column within the matrix.

For this tensor:

- $\mathbf{T}_{111} = 1$ (layer 1, row 1, column 1—top-left element of first matrix)
- $\mathbf{T}_{112} = 2$ (layer 1, row 1, column 2—top-right element of first matrix)
- $\mathbf{T}_{221} = 7$ (layer 2, row 2, column 1—bottom-left element of second matrix)
- $\mathbf{T}_{222} = 8$ (layer 2, row 2, column 2—bottom-right element of second matrix).

This indexing convention is standard in ML libraries like TensorFlow and PyTorch, where the first index typically represents batch/layer depth, the second represents spatial dimensions (height), and the third represents spatial dimensions (width) for image data, or time for sequences.

Python Example: 3D Tensor Indexing

```python
import numpy as np


T = np.array([[[1, 2], [3, 4]], [[5, 6], [7,
    8]]])

print(f"T[1,1,1] = {T[1,1,1]}")
print(f"T[0,0,0] = {T[0,0,0]}")
print(f"T[1,1,0] = {T[1,1,0]}")
```

Output:

```
T[1,1,1]  =  8
T[0,0,0]  =  1
T[1,1,1]  =  7
```

Note that in Python (NumPy), indexing starts from 0, so $\mathbf{T}_{111}$ corresponds to $\mathbf{T}[0, 0, 0]$, $\mathbf{T}_{221}$ corresponds to $\mathbf{T}[1, 1, 0]$, and so on.

Fourth-Order Tensor
A fourth-order tensor $\mathbf{S} \in \mathbb{R}^{2\times2\times2\times2}$ represents a collection of $2 \times 2 \times 2$ tensors, with elements $\mathbf{S}_{ijkl}$.
Fifth-Order Tensor
A fifth-order tensor $\mathbf{U} \in \mathbb{R}^{2\times2\times2\times2\times2}$ might represent a video dataset with dimensions (sample, height, width, channels, time), capturing complex spatiotemporal interactions.

4.1.1 Tensor Notation

Tensors are denoted by bold uppercase letters (e.g., $\mathbf{T}$), while their individual elements are accessed using multiple indices that correspond to each dimension. For a third-order tensor, elements are denoted as $\mathbf{T}_{ijk}$, where i, j, and k represent the indices for the first, second, and third dimensions respectively. This multi-index notation generalizes the single subscript for vectors and double subscripts for matrices.

The *Einstein summation convention*, widely used in tensor calculus and physics, provides a compact notation by implying summation over repeated indices. Consider matrix multiplication:

$$C_{ij} = \sum_{k=1}^{n} A_{ik} B_{kj}.$$

Under the Einstein convention, the explicit summation symbol is omitted, and repeated indices (here k) are automatically summed over:

$$C_{ij} = A_{ik} B_{kj},$$

where the repeated index k indicates summation from 1 to n.

For higher-order tensors, this convention extends naturally. For example, contracting a third-order tensor $\mathbf{T}_{ijk}$ with a vector $\mathbf{v}_k$ yields:

$$R_{ij} = \mathbf{T}_{ijk}\mathbf{v}_k = \sum_{k=1}^{n} T_{ijk} v_k,$$

where the repeated index k is summed over, producing a matrix result.

Example 4.2 For a fourth-order tensor $\mathbf{S}$ and vectors $\mathbf{u}, \mathbf{v}, \mathbf{w}, \mathbf{x}$, a quadruple contraction yields a scalar:

$$s = \mathbf{S}_{ijkl}\mathbf{u}_i\mathbf{v}_j\mathbf{w}_k\mathbf{x}_l.$$

Let $\mathbf{S}_{ijkl}$ be a $2\times2\times2\times2$ tensor where all entries are 1. Let the vectors be:

$$\mathbf{u} = [1, 2], \quad \mathbf{v} = [0, 1], \quad \mathbf{w} = [1, 1], \quad \mathbf{x} = [2, 1].$$

Using Einstein summation:

$$s = \sum_{i=1}^{2}\sum_{j=1}^{2}\sum_{k=1}^{2}\sum_{l=1}^{2}\mathbf{S}_{ijkl}\mathbf{u}_i\mathbf{v}_j\mathbf{w}_k\mathbf{x}_l.$$

Since $\mathbf{S}_{ijkl} = 1$ for all indices, this reduces to:

$$s = \sum_{i=1}^{2}\mathbf{u}_i\sum_{j=1}^{2}\mathbf{v}_j\sum_{k=1}^{2}\mathbf{w}_k\sum_{l=1}^{2}\mathbf{x}_l.$$

Computing each sum:

$$\sum\mathbf{u}_i = 1 + 2 = 3,$$

$$\sum\mathbf{v}_j = 0 + 1 = 1,$$

$$\sum\mathbf{w}_k = 1 + 1 = 2,$$

$$\sum\mathbf{x}_l = 2 + 1 = 3.$$

Thus,

$$s = 3 \times 1 \times 2 \times 3 = 18.$$

4.2 Tensor Operations

Tensor operations generalize vector and matrix operations. We include additional operations like mode-n product, tensor unfolding, tensor norms, and tensor transformations.

4.2.1 Tensor Addition and Subtraction

Tensor addition and subtraction are element-wise, requiring identical shapes. For $\mathbf{A}, \mathbf{B} \in \mathbb{R}^{m \times n \times p}$:

$$\mathbf{C}_{ijk} = \mathbf{A}_{ijk} + \mathbf{B}_{ijk},$$

$$\mathbf{D}_{ijk} = \mathbf{A}_{ijk} - \mathbf{B}_{ijk}.$$

Example 4.3 For:

$$\mathbf{A} = \left[\begin{bmatrix} 1 & 2 \\ 3 & 4 \end{bmatrix}, \begin{bmatrix} 5 & 6 \\ 7 & 8 \end{bmatrix} \right], \quad \mathbf{B} = \left[\begin{bmatrix} 0 & 1 \\ 1 & 0 \end{bmatrix}, \begin{bmatrix} 1 & 0 \\ 0 & 1 \end{bmatrix} \right],$$

we compute $\mathbf{A} + \mathbf{B}$ as follows.

$$\mathbf{C}_{111} = 1 + 0 = 1$$

$$\mathbf{C}_{112} = 2 + 1 = 3$$

$$\mathbf{C}_{121} = 3 + 1 = 4$$

$$\mathbf{C}_{122} = 4 + 0 = 4$$

$$\mathbf{C}_{211} = 5 + 1 = 6.$$

And so on, yielding $\mathbf{C} = \mathbf{A} + \mathbf{B}$:

$$\mathbf{C} = \left[\begin{bmatrix} 1 & 3 \\ 4 & 4 \end{bmatrix}, \begin{bmatrix} 6 & 6 \\ 7 & 9 \end{bmatrix} \right].$$

Subtraction follows similarly, $\mathbf{D} = \mathbf{A} - \mathbf{B}$:

$$\mathbf{D} = \left[\begin{bmatrix} 1 & 1 \\ 2 & 4 \end{bmatrix}, \begin{bmatrix} 4 & 6 \\ 7 & 7 \end{bmatrix} \right].$$

Python Example: Tensor Addition and Subtraction

```python
import numpy as np

A = np.array([[[1, 2], [3, 4]], [[5, 6], [7, 8]]])
B = np.array([[[0, 1], [1, 0]], [[1, 0], [0, 1]]])

```

```
 6  C = A + B
 7  print("Tensor A + B:\n", C)
 8
 9  D = A - B
10  print("Tensor A - B:\n", D)
```

Output:

```
Tensor A + B:
[[[1 3]
[4 4]]
[[6 6]
[7 9]]]
Tensor A - B:
[[[1 1]
[2 4]]
[[4 6]
[7 7]]]
```

Tensor addition combines data from multiple sensors (e.g., temperature, pressure, humidity) represented as third-order tensors to average or integrate measurements for ML model training.

Example 4.4 Consider the following tensors:

$$A = \left[\begin{bmatrix} 20 & 21 \\ 22 & 23 \end{bmatrix}, \begin{bmatrix} 24 & 25 \\ 26 & 27 \end{bmatrix} \right], \quad B = \left[\begin{bmatrix} 19 & 20 \\ 21 & 22 \end{bmatrix}, \begin{bmatrix} 23 & 24 \\ 25 & 26 \end{bmatrix} \right].$$

Their sum $A + B$ is:

$$C = \left[\begin{bmatrix} 39 & 41 \\ 43 & 45 \end{bmatrix}, \begin{bmatrix} 47 & 49 \\ 51 & 53 \end{bmatrix} \right].$$

4.2.2 Tensor-Scalar Multiplication

Scalar multiplication scales every element of a tensor by a constant $s \in \mathbb{R}$:

$$\mathbf{B}_{ijk} = s \cdot \mathbf{A}_{ijk}.$$

Example 4.5 For $s = 2$, and:

$$A = \left[\begin{bmatrix} 1 & 2 \\ 3 & 4 \end{bmatrix}, \begin{bmatrix} 5 & 6 \\ 7 & 8 \end{bmatrix} \right],$$

we compute $s\mathbf{A}$:

$$\mathbf{B}_{111} = 2 \cdot 1 = 2$$

$$\mathbf{B}_{112} = 2 \cdot 2 = 4.$$

And so on, yielding:

$$s\mathbf{A} = \left[\begin{bmatrix} 2 & 4 \\ 5 & 8 \end{bmatrix}, \begin{bmatrix} 10 & 12 \\ 14 & 16 \end{bmatrix}\right].$$

Python Example: Scalar Multiplication

```python
import numpy as np

A = np.array([[[1, 2], [3, 4]], [[5, 6], [7,
    8]]])
s = 2

B = s * A
print("Scalar multiplication s * A:\n", B)
```

Output:

```
Scalar multiplication s * A:
[[[ 2   4]
 [ 6   8]]
 [[10 12]
 [14 16]]]
```

In ML, scalar multiplication is used to rescale features during preprocessing, for example when adjusting feature magnitudes for neural network training.

Example 4.6 For a feature tensor $\mathbf{F} \in \mathbb{R}^{3 \times 3 \times 3}$, scaling by a factor $s = 1.5$ normalizes the feature magnitudes to improve model convergence.

4.2.3 *Tensor Product and Kronecker Product*

The tensor product (or outer product) of two tensors produces a higher-order tensor whose order is the sum of the orders of the operands. For vectors a in $\mathbb{R}^m$ and b in $\mathbb{R}^n$, the outer product $a \circ b$ yields a matrix in $\mathbb{R}^{m \times n}$. For matrices A in $\mathbb{R}^{m_1 \times n_1}$ and B in $\mathbb{R}^{m_2 \times n_2}$, the full outer product yields a fourth-order tensor in $\mathbb{R}^{m_1 \times n_1 \times m_2 \times n_2}$. A closely related and widely used operation is the Kronecker product, which arranges the scaled copies of B by entries of A into a single block matrix $A \otimes B$ in $\mathbb{R}^{(m_1 m_2) \times (n_1 n_2)}$, with elements:

$$(\mathbf{A} \otimes \mathbf{B})_{i_1 i_2, j_1 j_2} = \mathbf{A}_{i_1 j_1} \cdot \mathbf{B}_{i_2 j_2}.$$

Example 4.7 For:

$$\mathbf{A} = \begin{bmatrix} 1 & 2 \\ 3 & 4 \end{bmatrix}, \quad \mathbf{B} = \begin{bmatrix} 0 & 1 \\ 1 & 0 \end{bmatrix},$$

we compute $\mathbf{A} \otimes \mathbf{B}$:

$$\mathbf{A} \otimes \mathbf{B} = \begin{bmatrix} 1 \cdot \begin{bmatrix} 0 & 1 \\ 1 & 0 \end{bmatrix} & 2 \cdot \begin{bmatrix} 0 & 1 \\ 1 & 0 \end{bmatrix} \\ 3 \cdot \begin{bmatrix} 0 & 1 \\ 1 & 0 \end{bmatrix} & 4 \cdot \begin{bmatrix} 0 & 1 \\ 1 & 0 \end{bmatrix} \end{bmatrix} = \begin{bmatrix} \begin{bmatrix} 0 & 1 \\ 1 & 0 \end{bmatrix} & \begin{bmatrix} 0 & 2 \\ 2 & 0 \end{bmatrix} \\ \begin{bmatrix} 0 & 3 \\ 3 & 0 \end{bmatrix} & \begin{bmatrix} 0 & 4 \\ 4 & 0 \end{bmatrix} \end{bmatrix}.$$

Flattened, this is a 4×4 matrix:

$$\begin{bmatrix} 0 & 1 & 0 & 2 \\ 1 & 0 & 2 & 0 \\ 0 & 3 & 0 & 4 \\ 3 & 0 & 4 & 0 \end{bmatrix}.$$

Python Example: Tensor Product

```python
import numpy as np

A = np.array([[1, 2], [3, 4]])
B = np.array([[0, 1], [1, 0]])

C = np.kron(A, B)
print(C)
```

Output:

```
[[0 1 0 2]
 [1 0 2 0]
 [0 3 0 4]
 [3 0 4 0]]
```

In ML, the tensor product is used to construct rank-one components in tensor decompositions for feature interactions in recommendation systems.

4.2.4 Tensor Contraction

Tensor contraction reduces a tensor's order by summing over repeated indices. For a third-order tensor $\mathbf{T}_{ijk}$ and a vector $\mathbf{v}_k$:

$$\mathbf{R}_{ij} = \sum_k \mathbf{T}_{ijk}\mathbf{v}_k.$$

Example 4.8 For:

$$\mathbf{T} = \left[\begin{bmatrix} 1 & 1 & 1 \\ 1 & 1 & 1 \end{bmatrix}, \begin{bmatrix} 1 & 1 & 1 \\ 1 & 1 & 1 \end{bmatrix}\right], \quad \mathbf{v} = \begin{bmatrix} 1 \\ 2 \\ 3 \end{bmatrix},$$

contract over the third index:

$$\mathbf{R}_{11} = 1 \cdot 1 + 1 \cdot 2 + 1 \cdot 3 = 6$$

Similarly, $\mathbf{R}_{ij} = 6$, yielding:

$$\mathbf{R} = \begin{bmatrix} 6 & 6 \\ 6 & 6 \end{bmatrix}.$$

Python Example: Tensor Contraction

```python
import numpy as np

T = np.ones((2, 2, 3))
v = np.array([1, 2, 3])

R = np.tensordot(T, v, axes=([2], [0]))
print("Contraction result:\n", R)
```

Output:

```
[[6. 6.]
 [6. 6.]]
```

For a fourth-order tensor $\mathbf{S}_{ijkl}$ and matrix $\mathbf{A}_{kl}$:

$$\mathbf{R}_{ij} = \sum_k \sum_l \mathbf{S}_{ijkl}\mathbf{A}_{kl}.$$

Example 4.9 For:

$$\mathbf{S} = \text{all ones, shape } (2, 2, 2, 2), \quad \mathbf{A} = \begin{bmatrix} 1 & 2 \\ 3 & 4 \end{bmatrix},$$

compute the double contraction:

$$\mathbf{R}_{11} = \sum_{k} \sum_{l} \mathbf{S}_{11kl} \mathbf{A}_{kl} = 1 \cdot 1 + 1 \cdot 2 + 1 \cdot 3 + 1 \cdot 4 = 10.$$

Similarly, $\mathbf{R}_{ij} = 10$, yielding:

$$\mathbf{R} = \begin{bmatrix} 10 & 10 \\ 10 & 10 \end{bmatrix}.$$

Python Example: Double Contraction

```python
import numpy as np

S = np.ones((2, 2, 2, 2))    # Shape (2, 2, 2, 2)
A = np.array([[1, 2], [3, 4]])    # Shape (2, 2)

R = np.tensordot(S, A, axes=([2, 3], [0, 1]))
print("Double contraction result:\n", R)
```

Output:

```
[[10. 10.]
 [10. 10.]]
```

In ML, tensor contraction is used in attention mechanisms to compute weighted sums of features across multiple dimensions in transformer models.

4.2.5 Mode-n Product

The mode-n product generalizes matrix multiplication to higher-order tensors. The mode-n product multiplies a tensor by a matrix along its n-th mode. For $\mathbf{T} \in \mathbb{R}^{I_1 \times I_2 \times I_3}$ and $\mathbf{A} \in \mathbb{R}^{J \times I_n}$:

$$(\mathbf{T} \times_n \mathbf{A})_{i_1, \ldots, i_{n-1}, j, i_{n+1}, \ldots, i_N} = \sum_{i_n} \mathbf{T}_{i_1, \ldots, i_N} \mathbf{A}_{j, i_n}.$$

Example 4.10 For $\mathbf{T} \in \mathbb{R}^{2 \times 2 \times 2}$, $\mathbf{A} = \begin{bmatrix} 1 & 2 \\ 3 & 4 \end{bmatrix}$, the mode-1 product $\mathbf{T} \times_1 \mathbf{A}$ transforms the first mode, yielding a tensor of shape $(2, 2, 2)$.

Python Example: Mode-n Product

```python
import numpy as np
import tensorly as tl
#pip install tensorly (if not installed)

T = np.array([[[1, 2], [3, 4]], [[5, 6], [7,
    8]]])
A = np.array([[1, 2], [3, 4]])

result = tl.tenalg.mode_dot(T, A, mode=0)
print("Mode-1 product result shape:",
    result.shape)
```

Output:

```
(2, 2, 2)
```

In tensor regression, mode-n products are essential operations that allow models to incorporate and predict multidimensional outcomes while preserving the inherent structure of the data. Unlike traditional regression, which flattens multidimensional inputs into vectors or matrices, tensor regression leverages the full dimensionality of the data tensor by applying mode-wise transformations with corresponding coefficient matrices. This approach is particularly useful in fields like neuroimaging, where brain activity is naturally represented as high-order tensors (e.g., subject $\times$ region $\times$ time). By using mode-n products, tensor regression captures complex interactions across multiple modes, leading to more interpretable models and improved predictive performance in high-dimensional applications.

4.2.6 Tensor Unfolding

Tensor unfolding (matricization) rearranges a tensor into a matrix along one mode. For $\mathbf{T} \in \mathbb{R}^{I_1 \times I_2 \times I_3}$, the mode-1 unfolding is $\mathbf{T}_{(1)} \in \mathbb{R}^{I_1 \times (I_2 I_3)}$.

Example 4.11 For:

$$\mathbf{T} = \left[\begin{bmatrix} 1 & 2 \\ 3 & 4 \end{bmatrix}, \begin{bmatrix} 5 & 6 \\ 7 & 8 \end{bmatrix} \right],$$

mode-1 unfolding yields:

$$\mathbf{T}_{(1)} = \begin{bmatrix} 1\ 2\ 3\ 4 \\ 5\ 6\ 7\ 8 \end{bmatrix}.$$

Python Example: Tensor Unfolding

```python
import numpy as np
import tensorly as tl

T = np.array([[[1, 2], [3, 4]], [[5, 6], [7,
    8]]])

T1 = tl.unfold(T, mode=0)
print("Mode -1 unfolding:\n", T1)
```

Output:

```
[[1 2 3 4]
 [5 6 7 8]]
```

This transformation enables the application of powerful matrix techniques such as SVD (see Chap. 11 for more details), which are well-established for reducing dimensionality and extracting latent structures. By unfolding a tensor, one can isolate and analyze variations along particular dimensions, facilitating efficient compression and noise reduction in high-dimensional data. This approach is widely used in areas like computer vision and ML, where it helps simplify complex datasets while preserving essential information, thereby enabling more effective feature extraction and downstream analysis.

4.2.7   Tensor Norms

Tensor norms generalize vector and matrix norms to higher-dimensional arrays, providing essential measures of "magnitude" for multidimensional data structures. These norms are crucial for optimization, regularization, and error analysis in ML applications involving complex data like images, videos, and feature maps in deep neural networks. They quantify the overall "size" of a tensor while maintaining compatibility with lower-dimensional norms.

The most commonly used tensor norm is the **Frobenius norm**, which extends the Euclidean norm to tensors by treating the entire tensor as a flattened vector of all its elements:

$$\|\mathbf{T}\|_F = \sqrt{\sum_{i=1}^{I}\sum_{j=1}^{J}\sum_{k=1}^{K} \mathbf{T}_{ijk}^2},$$

where $\mathbf{T} \in \mathbb{R}^{I \times J \times K}$ is a third-order tensor. This norm is computationally efficient, invariant under orthogonal transformations, and corresponds to the Euclidean norm when the tensor is flattened into a vector, making it particularly suitable for gradient-based optimization and loss function design.

Example 4.12 For:

$$\mathbf{T} = \left[\begin{bmatrix} 1 & 2 \\ 3 & 4 \end{bmatrix}, \begin{bmatrix} 5 & 6 \\ 7 & 8 \end{bmatrix}\right],$$

$$\|\mathbf{T}\|_F = \sqrt{1^2 + 2^2 + 3^2 + 4^2 + 5^2 + 6^2 + 7^2 + 8^2} = \sqrt{204} \approx 14.28.$$

Python Example: Frobenius Norm

```python
import numpy as np

T = np.array([[[1, 2], [3, 4]], [[5, 6], [7, 8]]])

norm = np.linalg.norm(T)
print(f"Frobenius norm:{norm:.2f}")
```

Output:

```
Frobenius norm: 14.28
```

Besides the Frobenius norm, several other tensor norms exist, including Nuclear Norm, Spectral Norm, Entrywise ℓ_p Norms, Mode-n Norms, Spectral p-Norms, and Tensor p-Spectral Norm. Each norm serves different purposes in optimization, regularization, and analysis of multidimensional data. Each is discussed in turn below.

Nuclear Norm (Trace Norm)

The nuclear norm promotes low-rank solutions and serves as a convex relaxation of the tensor rank, commonly used in tensor completion and collaborative filtering. Singular values, obtained from the SVD of the unfolded tensor, represent the importance of each mode in capturing the tensor's structure. By summing these singular values, the nuclear norm encourages solutions that retain only the most significant components, effectively reducing dimensionality while preserving essential information. We will study singular values and SVD in Chap. 11 in detail.

$$\|\mathbf{T}\|_* = \sum_i \sigma_i(\mathbf{T}_{(1)}),$$

where $\sigma_i(\mathbf{T}_{(1)})$ are the singular values of the mode-1 unfolding.

Example 4.13 For a tensor $\mathbf{T} \in \mathbb{R}^{2 \times 2 \times 2}$ with singular values [3, 1] from mode-1 unfolding, $\|\mathbf{T}\|_* = 3 + 1 = 4$.

Spectral Norm

The spectral norm measures the maximum "gain" or amplification of the tensor as a linear operator, crucial for stability analysis and bounding error propagation.

$$\|\mathbf{T}\|_2 = \sigma_{max}(\mathbf{T}_{(1)}),$$

where $\sigma_{\max}(\mathbf{T}_{(1)})$ is the largest singular value of the mode-1 unfolding.

Example 4.14 For the same tensor with singular values [3, 1], $\|\mathbf{T}\|_2 = 3$, representing the maximum stretch factor.

Entrywise ℓ_p Norms

These generalize vector p-norms to tensors, with ℓ_1 promoting sparsity and ℓ_∞ measuring the largest entry magnitude.

$$\|\mathbf{T}\|_p = \left(\sum_{i,j,k} |\mathbf{T}_{ijk}|^p \right)^{1/p}, \quad p \geq 1.$$

Example 4.15 For $\mathbf{T} = [[[1, 2], [3, 4]], [[5, 6], [7, 8]]]$, the ℓ_1 norm is $\|\mathbf{T}\|_1 = |1| + |2| + |3| + |4| + |5| + |6| + |7| + |8| = 36$.

Mode-n Norms

Mode-n norms apply matrix norms to specific unfoldings, useful for structured sparsity and multi-task learning applications.

$$\|\mathbf{T}\|_{2,1} = \sum_j \|\mathbf{T}_{(1)[:,j,:]}\|_2,$$

summing the ℓ_2 norms of columns in the mode-1 unfolding.

Example 4.16 If mode-1 unfolding columns have ℓ_2 norms [2, 3], then $\|\mathbf{T}\|_{2,1} = 2 + 3 = 5$.

Spectral p-Norms

These generalize nuclear and spectral norms, providing flexible regularization for different sparsity patterns.

$$\|\mathbf{T}\|_p = \left(\sum_i \sigma_i(\mathbf{T}_{(1)})^p \right)^{1/p}.$$

Example 4.17 For singular values [3, 1] with $p = 2$, $\|\mathbf{T}\|_2 = \sqrt{3^2 + 1^2} = \sqrt{10} \approx 3.16$.

Tensor p-Spectral Norm

The tensor p-spectral norm combines entrywise and spectral properties for structured regularization in multilinear models.

$$\|\mathbf{T}\|_{p,q} = \left(\sum_i \left(\sum_j |\mathbf{T}_{ij}|^p \right)^{q/p} \right)^{1/q}.$$

For a matrix treated as 2D tensor with $p = 2, q = 1$, this reduces to the $\ell_{2,1}$ norm for group sparsity.

These norms extend matrix norm concepts to higher dimensions, each serving specialized roles in ML algorithms for dimensionality reduction, regularization, and robust optimization.

4.2.8 Tensor Transformations

Tensors transform under coordinate changes. For a second-order tensor $\mathbf{A}$, under a basis change $\mathbf{Q}$:

$$\mathbf{A}' = \mathbf{Q}^T \mathbf{A} \mathbf{Q}.$$

For a third-order tensor $\mathbf{T}$, with transformation matrices $\mathbf{Q}_1, \mathbf{Q}_2, \mathbf{Q}_3$:

$$\mathbf{T}'_{ijk} = \sum_{p,q,r} \mathbf{Q}_{1ip} \mathbf{Q}_{2jq} \mathbf{Q}_{3kr} \mathbf{T}_{pqr}.$$

Example 4.18 For a rotation matrix $\mathbf{Q} = \begin{bmatrix} \cos\theta & -\sin\theta \\ \sin\theta & \cos\theta \end{bmatrix}$, a second-order tensor $\mathbf{A}$ transforms as above, preserving data properties in ML applications.

Python Example: Tensor Transformation

```python
import numpy as np

A = np.array([[1, 2], [2, 3]])

theta = np.pi / 4
Q = np.array([
    [np.cos(theta), -np.sin(theta)],
    [np.sin(theta),  np.cos(theta)]
    ])

A_prime = Q.T @ A @ Q

np.set_printoptions(precision=2, suppress=True)
print("Transformed tensor:\n", A_prime)
```

Output:

```
[[ 4.  1. ]
 [ 1.  0.]]
```

Tensor transformations are crucial in computer vision for data augmentation techniques, such as rotation and scaling of images, which help improve the robustness and generalization of ML models.

4.3 Tensor Decompositions

Tensor decompositions factorize tensors into simpler components, reducing dimensionality and computational complexity.

4.3.1 CP Decomposition

The Canonical Polyadic (CP) decomposition, also known as CANDECOMP/-PARAFAC, approximates a tensor as a sum of rank-one tensors:

$$\mathcal{T}_{ijk} \approx \sum_{r=1}^{R} a_{ir}^{(1)} \, a_{jr}^{(2)} \, a_{kr}^{(3)},$$

where R is the rank of the decomposition, and $a_{ir}^{(n)}$ are the elements of the factor matrices along each mode n.

This dual nomenclature arises from its historical development: 'PARAFAC' (Parallel Factors) was introduced by Harshman in 1970, while 'CANDECOMP' (Canonical Decomposition) was independently proposed by Carroll and Chang in the same year, independently formalizing the same sum-of-rank-one model. The terms were unified as CP to acknowledge both origins, reflecting its foundational role in chemometrics and psychometrics.

Example 4.19 For a tensor $\mathbf{T} \in \mathbb{R}^{2 \times 2 \times 2}$:

$$\mathcal{T} \approx \mathbf{a}_1 \circ \mathbf{b}_1 \circ \mathbf{c}_1 + \mathbf{a}_2 \circ \mathbf{b}_2 \circ \mathbf{c}_2,$$

where $\circ$ denotes the outer product, yielding rank-one tensors summed to reconstruct the original data.

CP decomposition is a powerful tensor factorization technique widely used in ML to disentangle complex multi-way data, such as recommendation system interactions involving users, items, and contextual factors. By decomposing the tensor into rank-one components, CP isolates underlying latent patterns, enabling the identification of distinct user preferences or item characteristics. This enhances interpretability by separating overlapping signals and mitigating noise, leading to more accurate predictions and deeper insights into user behavior. Its efficacy with multidimensional datasets makes CP particularly valuable in personalized advertising, e-commerce, and content recommendation, where mixed signals are prevalent.

4.3.2   Tucker Decomposition

Tucker decomposition is an effective tensor factorization method used to compress high-dimensional data such as color images, which are naturally represented as three-dimensional tensors with dimensions corresponding to height, width, and color channels. By decomposing the image tensor into a smaller core tensor multiplied by factor matrices along each mode, Tucker decomposition achieves significant data reduction while preserving essential structural and color information. This compression reduces storage requirements and computational costs, making it valuable for image transmission and real-time processing applications in computer vision. Moreover, the flexible multilinear rank selection in Tucker decomposition allows for a tunable trade-off between compression ratio and reconstruction quality, enabling efficient encoding without severe loss of visual fidelity.

Tucker decomposition factorizes a tensor into a core tensor and factor matrices:

$$\mathbf{T}_{ijk} \approx \sum_{p,q,r} \mathbf{G}_{pqr} \mathbf{A}_{ip} \mathbf{B}_{jq} \mathbf{C}_{kr}.$$

For $\mathbf{T} \in \mathbb{R}^{2 \times 2 \times 2}$, Tucker decomposition involves a core tensor $\mathbf{G} \in \mathbb{R}^{2 \times 2 \times 2}$ and matrices $\mathbf{A}, \mathbf{B}, \mathbf{C} \in \mathbb{R}^{2 \times 2}$.

Python Example: Tucker Decomposition

```python
import numpy as np
import tensorly as tl
from tensorly.decomposition import tucker

# 3rd-order tensor
T = np.array([[[1, 2], [3, 4]], [[5, 6], [7,
    8]]])

# Tucker decomposition
core, factors = tucker(T, rank=[2, 2, 2])
reconstructed = tl.tucker_to_tensor((core,
    factors))
print("Reconstructed tensor shape:",
    reconstructed.shape)
```

Output:

```
          Reconstructed tensor shape: (2, 2, 2)
```

4.3.3 HOSVD

HOSVD decomposes a tensor into a core tensor and orthogonal factor matrices:

$$\mathbf{T}_{ijk} = \sum_{p,q,r} \mathbf{S}_{pqr} \mathbf{U}_{ip} \mathbf{V}_{jq} \mathbf{W}_{kr}.$$

For $\mathbf{T} \in \mathbb{R}^{3 \times 3 \times 3}$, HOSVD computes orthogonal matrices via SVD along each mode.

Python Example: HOSVD

```python
import numpy as np
import tensorly as tl
from tensorly.decomposition import tucker
```

```
4
5  # 3rd-order tensor
6  T = np.random.rand(3, 3, 3)
7
8  # HOSVD
9  core, factors = tucker(T, rank=[3, 3, 3])
10 reconstructed = tl.tucker_to_tensor((core,
       factors))
11 print("Reconstructed tensor shape:",
       reconstructed.shape)
```

Output:

```
        Reconstructed tensor shape: (3, 3, 3)
```

4.3.4 TT Decomposition

TT decomposition represents a tensor as a sequence of third-order tensors:

$$\mathbf{T}_{i_1 i_2 \ldots i_N} \approx \sum_{r_1, r_2, \ldots, r_{N-1}} \mathbf{G}^{(1)}_{r_1 i_1 r_2} \mathbf{G}^{(2)}_{r_2 i_2 r_3} \cdots \mathbf{G}^{(N)}_{r_{N-1} i_N r_N}.$$

For a fourth-order tensor $\mathbf{T} \in \mathbb{R}^{2 \times 2 \times 2 \times 2}$, TT decomposition yields a sequence of cores, reducing storage for high-dimensional tensors in deep learning.

4.3.5 Tensor Rank Estimation

Unlike matrix rank, tensor rank is not uniquely defined and is generally difficult to compute. Estimating the rank of a tensor (minimum number of rank-one tensors in CP decomposition) is challenging and often involves iterative optimization.
For a third-order tensor, numerical methods like alternating least squares (ALS) estimate the CP rank by minimizing the reconstruction error.

Python Example: CP Decomposition with Prescribed Rank

```
1  import numpy as np
2  import tensorly as tl
3  from tensorly.decomposition import parafac
```

```
 4
 5  # 3rd-order tensor
 6  T = np.random.rand(3, 3, 3)
 7
 8  # CP decomposition with rank estimation
 9  weights, factors = parafac(T, rank=3)
10  print("Estimated rank:", len(weights))
```

Output:

```
Estimated rank: 3
```

4.4 Properties of Tensors

The following properties generalize familiar matrix properties to higher-order tensors:

- **Symmetric Tensor**: A tensor is called symmetric if its entries remain unchanged under any permutation of its indices. For example, a third-order tensor $\mathbf{T}$ is symmetric if $T_{ijk} = T_{ikj} = T_{jik} = T_{jki} = T_{kij} = T_{kji}$ for all indices i, j, k. This property generalizes the concept of symmetric matrices to higher-order tensors and is important in applications such as kernel methods in ML where symmetry reflects data constraints.

  Consider a symmetric third-order tensor $\mathbf{T} \in \mathbb{R}^{2 \times 2 \times 2}$ where $T_{ijk} = T_{ikj} = T_{jik} = T_{jki} = T_{kij} = T_{kji}$. Define the tensor with the following independent elements:

$$\mathbf{T}_{111} = 1, \quad \mathbf{T}_{112} = \mathbf{T}_{121} = \mathbf{T}_{211} = 2,$$

$$\mathbf{T}_{122} = \mathbf{T}_{212} = \mathbf{T}_{221} = 3, \quad \mathbf{T}_{222} = 4.$$

  This creates a tensor with only 4 independent parameters instead of 8, demonstrating the efficiency of symmetry. The complete tensor structure becomes:

$$\mathbf{T} = \left[\begin{bmatrix} 1 & 2 \\ 2 & 3 \end{bmatrix}, \begin{bmatrix} 2 & 3 \\ 3 & 4 \end{bmatrix} \right],$$

  where each matrix layer respects the symmetry constraint. Verification shows $\mathbf{T}_{112} = \mathbf{T}_{121} = \mathbf{T}_{211} = 2$ and $\mathbf{T}_{122} = \mathbf{T}_{212} = \mathbf{T}_{221} = 3$, confirming the permutation invariance required for symmetric tensors.

- **Orthogonal Tensor**: For a second-order tensor $\mathbf{Q}$, orthogonality means that its transpose is also its inverse, i.e., $\mathbf{Q}^T \mathbf{Q} = \mathbf{I}$, where $\mathbf{I}$ is the identity matrix. Such tensors represent rotations or reflections that preserve lengths and angles in Euclidean space, and they are fundamental in data augmentation, computer vision, and neural network architectures.

Consider a 2×2 rotation matrix by 90 degrees counterclockwise:

$$\mathbf{Q} = \begin{bmatrix} 0 & -1 \\ 1 & 0 \end{bmatrix}.$$

Verify orthogonality by computing $\mathbf{Q}^T \mathbf{Q}$:

$$\mathbf{Q}^T = \begin{bmatrix} 0 & 1 \\ -1 & 0 \end{bmatrix},$$

$$\mathbf{Q}^T \mathbf{Q} = \begin{bmatrix} 0 & 1 \\ -1 & 0 \end{bmatrix} \begin{bmatrix} 0 & -1 \\ 1 & 0 \end{bmatrix} = \begin{bmatrix} 0 \cdot 0 + 1 \cdot 1 & 0 \cdot (-1) + 1 \cdot 0 \\ -1 \cdot 0 + 0 \cdot 1 & -1 \cdot (-1) + 0 \cdot 0 \end{bmatrix} = \begin{bmatrix} 1 & 0 \\ 0 & 1 \end{bmatrix} = \mathbf{I}.$$

The length preservation property can be verified with any vector. For $\mathbf{v} = [1, 0]^T$:

$$\|\mathbf{Q}\mathbf{v}\| = \left\| \begin{bmatrix} 0 & -1 \\ 1 & 0 \end{bmatrix} \begin{bmatrix} 1 \\ 0 \end{bmatrix} \right\| = \left\| \begin{bmatrix} 0 \\ 1 \end{bmatrix} \right\| = \sqrt{0^2 + 1^2} = 1 = \|\mathbf{v}\|.$$

- **Rank**: The rank of a tensor is defined as the minimum number of rank-one tensors (outer products of vectors) needed to express it exactly in a CP decomposition. Unlike matrix rank, tensor rank can be more complex and is crucial for understanding tensor complexity and compressibility, with applications in data analysis, ML, and model compression.

  Consider a third-order tensor $\mathbf{T} \in \mathbb{R}^{2 \times 2 \times 2}$ with CP rank 2, expressed as the sum of two rank-one tensors:

$$\mathbf{T} = \mathbf{u}_1 \circ \mathbf{v}_1 \circ \mathbf{w}_1 + \mathbf{u}_2 \circ \mathbf{v}_2 \circ \mathbf{w}_2,$$

where

$$\mathbf{u}_1 = \begin{bmatrix} 1 \\ 2 \end{bmatrix}, \quad \mathbf{v}_1 = \begin{bmatrix} 3 \\ 4 \end{bmatrix}, \quad \mathbf{w}_1 = \begin{bmatrix} 5 \\ 6 \end{bmatrix},$$

$$\mathbf{u}_2 = \begin{bmatrix} 1 \\ 1 \end{bmatrix}, \quad \mathbf{v}_2 = \begin{bmatrix} 1 \\ 1 \end{bmatrix}, \quad \mathbf{w}_2 = \begin{bmatrix} 1 \\ 1 \end{bmatrix}.$$

The first rank-one tensor is:
Given

$$\mathbf{T} = \mathbf{u}_1 \circ \mathbf{v}_1 \circ \mathbf{w}_1 + \mathbf{u}_2 \circ \mathbf{v}_2 \circ \mathbf{w}_2,$$

with

$$\mathbf{u}_1 = \begin{bmatrix} 1 \\ 2 \end{bmatrix}, \quad \mathbf{v}_1 = \begin{bmatrix} 3 \\ 4 \end{bmatrix}, \quad \mathbf{w}_1 = \begin{bmatrix} 5 \\ 6 \end{bmatrix},$$

and

$$\mathbf{u}_2 = \begin{bmatrix} 1 \\ 1 \end{bmatrix}, \quad \mathbf{v}_2 = \begin{bmatrix} 1 \\ 1 \end{bmatrix}, \quad \mathbf{w}_2 = \begin{bmatrix} 1 \\ 1 \end{bmatrix}.$$

The tensor entries are computed as

$$T_{ijk} = u_i v_j w_k.$$

The first rank-one tensor is

$$\mathcal{T}^{(1)} = \mathbf{u}_1 \circ \mathbf{v}_1 \circ \mathbf{w}_1.$$

For $k = 1$ ($w_1 = 5$),

$$\mathcal{T}^{(1)}(:, :, 1) = 5 \begin{bmatrix} 1 \cdot 3 & 1 \cdot 4 \\ 2 \cdot 3 & 2 \cdot 4 \end{bmatrix} = \begin{bmatrix} 15 & 20 \\ 30 & 40 \end{bmatrix}.$$

For $k = 2$ ($w_2 = 6$),

$$\mathcal{T}^{(1)}(:, :, 2) = 6 \begin{bmatrix} 1 \cdot 3 & 1 \cdot 4 \\ 2 \cdot 3 & 2 \cdot 4 \end{bmatrix} = \begin{bmatrix} 18 & 24 \\ 36 & 48 \end{bmatrix}.$$

Thus,

$$\mathcal{T}^{(1)} = \begin{bmatrix} 15 & 20 & 18 & 24 \\ 30 & 40 & 36 & 48 \end{bmatrix}.$$

The second rank-one tensor is

$$\mathcal{T}^{(2)} = \mathbf{u}_2 \circ \mathbf{v}_2 \circ \mathbf{w}_2.$$

Since all entries are 1,

$$\mathcal{T}^{(2)}(:, :, 1) = \begin{bmatrix} 1 & 1 \\ 1 & 1 \end{bmatrix}, \qquad \mathcal{T}^{(2)}(:, :, 2) = \begin{bmatrix} 1 & 1 \\ 1 & 1 \end{bmatrix}.$$

Hence,

$$\mathcal{T}^{(2)} = \begin{bmatrix} 1 & 1 & 1 & 1 \\ 1 & 1 & 1 & 1 \end{bmatrix}.$$

Finally, their sum is

$$\mathbf{T} = \mathcal{T}^{(1)} + \mathcal{T}^{(2)}.$$

For $k = 1$,

$$\mathbf{T}(:,:,1) = \begin{bmatrix} 15 & 20 \\ 30 & 40 \end{bmatrix} + \begin{bmatrix} 1 & 1 \\ 1 & 1 \end{bmatrix} = \begin{bmatrix} 16 & 21 \\ 31 & 41 \end{bmatrix}.$$

For $k = 2$,

$$\mathbf{T}(:,:,2) = \begin{bmatrix} 18 & 24 \\ 36 & 48 \end{bmatrix} + \begin{bmatrix} 1 & 1 \\ 1 & 1 \end{bmatrix} = \begin{bmatrix} 19 & 25 \\ 37 & 49 \end{bmatrix}.$$

Therefore,

$$\mathbf{T} = \begin{bmatrix} 16 & 21 & 19 & 25 \\ 31 & 41 & 37 & 49 \end{bmatrix}.$$

This decomposition shows that the CP rank is at most 2, as fewer than two rank-one tensors cannot exactly reconstruct $\mathbf{T}$.

Example 4.20 For:

$$\mathbf{Q} = \begin{bmatrix} \cos\theta & -\sin\theta \\ \sin\theta & \cos\theta \end{bmatrix},$$

$$\mathbf{Q}^T\mathbf{Q} = \begin{bmatrix} 1 & 0 \\ 0 & 1 \end{bmatrix}.$$

since Q is an orthogonal (rotation) matrix whose columns are orthonormal.
A third-order tensor $\mathbf{T}$ is symmetric if $\mathbf{T}_{ijk} = \mathbf{T}_{\pi(i,j,k)}$ for all permutations π. For example:

$$\mathbf{T}_{ijk} = i + j + k,$$

is symmetric since the value depends only on the sum of indices.

4.5 Tensors in ML

Tensors provide a natural and efficient way to represent both data and model parameters in ML, especially when dealing with multidimensional structured data. Their multi-way structure allows models to capture complex relationships across different modes or dimensions simultaneously.

- **Images**: Images are commonly represented as third-order tensors where the dimensions correspond to height, width, and color channels (e.g., RGB). This tensor representation preserves spatial and color information, enabling Convolutional Neural Networks (CNNs) to learn hierarchical features effectively by exploiting local correlations in the data.

- **Videos**: Videos extend image data by adding a temporal dimension, resulting in a fourth-order tensor with height, width, time (frames), and color channels. This structure captures spatial, temporal, and color information jointly, allowing models such as 3D CNNs or tensor-based recurrent networks to analyze motion patterns and temporal dependencies crucial for action recognition, video summarization, and surveillance.
- **Word Embeddings**: In NLP, word embeddings can be organized into a third-order tensor where the dimensions represent vocabulary size, embedding dimension, and context or positional information. This multi-way structure facilitates capturing semantic relationships, word sense disambiguation, and context-dependent meanings, enabling more expressive language models and improving tasks like machine translation and sentiment analysis.

Python Example: Image as a Tensor

```python
import numpy as np

# Simulate a 2x2 RGB image
image = np.array([[[255, 0, 0], [0, 255, 0]],
[[0, 0, 255], [255, 255, 255]]])

print("Image tensor shape:", image.shape)
```

Output:

```
Image tensor shape: (2, 2, 3)
```

- **CNNs:** In CNNs, the weights of convolutional layers are naturally represented as fourth-order tensors, with dimensions corresponding to the number of input channels, output channels (or filters), kernel height, and kernel width. This multidimensional structure enables the network to learn spatially localized patterns across multiple input feature maps simultaneously, capturing hierarchical features such as edges, textures, and complex shapes. By organizing weights in this tensor form, CNNs efficiently perform convolution operations that slide the kernels over the input data, producing output feature maps that retain spatial relationships. The tensor representation also facilitates techniques like tensor factorization for model compression and acceleration, which are crucial for deploying CNNs in resource-constrained environments.

Python Example: CNN Weights

```python
import numpy as np

# Simulate CNN weights
weights = np.random.rand(4, 8, 3, 3)
print("CNN weights shape:", weights.shape)
```

Output:

```
CNN weights shape: (4, 8, 3, 3)
```

- **Feature Extraction**: Multi-scale feature maps extracted by CNNs or other feature extractors are naturally represented as third-order tensors, with dimensions corresponding to height, width, and the number of feature channels. These tensors encapsulate rich hierarchical information at different resolutions, allowing models to detect edges, textures, and object parts effectively. The tensor format facilitates operations like multi-scale fusion and attention mechanisms to enhance recognition and segmentation tasks.
- **Motion Analysis**: Videos and dynamic scenes involve spatiotemporal data that can be represented as fourth-order tensors, incorporating height, width, time (frame sequence), and feature channels. This representation allows for the simultaneous analysis of spatial structure and temporal evolution, which is essential for tasks such as action recognition, tracking, and behavior analysis. Tensor-based methods enable efficient modeling of motion patterns and temporal dependencies while preserving spatial details.

Python Example: Feature Map Tensor

```python
import numpy as np

# Simulate a feature map tensor
feature_map = np.random.rand(64, 64, 128)
print("Feature map tensor shape:",
    feature_map.shape)
```

Output:

```
Feature map tensor shape: (64, 64, 128)
```

- **Electroencephalography (EEG) Analysis**: Third-order tensor (channels, time, trials) for brain-computer interfaces and neural signal processing in ML applications.
- **Audio Processing**: Third-order tensor (frequency, time, channels) for speech recognition and audio classification tasks.

Python Example: EEG Tensor

```python
import numpy as np
import tensorly as tl
from tensorly.decomposition import parafac

# Simulate EEG data (channels, time, trials)
eeg = np.random.rand(32, 1000, 10)

# CP decomposition
weights, factors = parafac(eeg, rank=5)
print("Channel factor matrix shape:",
    factors[0].shape)
```

Output:

```
Channel factor matrix shape: (32, 5)
```

- **Magnetic Resonance Imaging (MRI) and Computed Tomography (CT) Fusion**: Third-order tensors (height, width, modalities) for multi-modal medical image analysis in diagnostic ML.
- **Dynamic MRI**: Fourth-order tensors (height, width, time, modalities) for temporal analysis in medical imaging applications.

Python Example: MRI Tensor

```python
import numpy as np

# Simulate MRI data (height, width, modalities)
mri = np.random.rand(256, 256, 3)
print("MRI tensor shape:", mri.shape)
```

Output:

```
MRI tensor shape: (256, 256, 3)
```

- **Social Networks**: Third-order tensor (nodes, nodes, time) for dynamic network analysis and link prediction in graph neural networks.
- **Traffic Networks**: Third-order tensor (locations, locations, time) for spatiotemporal forecasting in transportation ML.

Python Example: Social Network Tensor

```python
import numpy as np
import tensorly as tl
from tensorly.decomposition import parafac

# Simulate a social network tensor
network = np.random.randint(0, 2, size=(50, 50,
    10))

# CP decomposition
weights, factors = parafac(network, rank=5)
print("Node factor matrix shape:",
    factors[0].shape)
```

Output:

```
Node factor matrix shape: (50, 5)
```

Chapter Summary

This chapter introduced tensors as powerful generalizations of scalars, vectors, and matrices, extending linear algebra to multidimensional structures essential for handling complex data in ML and related fields. It began with the formal definition of tensors by order and shape, where zeroth-order tensors are scalars, first-order are vectors, second-order are matrices, and higher-order tensors organize data in three or more dimensions, as visualized in Fig. 4.1. Examples included a third-order tensor $\mathbf{T} \in \mathbb{R}^{2 \times 2 \times 2}$ with elements accessed via indices $\mathbf{T}_{ijk}$, and higher-order cases like fourth- and fifth-order tensors for videos and spatiotemporal data. Tensor notation, including the Einstein summation convention for implied contractions like $C_{ij} = A_{ik} B_{kj}$, was explained with examples, emphasizing multi-index access and summation over repeated indices.

Section 4.2 explored tensor operations, starting with element-wise addition and subtraction for identical shapes, as in the sum yielding $\mathbf{C} = \begin{bmatrix} \begin{bmatrix} 1 & 3 \\ 4 & 4 \end{bmatrix}, \begin{bmatrix} 6 & 6 \\ 7 & 9 \end{bmatrix} \end{bmatrix}$, and scalar multiplication for uniform scaling. The tensor product (outer product) combined lower-order tensors into higher-order ones, demonstrated by the Kronecker product $\mathbf{A} \otimes \mathbf{B}$ flattening to a 4×4 matrix. Tensor contraction reduced order via summation, such as contracting a third-order tensor with a vector to produce a matrix, with double contractions yielding scalars. The mode-n product transformed specific dimensions, unfolding (matricization) rearranged tensors into matrices for SVD application, and norms like the Frobenius $\|\mathbf{T}\|_F = \sqrt{\sum \mathbf{T}_{ijk}^2} \approx 14.28$ were computed, alongside nuclear, spectral, and ℓ_p variants for regularization. Tensor transformations under coordinate changes, such as $\mathbf{A}' = \mathbf{Q}^T \mathbf{A} \mathbf{Q}$ for rotations, were illustrated.

Tensor decompositions in Sect. 4.3 factored higher-order structures for compression and analysis: Canonical Polyadic (CP) as sums of rank-one tensors $\mathbf{T} \approx \sum_r \mathbf{a}_r^{(1)} \circ \mathbf{a}_r^{(2)} \circ \mathbf{a}_r^{(3)}$, Tucker into a core and factor matrices, Higher-Order SVD (HOSVD) with orthogonal factors, andTT as sequential third-order cores for high-dimensional efficiency. Rank estimation via alternating least squares (ALS) minimized reconstruction error. Properties in Sect. 4.4 included symmetric tensors invariant under index permutation, orthogonal tensors generalizing rotations with $\mathbf{Q}^T \mathbf{Q} = \mathbf{I}$, and tensor rank as the minimal number of rank-one components in CP decomposition.

Section 4.5 highlighted applications in ML: third-order tensors for images ($height \times width \times channels$), fourth-order for videos (adding time), word embeddings ($vocabulary \times dimension \times context$), feature maps, motion analysis, electroencephalography (EEG) (channels$\times$time$\times$trials), audio, MRI or CT fusion, and social or traffic networks, with Python examples using NumPy and TensorLy for decompositions like CP and Tucker.

Takeaways

Key takeaways from the chapter included:

- Tensors as multidimensional arrays defined by order and shape, with notation enabling efficient access and operations like Einstein summation for contractions.
- Core operations, addition, scalar multiplication, tensor products, contractions, mode-n products, unfolding, norms (Frobenius, nuclear, spectral), and transformations, generalizing vector or matrix tools for high-dimensional data processing.
- Decompositions such as CP, Tucker, HOSVD, and TT for dimensionality reduction, compression, and latent structure extraction, with rank estimation via optimization.
- Properties like symmetry, orthogonality, and rank providing structural insights for efficient representations and analysis.
- Practical implementations in NumPy or TensorLy for computations and applications in ML domains including computer vision, NLP, signal processing, medical imaging, and graph analysis.

By exploring these concepts, readers gained the ability to model and manipulate multidimensional data, laying the groundwork for advanced topics like linear systems and transformations in subsequent chapters.

Exercises

1. Compute $\mathbf{A} + \mathbf{B}$ for:

$$\mathbf{A} = \left[\begin{bmatrix} 1 & 2 \\ 3 & 4 \end{bmatrix}, \begin{bmatrix} 5 & 6 \\ 7 & 8 \end{bmatrix}\right], \quad \mathbf{B} = \left[\begin{bmatrix} 1 & 0 \\ 0 & 1 \end{bmatrix}, \begin{bmatrix} 0 & 1 \\ 1 & 0 \end{bmatrix}\right].$$

2. Compute $3 \cdot \mathbf{A}$ using the tensor defined in Exercise 1.
3. Compute $\mathbf{A} \otimes \mathbf{B}$ for:

$$\mathbf{A} = \begin{bmatrix} 1 & 2 \\ 3 & 4 \end{bmatrix}, \quad \mathbf{B} = \begin{bmatrix} 0 & 1 \\ 1 & 0 \end{bmatrix}.$$

4. Perform the mode-3 contraction of $\mathbf{T} \in \mathbb{R}^{2 \times 2 \times 3}$ with $\mathbf{v} \in \mathbb{R}^3$, i.e.,

$$(\mathbf{T} \times_3 \mathbf{v})_{ij} = \sum_{k=1}^{3} T_{ijk} v_k.$$

$$\mathbf{T} = \left[\begin{bmatrix} 1 & 2 & 3 \\ 4 & 5 & 6 \end{bmatrix}, \begin{bmatrix} 7 & 8 & 9 \\ 10 & 11 & 12 \end{bmatrix}\right], \quad \mathbf{v} = \begin{bmatrix} 1 \\ 2 \\ 3 \end{bmatrix}.$$

5. Verify the symmetry of $\mathbf{A} = \begin{bmatrix} 1 & 2 \\ 2 & 3 \end{bmatrix}$.
6. Compute the double contraction $\sum_{k,l} S_{ijkl} A_{kl}$, where $\mathbf{S} \in \mathbb{R}^{2 \times 2 \times 2 \times 2}$ has all entries equal to 1.

$$\mathbf{S} = \text{all ones, shape } (2, 2, 2, 2), \quad \mathbf{A} = \begin{bmatrix} 1 & 2 \\ 3 & 4 \end{bmatrix}.$$

7. Generate a random tensor $\mathbf{T} \in \mathbb{R}^{3 \times 3 \times 3}$, perform CP decomposition with rank $R = 2$, and verify that the reconstructed tensor has shape $3 \times 3 \times 3$.
8. Compute the mode-1 product $\mathbf{Y} = \mathbf{T} \times_1 \mathbf{U}$, where $\mathbf{T} \in \mathbb{R}^{3 \times 3 \times 3}$ and $\mathbf{U} \in \mathbb{R}^{2 \times 3}$.
9. Let $\mathbf{T} \in \mathbb{R}^{2 \times 2 \times 2}$ with slices:

$$\mathbf{T} = \left[\begin{bmatrix} 1 & 2 \\ 3 & 4 \end{bmatrix}, \begin{bmatrix} 5 & 6 \\ 7 & 8 \end{bmatrix}\right].$$

Compute its Frobenius norm.
10. Let $\mathbf{Q}$ be the 2D rotation matrix with $\theta = \pi/4$. Compute the transformed tensor:

$$\mathbf{A}' = \mathbf{Q}^T \mathbf{A} \mathbf{Q}.$$

11. Generate a random tensor $\mathbf{T} \in \mathbb{R}^{2 \times 2 \times 2 \times 2}$, perform TT decomposition, reconstruct the tensor, and verify that the reconstructed tensor has the same shape.

12. Using CP decomposition, approximate a $4 \times 4 \times 4$ random tensor with increasing rank R and determine the smallest R that achieves a low reconstruction error.

Chapter 5
Linear Systems

Introduction

Linear systems are the backbone of linear algebra, providing a structured way to model and solve a wide array of problems across mathematics, physics, engineering, and computer science. At their core, they represent a set of linear equations that must be satisfied simultaneously, encoding relationships between variables through coefficients and constants. These systems are not merely abstract constructs but are pivotal in applications ranging from optimizing ML models to simulating physical systems and solving real-world problems like traffic flow or resource allocation.

Far from being just academic exercises, systems of linear equations orchestrate a vast range of technologies that shape our world, from physics simulations and structural engineering to optimization routines in ML models. They encode balance, constraint, and structure, forming the skeleton of more complex mathematical models used in ML and data science today. This chapter aims to demystify linear systems by exploring their algebraic structure, geometric interpretations, and computational solutions, with a particular emphasis on their relevance to ML.

In this chapter, we will begin with the anatomy of a linear system: what it means for equations to be consistent or inconsistent, how solution sets behave geometrically, and why seemingly simple changes in coefficients can lead to profound shifts in outcomes. These are not just curiosities; they're the difference between stable predictions and numerical failure in ML pipelines. From there, we'll explore methods for solving systems, such as substitution, elimination, and matrix-based techniques like Gaussian elimination. We will also examine fundamental concepts like linear combinations, span, linear dependence, and matrix rank, which determine the nature of solutions, unique, infinite, or none.

To ground this theory in practice, we'll provide visualizations with real Python implementations using NumPy, illustrating how these ideas map directly onto ML methods like LR. The chapter is designed to be accessible to beginners, with clear, concrete, and gradually layered complexity, while retaining enough depth to intrigue

© The Author(s), under exclusive license to Springer Nature Singapore Pte Ltd. 2026

Md. Jalil Piran, *Linear Algebra with Applications in Machine Learning*,

https://doi.org/10.1007/978-981-95-5167-5_5

advanced readers. Each section builds on the previous one, accompanied by figures to illustrate geometric interpretations and code snippets to demonstrate computational approaches.

The final sections explicitly connect these linear algebraic principles to ML, showing how linear systems underpin algorithms like LR and feature engineering. By the end, you will not only be able to solve linear systems but also understand their geometric significance, develop an intuition for their structure, and appreciate their critical role in ML and data science.

Topics Covered

This chapter is organized as follows:

- **Section** 5.1 **Systems of Linear Equations; Definition and Basics**: Introduces the definition, matrix representation, and geometric interpretation of linear systems, including unique, infinite, and no-solution cases.
- **Section** 5.2 **Linear Combinations**: Explores linear combinations and their role in defining solution spaces.
- **Section** 5.3 **Span**: Defines the span of vectors and its geometric interpretation as lines, planes, or higher-dimensional subspaces.
- **Section** 5.4 **Linear Dependence and Independence**: Discusses the concepts of linear dependence and independence and their impact on solution sets.
- **Section** 5.5 **Solving Methods for Linear Systems**: Covers methods like substitution, elimination, and Gaussian elimination, using augmented matrices and row operations.
- **Section** 5.6 **Matrix Rank and Solutions**: Examines matrix rank, nullity, and their role in determining solution existence and uniqueness.
- **Section** 5.7 **Linear Systems in ML**: Connects linear systems to ML applications, such as LR and feature engineering.
- **Summary and Exercises**: Recaps key concepts and provides practical and theoretical problems to reinforce concepts.

All code accompanying the chapter is publicly available at https://github.com/jalil-piran/Linear-Algebra-with-Applications-in-Machine-Learning/blob/main/Chapter_05_Linear_Systems.ipynb.

This chapter invites you not only to solve equations but to **see** them, to develop an intuition for their structure, an eye for their geometric behavior, and a feel for the implications they hold in ML contexts.

5.1   Systems of Linear Equations: Definition and Basics

A system of linear equations provides a structured way to model real-world problems, such as networks, resource allocation, and optimization, by expressing multiple linear relationships simultaneously.

Definition

A **system of linear equations** consists of multiple linear equations involving unknowns $x_1, x_2, \ldots, x_n$. A general system with m equations and n unknowns is written as:

$$\begin{cases} a_{11}x_1 + a_{12}x_2 + \cdots + a_{1n}x_n = b_1 \\ a_{21}x_1 + a_{22}x_2 + \cdots + a_{2n}x_n = b_2 \\ \vdots \\ a_{m1}x_1 + a_{m2}x_2 + \cdots + a_{mn}x_n = b_m \end{cases}$$

This can be compactly represented in matrix form as:

$$A\mathbf{x} = \mathbf{b},$$

where:

- $A \in \mathbb{R}^{m \times n}$: the coefficient matrix with entries a_{ij},
- $\mathbf{x} \in \mathbb{R}^n$: the vector of unknowns $[x_1, x_2, \ldots, x_n]^T$,
- $\mathbf{b} \in \mathbb{R}^m$: the constant vector $[b_1, b_2, \ldots, b_m]^T$.

The **solution set** comprises all vectors $\mathbf{x}$ that satisfy $A\mathbf{x} = \mathbf{b}$. A system is classified as:

- **Consistent**: At least one solution exists (unique or infinitely many).
- **Inconsistent**: No solution exists.

Geometrically, each equation represents a hyperplane in $\mathbb{R}^n$. The solution set is the intersection of these hyperplanes:

- In $\mathbb{R}^2$, equations are lines, and the solution may be a point (unique), a line (infinitely many), or empty (no solution).
- In $\mathbb{R}^3$, equations are planes, and the solution may be a point, line, plane, or empty set.

5.1.1 Unique Solution

A system of linear equations has a unique solution when the coefficient matrix $A \in \mathbb{R}^{m \times n}$ is square ($m = n$) and invertible, meaning it has full rank (i.e., $\text{rank}(A) = n$). In geometric terms, this corresponds to the case where each equation represents a hyperplane in $\mathbb{R}^n$, and all hyperplanes intersect at exactly one common point. This unique point of intersection represents the single solution to the system. Algebraically, the solution can be found by computing $\mathbf{x} = A^{-1}\mathbf{b}$, where $\mathbf{b}$ is the right-hand side vector.

Existence and uniqueness are guaranteed by the Invertible Matrix Theorem, which states that a square matrix is invertible if and only if its determinant is non-zero. In practical applications, such systems arise in areas like circuit analysis, mechanical equilibrium problems, and control systems, where exact solutions are necessary for stability and predictability.

Example 5.1 Consider the system:

$$\begin{cases} 3x_1 - x_2 = 10 \\ x_1 - 3x_2 = 0 \end{cases}$$

In matrix form:

$$A = \begin{bmatrix} 3 & -1 \\ 1 & -3 \end{bmatrix}, \quad \mathbf{x} = \begin{bmatrix} x_1 \\ x_2 \end{bmatrix}, \quad \mathbf{b} = \begin{bmatrix} 10 \\ 0 \end{bmatrix}.$$

We can solve it using **substitution method**. This method involves solving one equation for one variable and substituting the result into the other equation. In this example, we first express x_1 in terms of x_2 using the second equation. Then, substituting this expression into the first equation allows us to solve for x_2 directly. Once x_2 is known, we substitute it back to find x_1. This approach reduces a system of equations to a single-variable equation and is especially effective for small systems where at least one equation can be easily rearranged.

1. From the second equation: $x_1 = 3x_2$.
2. Substitute into the first equation: $3(3x_2) - x_2 = 10 \implies 9x_2 - x_2 = 10 \implies 8x_2 = 10 \implies x_2 = \frac{5}{4}$.
3. Then: $x_1 = 3 \cdot \frac{5}{4} = \frac{15}{4}$.

The solution is $\mathbf{x} = \begin{bmatrix} \frac{15}{4} \\ \frac{5}{4} \end{bmatrix}$.

We can verify the solution as follows:

- First equation: $3 \cdot \frac{15}{4} - \frac{5}{4} = \frac{45}{4} - \frac{5}{4} = 10$.
- Second equation: $\frac{15}{4} - 3 \cdot \frac{5}{4} = \frac{15}{4} - \frac{15}{4} = 0$.

Geometrically, as shown in Fig. 5.1, the lines intersect at $\left(\frac{15}{4}, \frac{5}{4}\right)$.

Python Example: System of Linear Equations

```python
import numpy as np

A = np.array([[3, -1], [1, -3]])
b = np.array([10, 0])

solution = np.linalg.solve(A, b)
print("Solution:", solution)
```

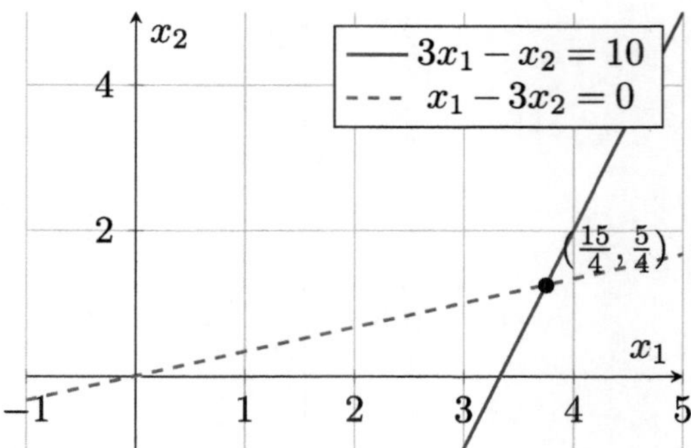

Fig. 5.1 Unique solution at the intersection of two lines in $\mathbb{R}^2$

Output:

```
Solution: [3.75 1.25]
```

Example 5.2 Consider the system:

$$\begin{cases} 2x_1 - 3x_2 = 1, \\ x_1 = 1. \end{cases}$$

Substituting $x_1 = 1$ into the first equation:

$$2(1) - 3x_2 = 1 \Rightarrow -3x_2 = -1 \Rightarrow x_2 = \frac{1}{3}.$$

As shown in Fig. 5.2, the solution is:

$$\boxed{(x_1, x_2) = \left(1, \frac{1}{3}\right)}$$

5.1.2 No Solution

A system of linear equations has no solution when it is inconsistent, meaning that the equations contradict one another and there is no set of values for the variables that satisfies all equations simultaneously. This typically occurs when the coefficient matrix leads to parallel hyperplanes that never intersect, such as two lines in $\mathbb{R}^2$ with the same slope but different intercepts. Algebraically, inconsistency arises when row reduction [Sect. 5.5.3 leads to a row of zeros in the coefficient matrix paired with

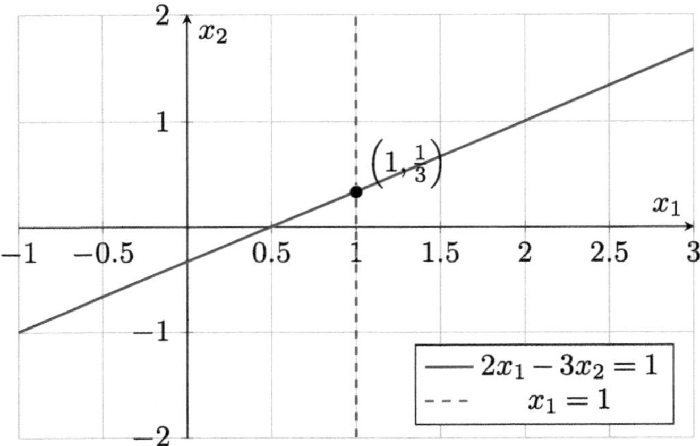

Fig. 5.2 Graphical solution of the system $2x_1 - 3x_2 = 1$ and $x_1 = 1$. The intersection point $(1, \frac{1}{3})$ satisfies both equations, representing the unique solution

a nonzero entry in the augmented column, indicating an impossible equation (e.g., $0 = 5$)]. Such systems are said to be overdetermined, often containing redundant or conflicting constraints. In practical applications, inconsistency may result from data errors, incompatible boundary conditions, or conflicting measurements, and detecting it is essential before attempting to solve the system.

Example 5.3 Consider:

$$\begin{cases} 3x_1 + x_2 = 10, \\ 6x_1 + 2x_2 = 0. \end{cases}$$

In matrix form:

$$A = \begin{bmatrix} 3 & 1 \\ 6 & 2 \end{bmatrix}, \quad \mathbf{x} = \begin{bmatrix} x_1 \\ x_2 \end{bmatrix}, \quad \mathbf{b} = \begin{bmatrix} 10 \\ 0 \end{bmatrix}.$$

Notice that the second equation is twice the first but with a different constant: $6x_1 + 2x_2 = 2(3x_1 + x_2) = 0$, while $3x_1 + x_2 = 10$. Scaling the first equation by 2 gives $6x_1 + 2x_2 = 20$, contradicting $6x_1 + 2x_2 = 0$. Thus, no solution exists. Figure 5.3 shows the parallel lines $x_2 = -3x_1 + 10$ and $x_2 = -3x_1$, which never intersect due to identical slopes but different intercepts.

Fig. 5.3 Parallel lines indicate an inconsistent system with no solution

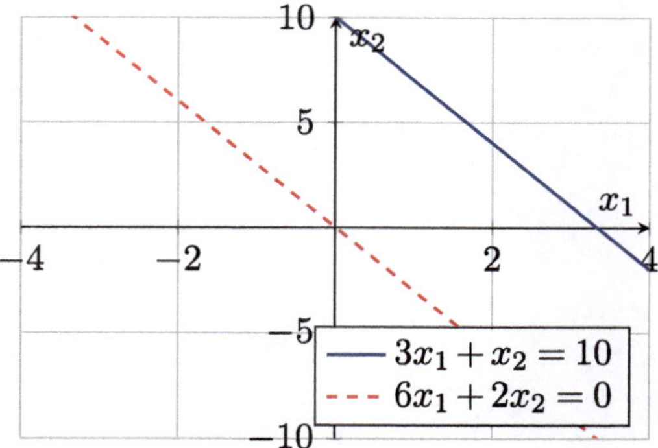

Python Example: System of Linear Equations

```python
import numpy as np

A = np.array([[3, 1], [6, 2]])
b = np.array([10, 0])

try:
    solution = np.linalg.solve(A, b)
except np.linalg.LinAlgError:
    print("System is inconsistent (no
        solution).")
```

Output:

```
System is inconsistent (no solution).
```

Example 5.4 Consider the system:

$$\begin{cases} 3x_1 + 6x_2 = 3 \\ 2x_1 + 4x_2 = 4 \end{cases}$$

We divide both equations:

$$\begin{cases} \frac{1}{3}(3x_1 + 6x_2) = 1 \Rightarrow x_1 + 2x_2 = 1 \\ \frac{1}{2}(2x_1 + 4x_2) = 2 \Rightarrow x_1 + 2x_2 = 2 \end{cases}$$

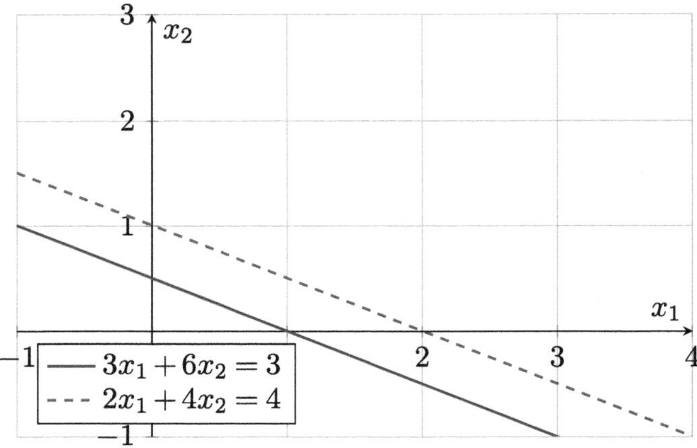

Fig. 5.4 Graphical representation of the system $3x_1 + 6x_2 = 3$ and $2x_1 + 4x_2 = 4$. The lines are parallel and do not intersect, indicating an inconsistent system with no solution

This leads to:

$$\begin{cases} x_1 + 2x_2 = 1 \\ x_1 + 2x_2 = 2 \Rightarrow \text{Contradiction} \end{cases}$$

The system has no solution. The lines are parallel and never intersect, as shown in Fig. 5.4.

5.1.3 Infinitely Many Solutions

A system of linear equations has infinitely many solutions when it is underdetermined, meaning that there are fewer independent equations than unknowns ($m < n$), or when the system contains linearly dependent equations that do not provide additional constraints. In such cases, some variables can take on arbitrary values, referred to as free variables, while others (dependent variables) are expressed in terms of them. This results in a solution space that forms a line, plane, or higher-dimensional subspace, depending on the number of free variables. Geometrically, the equations represent hyperplanes that intersect along a line or plane rather than at a single point. Row reduction of the augmented matrix typically reveals at least one row of zeros, indicating a dependency among the equations (see Sect. 5.5.3 for more details). Infinitely many solutions are common in parameterized models, constraint relaxation, and systems where symmetry or redundancy is present. Describing the solution set requires expressing variables in terms of parameters that span (Sect. 5.3) the null space of the coefficient matrix.

Example 5.5 Consider:

$$\begin{cases} 2x_1 - 3x_2 + x_3 = -10, \\ x_1 + x_3 = 3. \end{cases}$$

In matrix form:

$$A = \begin{bmatrix} 2 & -3 & 1 \\ 1 & 0 & 1 \end{bmatrix}, \quad \mathbf{x} = \begin{bmatrix} x_1 \\ x_2 \\ x_3 \end{bmatrix}, \quad \mathbf{b} = \begin{bmatrix} -10 \\ 3 \end{bmatrix}.$$

To solve it, we follow three steps:

Step 1- From the second equation: $x_1 + x_3 = 3 \implies x_3 = 3 - x_1$.

Step 2- Substitute into the first: $2x_1 - 3x_2 + (3 - x_1) = -10 \implies x_1 - 3x_2 + 3 = -10 \implies x_1 - 3x_2 = -13 \implies x_2 = \frac{x_1+13}{3}$.

Step 3- Parameterize: if we consider $x_1 = t$, then $x_2 = \frac{t+13}{3}, x_3 = 3 - t$.

The solution set is:

$$\mathbf{x} = \begin{bmatrix} t \\ \frac{t+13}{3} \\ 3 - t \end{bmatrix}, \quad t \in \mathbb{R}.$$

For $t = 2$, we get $\mathbf{x} = \begin{bmatrix} 2 \\ 5 \\ 1 \end{bmatrix}$.

For $t = -4$, $\mathbf{x} = \begin{bmatrix} -4 \\ 3 \\ 7 \end{bmatrix}$.

Figure 5.5 illustrates the two planes intersecting along a line in $\mathbb{R}^3$.

Python Example: System of Linear Equations

```python
import numpy as np

A = np.array([[2, -3, 1], [1, 0, 1]])
b = np.array([-10, 3])

t = 2
x_2 = (t + 13) / 3
x_3 = 3 - t
solution = [t, x_2, x_3]
print("One solution:", solution)
```

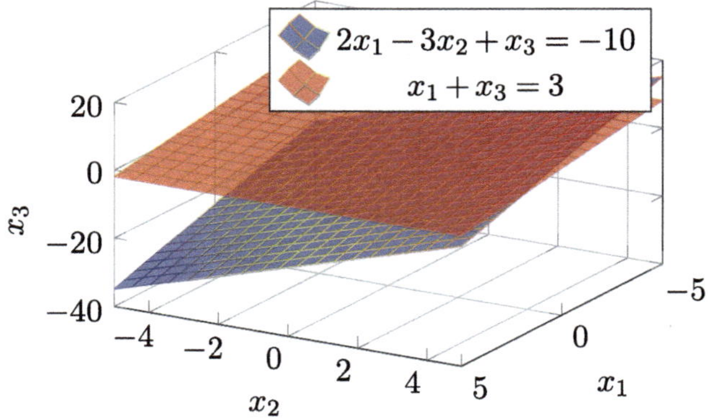

Fig. 5.5 Intersection of two planes forms a line, indicating infinitely many solutions

Output:

```
One solution: [2, 5.0, 1]
```

5.2 Linear Combinations

A linear combination expresses how different vectors can be scaled and added together to form new vectors, making it a key tool for understanding span, subspaces, and vector dependence.

Definition
A **linear combination** of vectors $\mathbf{v}_1, \mathbf{v}_2, \ldots, \mathbf{v}_n \in \mathbb{R}^m$ with real-valued scalars $c_1, c_2, \ldots, c_n \in \mathbb{R}$ is defined as:

$$c_1\mathbf{v}_1 + c_2\mathbf{v}_2 + \cdots + c_n\mathbf{v}_n.$$

This operation produces another vector in $\mathbb{R}^m$, formed by scaling each vector $\mathbf{v}_i$ by its corresponding scalar c_i and then summing the results. The concept of linear combination is a cornerstone of linear algebra, as it captures the idea of constructing new vectors from existing ones through weighted addition.

Geometrically, a linear combination of vectors represents a position in space that can be reached by moving along the directions of $\mathbf{v}_1, \ldots, \mathbf{v}_n$, scaled and summed together. For instance, if two vectors lie in a plane, then all their linear combinations will also lie in that same plane.

When the vectors $v_1, \ldots, v_n$ are placed as columns of a matrix $A \in \mathbb{R}^{m \times n}$, any linear combination of these vectors can be expressed as a matrix-vector product:

$$A\mathbf{x} = c_1 \mathbf{v}_1 + c_2 \mathbf{v}_2 + \cdots + c_n \mathbf{v}_n, \quad \text{where} \quad \mathbf{x} = \begin{bmatrix} c_1 \\ c_2 \\ \vdots \\ c_n \end{bmatrix}.$$

This formulation provides a compact and powerful way to represent linear combinations using matrix multiplication.

Furthermore, the linear system $A\mathbf{x} = \mathbf{b}$ is said to be *consistent* if and only if the vector $\mathbf{b}$ can be written as a linear combination of the columns of A. In other words, $\mathbf{b}$ must lie in the column space of A, denoted $\text{Col}(A)$. If no such combination exists, then the system has no solution.

Example 5.6 Let

$$A = \begin{bmatrix} 1 & 2 \\ 3 & 5 \end{bmatrix}, \quad \mathbf{x} = \begin{bmatrix} 2 \\ -1 \end{bmatrix}.$$

Then the linear combination is:

$$A\mathbf{x} = 2 \begin{bmatrix} 1 \\ 3 \end{bmatrix} + (-1) \begin{bmatrix} 2 \\ 5 \end{bmatrix} = \begin{bmatrix} 2 \\ 6 \end{bmatrix} + \begin{bmatrix} -2 \\ -5 \end{bmatrix} = \begin{bmatrix} 0 \\ 1 \end{bmatrix}.$$

Hence, $\mathbf{b} = \begin{bmatrix} 0 \\ 1 \end{bmatrix}$ is a linear combination of the columns of A.

Example 5.7 Given vectors $\mathbf{v}_1 = \begin{bmatrix} 1 \\ 1 \end{bmatrix}, \mathbf{v}_2 = \begin{bmatrix} 1 \\ 3 \end{bmatrix}, \mathbf{v}_3 = \begin{bmatrix} 1 \\ -1 \end{bmatrix}$ with coefficients $c_1 = -3, c_2 = 4, c_3 = 1$, compute:

$$-3\mathbf{v}_1 + 4\mathbf{v}_2 + 1\mathbf{v}_3 = -3 \begin{bmatrix} 1 \\ 1 \end{bmatrix} + 4 \begin{bmatrix} 1 \\ 3 \end{bmatrix} + \begin{bmatrix} 1 \\ -1 \end{bmatrix}.$$

- First component: $-3 \cdot 1 + 4 \cdot 1 + 1 \cdot 1 = -3 + 4 + 1 = 2$.
- Second component: $-3 \cdot 1 + 4 \cdot 3 + 1 \cdot (-1) = -3 + 12 - 1 = 8$.

Thus:

$$\begin{bmatrix} 2 \\ 8 \end{bmatrix}.$$

Figure 5.6 illustrates a linear combination of vectors $\mathbf{v}_1$, $\mathbf{v}_2$, and $\mathbf{v}_3$ in $\mathbb{R}^2$, resulting in the vector $\begin{bmatrix} 2 \\ 8 \end{bmatrix}$. Each colored arrow represents a contributing vector, and the black

Fig. 5.6 Linear combination of vectors resulting in $\begin{bmatrix} 2 \\ 8 \end{bmatrix}$, with projections on axes

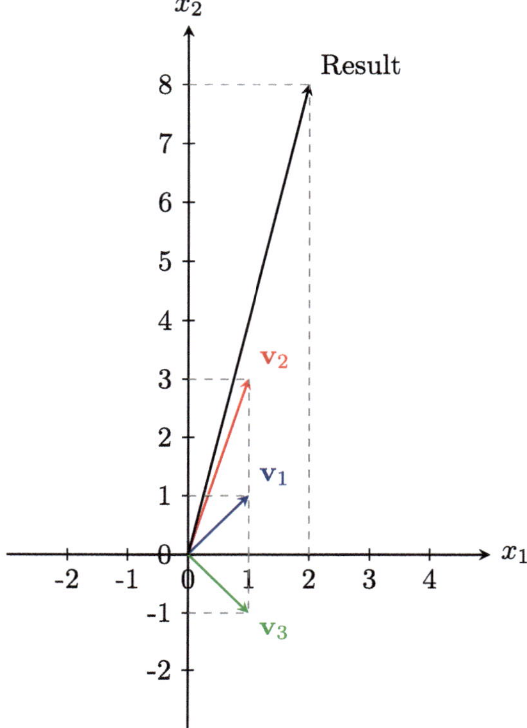

arrow labeled "Result" shows their combined effect. The gray dashed lines indicate the projections of each vector's components onto the x_1 and x_2 axes, helping to visualize how the individual vectors add up component-wise to form the final output.

Python Example: Linear Combinations

```python
import numpy as np

v1 = np.array([1, 1])
v2 = np.array([1, 3])
v3 = np.array([1, -1])
coeffs = np.array([-3, 4, 1])

result = coeffs[0]*v1 + coeffs[1]*v2 + coeffs[2]*v3
print("Result:", result)
```

Output:

```
Result: [2 8]
```

5.2.1 Linear Combination and Inconsistency

Example 5.8 Consider vectors:

$$\mathbf{v}_1 = \begin{bmatrix} 3 \\ 2 \end{bmatrix}, \quad \mathbf{v}_2 = \begin{bmatrix} 6 \\ 4 \end{bmatrix}, \quad \mathbf{b} = \begin{bmatrix} 3 \\ 4 \end{bmatrix}.$$

Can $\mathbf{b}$ be expressed as a linear combination: $x_1\mathbf{v}_1 + x_2\mathbf{v}_2 = \mathbf{b}$?
    This corresponds to:

$$\begin{cases} 3x_1 + 6x_2 = 3 \\ 2x_1 + 4x_2 = 4 \end{cases}, \quad A = \begin{bmatrix} 3 & 6 \\ 2 & 4 \end{bmatrix}, \quad \mathbf{x} = \begin{bmatrix} x_1 \\ x_2 \end{bmatrix}, \quad \mathbf{b} = \begin{bmatrix} 3 \\ 4 \end{bmatrix}.$$

Notice: $\mathbf{v}_2 = 2\mathbf{v}_1$, so the columns are linearly dependent. The second equation should
be $2x_1 + 4x_2 = 2$ since it is obtained by dividing the first equation by $3/2$, but the
system requires $2x_1 + 4x_2 = 4$. Hence, the system is inconsistent. Thus, the system is
inconsistent, and $\mathbf{b}$ is not in the span of $\mathbf{v}_1$, $\mathbf{v}_2$. This implies that no linear combination
of $\mathbf{v}_1$ and $\mathbf{v}_2$ can produce $\mathbf{b}$, as the row-reduced augmented matrix has a pivot in the
augmented column, indicating no solution exists.

Figure 5.7 illustrates the concept of vector span and inconsistency in a linear
system. The blue vectors $\begin{bmatrix} 3 \\ 2 \end{bmatrix}$ and $\begin{bmatrix} 6 \\ 4 \end{bmatrix}$ lie on the same dotted line, representing the

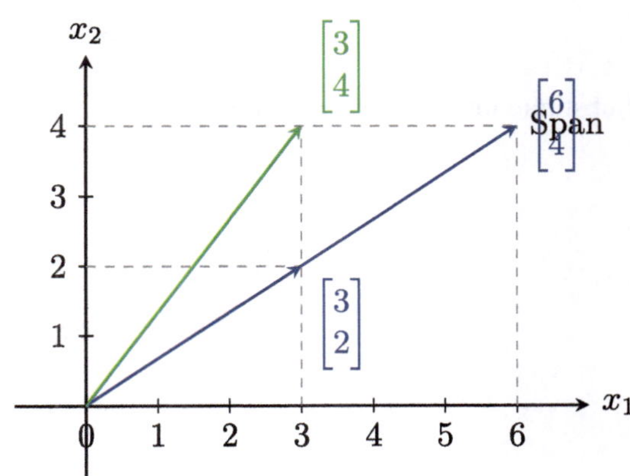

Fig. 5.7 Vector $\begin{bmatrix} 3 \\ 4 \end{bmatrix}$ (green) lies outside the span (dotted line), with projections shown for all vectors

span of a single direction in $\mathbb{R}^2$. In contrast, the green vector $\begin{bmatrix} 3 \\ 4 \end{bmatrix}$ lies outside this span, indicating that it cannot be expressed as a linear combination of the blue vectors.

5.2.2   Linear Combination and Existence of Solution

Example 5.9 Consider the vectors:

$$\mathbf{v}_1 = \begin{bmatrix} 2 \\ 1 \end{bmatrix}, \quad \mathbf{v}_2 = \begin{bmatrix} 6 \\ 3 \end{bmatrix}, \quad \mathbf{b} = \begin{bmatrix} -4 \\ -2 \end{bmatrix}.$$

We want to determine whether $\mathbf{b}$ is a linear combination of $\mathbf{v}_1$ and $\mathbf{v}_2$ (See Fig. 5.8). That is, does there exist a solution $(x_1, x_2) \in \mathbb{R}^2$ such that:

$$x_1 \mathbf{v}_1 + x_2 \mathbf{v}_2 = \mathbf{b}.$$

This can be written explicitly as a system of linear equations:

$$x_1 \begin{bmatrix} 2 \\ 1 \end{bmatrix} + x_2 \begin{bmatrix} 6 \\ 3 \end{bmatrix} = \begin{bmatrix} -4 \\ -2 \end{bmatrix} \quad \Longleftrightarrow \quad \begin{bmatrix} 2 & 6 \\ 1 & 3 \end{bmatrix} \begin{bmatrix} x_1 \\ x_2 \end{bmatrix} = \begin{bmatrix} -4 \\ -2 \end{bmatrix}.$$

This leads to the system of equations:

$$\begin{cases} 2x_1 + 6x_2 = -4, \\ x_1 + 3x_2 = -2. \end{cases}$$

Note that $\mathbf{v}_2 = 3\mathbf{v}_1$, i.e., the second vector is a scalar multiple of the first. This tells us that the two vectors are *linearly dependent* and span a one-dimensional subspace (a line) in $\mathbb{R}^2$. Therefore, any vector $\mathbf{b}$ that lies on this line can be expressed as a linear combination of $\mathbf{v}_1$ and $\mathbf{v}_2$.

We now solve the second equation:

$$x_1 + 3x_2 = -2 \quad \Rightarrow \quad x_1 = -2 - 3x_2.$$

Substitute into the first equation:

$$2x_1 + 6x_2 = -4,$$

$$2(-2 - 3x_2) + 6x_2 = -4,$$

$$-4 - 6x_2 + 6x_2 = -4,$$

$$-4 = -4.$$

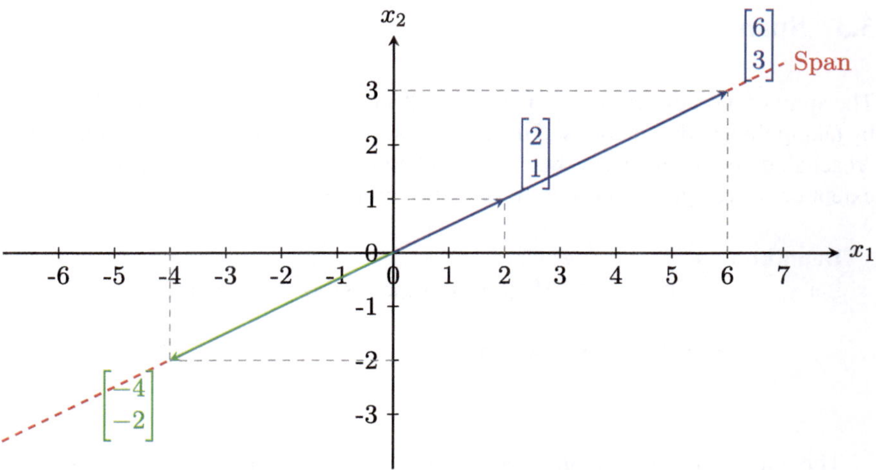

Fig. 5.8 All vectors lie on the extended span line (dashed red). The line continues beyond the vectors to clearly show the linear relationship $x_2 = 0.5x_1$

This simplification leads to a true statement that holds for all values of x_2, meaning the system has infinitely many solutions. The solution set forms a line in the (x_1, x_2)-plane.

Since the equation holds for all x_2, we can choose any value for x_2 and compute the corresponding x_1. For instance, let us choose:

$$x_2 = 0 \quad \Rightarrow \quad x_1 = -2 - 3(0) = -2.$$

So one particular solution is:

$$\mathbf{x} = \begin{bmatrix} -2 \\ 0 \end{bmatrix}.$$

The vector **b** is indeed a linear combination of $\mathbf{v}_1$ and $\mathbf{v}_2$, and the solution is not unique because the two vectors are linearly dependent. Any combination satisfying $x_1 = -2 - 3x_2$ gives a valid solution.

Consider the linear system $Ax = b$, where $A \in \mathbb{R}^{m \times n}$. The nature of the solutions is determined by comparing the rank of A and the augmented matrix $A \mid b$.

$$\text{Unique solution} \iff \text{rank}(A) = \text{rank}([A \mid b]) = n$$
$$\text{Infinite solutions} \iff \text{rank}(A) = \text{rank}([A \mid b]) < n$$
$$\text{No solution} \iff \text{rank}(A) < \text{rank}([A \mid b])$$

5.3　Span

The **span** of a set of vectors refers to the collection of all vectors that can be formed by taking linear combinations of those vectors. Intuitively, the span represents all the "reachable" points in space by scaling and adding the given vectors. It tells us the extent or "coverage" of those vectors in the ambient space.

> **Definition**
> Let $\mathbf{v}_1, \mathbf{v}_2, \ldots, \mathbf{v}_n \in \mathbb{R}^m$. The **span** of these vectors is defined as:
>
> $$\mathrm{Span}(\mathbf{v}_1, \ldots, \mathbf{v}_n) = \{c_1\mathbf{v}_1 + c_2\mathbf{v}_2 + \cdots + c_n\mathbf{v}_n \mid c_i \in \mathbb{R}\}.$$

This set contains all possible vectors that can be formed by applying real-valued weights c_i to the original vectors and summing the results.

The span is always a **subspace** of $\mathbb{R}^m$, a set that is:

- Closed under addition: the sum of any two vectors in the span is still in the span.
- Closed under scalar multiplication: scaling any vector in the span keeps it within the span.
- Contains the zero vector: achieved by setting all scalars $c_i = 0$.

If the vectors $\mathbf{v}_1, \ldots, \mathbf{v}_n$ are placed as columns in a matrix $A \in \mathbb{R}^{m \times n}$, then:

$$\mathrm{Span}(\mathbf{v}_1, \ldots, \mathbf{v}_n) = \mathrm{Col}(A),$$

where $\mathrm{Col}(A)$, the **column space** of A, consists of all vectors $\mathbf{b} \in \mathbb{R}^m$ for which the system $A\mathbf{x} = \mathbf{b}$ has a solution.

> **Remark**
> The span of any set of vectors always forms a subspace. Specifically, it:
>
> - Includes the zero vector.
> - Is closed under vector addition.
> - Is closed under scalar multiplication.

Special Case: Span of the Zero Vector

$$\mathrm{Span}(\mathbf{0}) = \{\mathbf{0}\}$$

This is the smallest possible subspace, also called the **trivial subspace**, containing only the zero vector.

Types of Span

There are several types of span as follows.

1. **Span of a Single Vector**

   Let $\mathbf{v} = \begin{bmatrix} 1 \\ 2 \end{bmatrix}$. Then:

   $$\mathrm{Span}(\mathbf{v}) = \left\{ c \begin{bmatrix} 1 \\ 2 \end{bmatrix} \mid c \in \mathbb{R} \right\}$$

   This set forms a line through the origin in the direction of $\mathbf{v}$. Every point on this line is a scaled version of $\mathbf{v}$, and this 1D subspace demonstrates how even a single vector can generate infinitely many directions if scaling is allowed.

   Figure 5.9 illustrates the vector $\mathbf{v} = \begin{bmatrix} 1 \\ 2 \end{bmatrix}$ and the one-dimensional subspace it spans in $\mathbb{R}^2$, represented by the dashed red line corresponding to the equation $y = 2x$. This line contains all scalar multiples of $\mathbf{v}$, highlighting how a single nonzero vector defines a line through the origin.

2. **Span of Two Linearly Independent Vectors**

   Let $\mathbf{v} = \begin{bmatrix} 1 \\ 0 \end{bmatrix}$, $\mathbf{u} = \begin{bmatrix} 0 \\ 1 \end{bmatrix}$. Then:

   $$\mathrm{Span}(\mathbf{v}, \mathbf{u}) = \left\{ a\mathbf{v} + b\mathbf{u} = \begin{bmatrix} a \\ b \end{bmatrix} \mid a, b \in \mathbb{R} \right\} = \mathbb{R}^2.$$

   These vectors are linearly independent and point along orthogonal axes, so together they span the entire 2D plane.

   Figure 5.10 shows the standard basis vectors $\mathbf{v} = \begin{bmatrix} 1 \\ 0 \end{bmatrix}$ and $\mathbf{u} = \begin{bmatrix} 0 \\ 1 \end{bmatrix}$, which span the entire two-dimensional space $\mathbb{R}^2$. The shaded blue region represents all possible linear combinations of $\mathbf{v}$ and $\mathbf{u}$, forming a complete plane through the origin. Since the vectors are linearly independent and point in orthogonal directions, they serve as a minimal basis for $\mathbb{R}^2$, demonstrating how two directions are sufficient to reach any point in the plane.

3. **Span in $\mathbb{R}^3$**

   Consider the standard basis vectors:

   $$\mathbf{e}_1 = \begin{bmatrix} 1 \\ 0 \\ 0 \end{bmatrix}, \quad \mathbf{e}_2 = \begin{bmatrix} 0 \\ 1 \\ 0 \end{bmatrix}, \quad \mathbf{e}_3 = \begin{bmatrix} 0 \\ 0 \\ 1 \end{bmatrix}.$$

   Then:

   $$\mathrm{Span}(\mathbf{e}_1, \mathbf{e}_2, \mathbf{e}_3) = \mathbb{R}^3.$$

Fig. 5.9 A vector **v** and the one-dimensional subspace it generates. The dashed line represents all scalar multiples of **v**, i.e., span{**v**} $= \{t(1, 2)/t \in R\}$

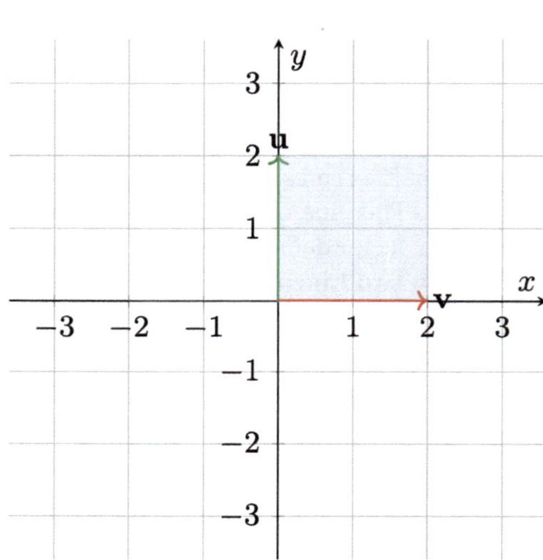

Fig. 5.10 The span of **v** $= (1, 0)$ and **u** $= (0, 1)$ is the full plane $\mathbb{R}^2$ the shaded region illustrates part of $\mathbb{R}^2$

Any 3D vector $\mathbf{x} = \begin{bmatrix} a \\ b \\ c \end{bmatrix}$ can be expressed as:

$$a\mathbf{e}_1 + b\mathbf{e}_2 + c\mathbf{e}_3.$$

The geometric nature of a span depends on the number and independence of the vectors:

- One nonzero vector spans a **line** (1D).
- Two linearly independent vectors span a **plane** (2D).
- Three linearly independent vectors in $\mathbb{R}^3$ span the entire **3D space**.

Connection to Basis and Dimension

The span of a set of vectors tells us what subspace they generate. If the vectors are linearly independent and minimal in number, they form a **basis** of that subspace. The number of vectors in any basis of a span is called its **dimension**:

$$\dim(\mathrm{Span}(\mathbf{v}_1, \ldots, \mathbf{v}_k)) = \text{number of linearly independent vectors among } \mathbf{v}_1, \ldots, \mathbf{v}_k.$$

This bridges the concepts of span, linear independence, basis, and dimension in a unified framework.

Example 5.10 Let:

$$S_1 = \left\{ \begin{bmatrix} 1 \\ -1 \end{bmatrix} \right\}, \quad S_2 = \left\{ \begin{bmatrix} 1 \\ -1 \end{bmatrix}, \begin{bmatrix} -2 \\ 2 \end{bmatrix} \right\}.$$

Observe that the second vector in S_2 is a scalar multiple of the first:

$$\begin{bmatrix} -2 \\ 2 \end{bmatrix} = -2 \cdot \begin{bmatrix} 1 \\ -1 \end{bmatrix}.$$

Hence, both sets span the same line through the origin:

$$\mathrm{Span}(S_1) = \mathrm{Span}(S_2) = \left\{ t \begin{bmatrix} 1 \\ -1 \end{bmatrix} \mid t \in \mathbb{R} \right\}.$$

Example 5.11 Let:

$$S_3 = \left\{ \begin{bmatrix} 1 \\ -1 \end{bmatrix}, \begin{bmatrix} -2 \\ 2 \end{bmatrix}, \begin{bmatrix} 2 \\ 1 \end{bmatrix} \right\}.$$

The first two vectors are linearly dependent, but the third vector is not a linear combination of them. Hence, the set S_3 spans all of $\mathbb{R}^2$. This is an example of redundancy in spanning sets: more vectors than needed, but still generating the full space.

5.4 Linear Dependence and Independence

Understanding whether a set of vectors is linearly independent or dependent is essential, as it directly relates to the structure and dimension of the subspace they span.

> **Definition**
> A set of vectors $\mathbf{v}_1, \mathbf{v}_2, \ldots, \mathbf{v}_n \in \mathbb{R}^m$ is said to be **linearly independent** if the only solution to the equation:

$$c_1\mathbf{v}_1 + c_2\mathbf{v}_2 + \cdots + c_n\mathbf{v}_n = \mathbf{0}$$

is the trivial solution:

$$c_1 = c_2 = \cdots = c_n = 0.$$

In other words, no vector in the set can be written as a linear combination of the others. Each vector contributes a new direction or dimension to the span.

Definition
A set of vectors is **linearly dependent** if there exists a nontrivial solution (i.e., at least one $c_i \neq 0$) such that:

$$c_1\mathbf{v}_1 + c_2\mathbf{v}_2 + \cdots + c_n\mathbf{v}_n = \mathbf{0}.$$

This implies that at least one of the vectors can be written as a linear combination of the others. In this case, the set is redundant in describing the span, and the vectors do not contribute additional dimensions to the subspace.

From geometric point of view:

- In $\mathbb{R}^2$: Two vectors are linearly dependent if they lie on the same line (i.e., are scalar multiples of each other).
- In $\mathbb{R}^3$: Three vectors are linearly dependent if they lie in the same plane or are coplanar (e.g., one vector lies in the plane formed by the other two).
- More generally, if the span of n vectors has dimension less than n, then the vectors are linearly dependent.

To test linear dependence, construct the homogeneous system:

$$A\mathbf{c} = \mathbf{0}, \quad \text{where } A = [\mathbf{v}_1 \ \mathbf{v}_2 \ \cdots \ \mathbf{v}_n].$$

Then:

- If the only solution is $\mathbf{c} = \mathbf{0}$, the columns are linearly independent.
- If a nontrivial solution exists, the columns are linearly dependent.

This can be checked using Gaussian elimination or computing the rank of the matrix.

Example 5.12 Let:

$$\mathbf{v}_1 = \begin{bmatrix} 2 \\ 3 \end{bmatrix}, \quad \mathbf{v}_2 = \begin{bmatrix} -4 \\ -6 \end{bmatrix}.$$

We observe that $\mathbf{v}_2 = -2\mathbf{v}_1$. Therefore, $\mathbf{v}_1$ and $\mathbf{v}_2$ are linearly dependent. They lie on the same line and span a one-dimensional subspace.

Example 5.13 Let:

$$\mathbf{v}_1 = \begin{bmatrix} 1 \\ 0 \end{bmatrix}, \quad \mathbf{v}_2 = \begin{bmatrix} 0 \\ 1 \end{bmatrix}.$$

These vectors point in orthogonal directions, and no scalar multiple of one can produce the other. They are linearly independent and span the entire plane $\mathbb{R}^2$.

Example 5.14 Let:

$$\mathbf{v}_1 = \begin{bmatrix} 1 \\ 0 \\ 0 \end{bmatrix}, \quad \mathbf{v}_2 = \begin{bmatrix} 0 \\ 1 \\ 0 \end{bmatrix}, \quad \mathbf{v}_3 = \begin{bmatrix} 1 \\ 1 \\ 0 \end{bmatrix}.$$

We see that $\mathbf{v}_3 = \mathbf{v}_1 + \mathbf{v}_2$. Therefore, these vectors are linearly dependent, and they span only the xy-plane in $\mathbb{R}^3$, not the full 3D space.

Example 5.15 Given:

$$\mathbf{v}_1 = \begin{bmatrix} 1 \\ 1 \\ 1 \end{bmatrix}, \quad \mathbf{v}_2 = \begin{bmatrix} 1 \\ -1 \\ 2 \end{bmatrix}, \quad \mathbf{v}_3 = \begin{bmatrix} 3 \\ 1 \\ 4 \end{bmatrix}.$$

Solve: $c_1\mathbf{v}_1 + c_2\mathbf{v}_2 + c_3\mathbf{v}_3 = \mathbf{0}$, or:

$$\begin{bmatrix} 1 & 1 & 3 \\ 1 & -1 & 1 \\ 1 & 2 & 4 \end{bmatrix} \begin{bmatrix} c_1 \\ c_2 \\ c_3 \end{bmatrix} = \begin{bmatrix} 0 \\ 0 \\ 0 \end{bmatrix}.$$

To determine whether the vectors are linearly dependent, we perform row reduction on the augmented matrix formed by placing the vectors as columns. Through a sequence of elementary row operations (including row swapping, scaling, and replacement), the matrix is transformed into Row Echelon Form (REF) (see Sect. 5.4.1 for more details). In this case, the final row of zeros indicates that one equation is redundant, revealing a rank of 2 for a system with 3 unknowns. This confirms the existence of a free variable, meaning the solution space is infinite and the vectors are linearly dependent.

$$\begin{bmatrix} 1 & 1 & 3 \\ 1 & -1 & 1 \\ 1 & 2 & 4 \end{bmatrix} \xrightarrow{R_2 - R_1} \begin{bmatrix} 1 & 1 & 3 \\ 0 & -2 & -2 \\ 1 & 2 & 4 \end{bmatrix} \xrightarrow{R_3 - R_1} \begin{bmatrix} 1 & 1 & 3 \\ 0 & -2 & -2 \\ 0 & 1 & 1 \end{bmatrix} \xrightarrow{R_3 + \frac{1}{2}R_2} \begin{bmatrix} 1 & 1 & 3 \\ 0 & -2 & -2 \\ 0 & 0 & 0 \end{bmatrix}.$$

The rank is 2, indicating a free variable: $c_3 = t$, $c_2 = -t$, $c_1 = -2t$.

Thus, the vectors are linearly dependent.

Example 5.16 Consider the set of vectors:

$$\mathbf{v}_1 = \begin{bmatrix} 1 \\ 1 \\ -2 \end{bmatrix}, \quad \mathbf{v}_2 = \begin{bmatrix} 1 \\ -1 \\ 2 \end{bmatrix}, \quad \mathbf{v}_3 = \begin{bmatrix} 3 \\ 1 \\ 4 \end{bmatrix}$$

We want to determine whether this set is linearly dependent or independent. To do this, we solve:

$$x_1 \mathbf{v}_1 + x_2 \mathbf{v}_2 + x_3 \mathbf{v}_3 = \mathbf{0}.$$

This gives the system:

$$x_1 \begin{bmatrix} 1 \\ 1 \\ -2 \end{bmatrix} + x_2 \begin{bmatrix} 1 \\ -1 \\ 2 \end{bmatrix} + x_3 \begin{bmatrix} 3 \\ 1 \\ 4 \end{bmatrix} = \begin{bmatrix} 0 \\ 0 \\ 0 \end{bmatrix},$$

Writing this as an augmented matrix:

$$\begin{bmatrix} 1 & 1 & 3 \\ 1 & -1 & 1 \\ -2 & 2 & 4 \end{bmatrix} \rightarrow \text{Row Reduce.}$$

Row Reduction Steps

- Step 1- Use the first row as the pivot.
  Replace $R_2 \leftarrow R_2 - R_1$: This operation subtracts the first row from the second row, eliminating the leading entry in R_2 to simplify the system.
  Replace $R_3 \leftarrow R_3 + 2R_1$: This operation adds twice the first row to the third row, helping to create a zero in a desired position of R_3.

$$\begin{bmatrix} 1 & 1 & 3 \\ 0 & -2 & -2 \\ 0 & 4 & 10 \end{bmatrix}.$$

- Step 2- Normalize the second row: $R_2 \leftarrow \frac{1}{-2} R_2$

$$\begin{bmatrix} 1 & 1 & 3 \\ 0 & 1 & 1 \\ 0 & 4 & 10 \end{bmatrix}.$$

- Step 3- Eliminate entries above and below the pivot in column 2: $R_3 \leftarrow R_3 - 4R_2$, $R_1 \leftarrow R_1 - R_2$

$$\begin{bmatrix} 1 & 0 & 2 \\ 0 & 1 & 1 \\ 0 & 0 & 6 \end{bmatrix}.$$

- Step 4- Normalize third row: $R_3 \leftarrow \frac{1}{6} R_3$

$$\begin{bmatrix} 1 & 0 & 2 \\ 0 & 1 & 1 \\ 0 & 0 & 1 \end{bmatrix}$$

- Step 5- Eliminate above the pivot in column 3:
$R_2 \leftarrow R_2 - R_3$,
$R_1 \leftarrow R_1 - 2R_3$

$$\begin{bmatrix} 1 & 0 & 0 \\ 0 & 1 & 0 \\ 0 & 0 & 1 \end{bmatrix}.$$

This is the identity matrix, i.e., the Reduced Row Echelon Form (RREF) (see the next subsection, e.g., 5.4.1).

Since the only solution is the trivial one:

$$x_1 = x_2 = x_3 = 0.$$

We conclude that the set $\{v_1, v_2, v_3\}$ is **linearly independent**.

5.4.1 RREF

The RREF of a matrix is a special form that a matrix can be transformed into using elementary row operations.

> **Definition**
> A matrix is in **RREF** if it satisfies the following conditions:
> 1. All nonzero rows are above any rows of all zeros.
> 2. The leading entry (first nonzero number from the left) of each nonzero row is 1, called a *leading 1*.
> 3. Each leading 1 is the only nonzero entry in its column.
> 4. The leading 1 of a lower row appears to the right of the leading 1 in the row above.

RREF is unique for any given matrix and is often used to solve systems of linear equations, determine the rank of a matrix, and find bases for row and column spaces.

5.4.2   *Connection to Span, Basis, and Dimension*

Linear independence plays a crucial role in defining a **basis**. A basis is a minimal set of linearly independent vectors that spans a vector space. The number of vectors in the basis equals the **dimension** of the space:

$$\text{If } \{\mathbf{v}_1, \ldots, \mathbf{v}_k\} \text{ is a basis for } V, \text{ then } \dim(V) = k.$$

- A set of vectors is linearly independent if no vector can be expressed as a combination of the others.
- A set is linearly dependent if at least one vector depends on the others.
- Independence implies uniqueness in solutions; dependence implies redundancy.
- The concept is foundational for understanding dimension, span, null space, and basis.

Example 5.17 Consider the following vectors in $\mathbb{R}^3$:

$$\mathbf{u} = \begin{bmatrix} 1 \\ -1 \\ 0 \end{bmatrix}, \quad \mathbf{v} = \begin{bmatrix} 0 \\ 1 \\ -1 \end{bmatrix}, \quad \mathbf{w} = \begin{bmatrix} 0 \\ 0 \\ 1 \end{bmatrix}, \quad \mathbf{w}^* = \begin{bmatrix} -1 \\ 0 \\ -1 \end{bmatrix}.$$

We investigate whether the sets $\{\mathbf{u}, \mathbf{v}, \mathbf{w}\}$ and $\{\mathbf{u}, \mathbf{v}, \mathbf{w}^*\}$ are linearly independent or dependent by solving:

$$x\mathbf{u} + y\mathbf{v} + z\mathbf{w} = \mathbf{0}, \quad \text{and} \quad x\mathbf{u} + y\mathbf{v} + z\mathbf{w}^* = \mathbf{0}.$$

Case 1: Set $\{\mathbf{u}, \mathbf{v}, \mathbf{w}\}$
Construct the matrix $A = [\mathbf{u} \ \mathbf{v} \ \mathbf{w}]$:

$$A = \begin{bmatrix} 1 & 0 & 0 \\ -1 & 1 & 0 \\ 0 & -1 & 1 \end{bmatrix}.$$

We check the rank of A. Since it is lower triangular with non-zero diagonal entries, it has full rank (rank 3), indicating that the only solution to $A \begin{bmatrix} x \\ y \\ z \end{bmatrix} = \mathbf{0}$ is the trivial solution $x = y = z = 0$. Therefore, the vectors are linearly independent and span the full $\mathbb{R}^3$ space.

Case 2: Set $\{\mathbf{u}, \mathbf{v}, \mathbf{w}^*\}$
Observe directly:

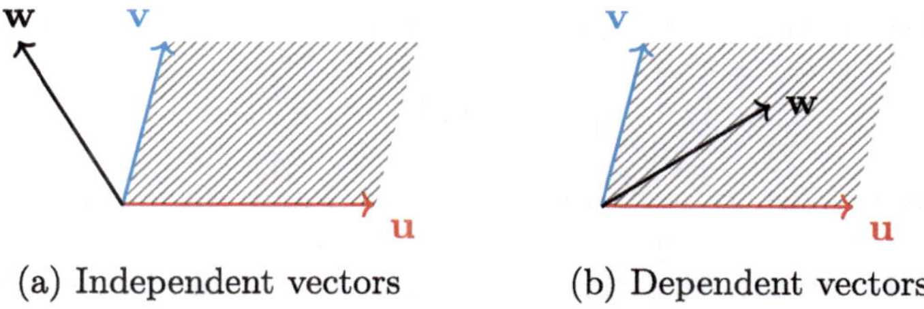

Fig. 5.11 Visualization of three linear independent (left) and dependent (right) vectors in 3-dimensions

$$\mathbf{u} + \mathbf{v} + \mathbf{w}^* = \begin{bmatrix} 1 \\ -1 \\ 0 \end{bmatrix} + \begin{bmatrix} 0 \\ 1 \\ -1 \end{bmatrix} + \begin{bmatrix} -1 \\ 0 \\ -1 \end{bmatrix} = \begin{bmatrix} 0 \\ 0 \\ -2 \end{bmatrix} \neq \mathbf{0}.$$

However, we test linear dependence:

$$\mathbf{w}^* = -\mathbf{u} - \mathbf{v},$$

which implies that $\mathbf{w}^*$ can be written as a linear combination of $\mathbf{u}$ and $\mathbf{v}$, making the set linearly dependent. Geometrically, all three vectors lie on the same plane in $\mathbb{R}^3$ (Fig. 5.11).

In the left subfigure above, vectors $\mathbf{u}$, $\mathbf{v}$, $\mathbf{w}$ are linearly independent, meaning they span the entire three-dimensional space. In the right subfigure, $\mathbf{w}$ lies in the plane formed by $\mathbf{u}$ and $\mathbf{v}$, making the set linearly dependent and reducing the span to a two-dimensional subspace.

Matrix Interpretation with Solution Behavior

Example 5.18 Consider the matrix equation:

$$A\mathbf{x} = \mathbf{b},$$

where

$$A = \begin{bmatrix} 6 & 1 & 7 \\ 3 & 8 & 11 \\ 3 & 3 & 6 \end{bmatrix}, \quad \mathbf{b} = \begin{bmatrix} 14 \\ 22 \\ 12 \end{bmatrix}.$$

Solving this system:

$$\begin{bmatrix} 6 & 1 & 7 \\ 3 & 8 & 11 \\ 3 & 3 & 6 \end{bmatrix} \begin{bmatrix} x_1 \\ x_2 \\ x_3 \end{bmatrix} = \begin{bmatrix} 14 \\ 22 \\ 12 \end{bmatrix} \Rightarrow \begin{bmatrix} x_1 \\ x_2 \\ x_3 \end{bmatrix} = \begin{bmatrix} 1 \\ 1 \\ 1 \end{bmatrix}.$$

Observe that the third column of A is the sum of the first two columns:

$$[7\ 11\ 6]^T = [6\ 3\ 3]^T + [1\ 8\ 3]^T.$$

This means the columns satisfy $1 \cdot a_1 + 1 \cdot a_2 - 1 \cdot a_3 = 0$, a nontrivial solution to the homogeneous system $Ac = 0$. Therefore, the column vectors of A are linearly dependent. The nontrivial null vector $c = [1\ 1\ -1]^T$ confirms that the system $Ax = b$, while solvable, does not have a unique solution.

5.5 Solving Methods for Linear Systems

In Sect. 5.1, we introduced systems of linear equations. In this section, we explore additional methods that are useful for solving such systems.

Solving a system of linear equations means finding the values of variables that satisfy all equations simultaneously. In matrix form, this is expressed as

$$A\mathbf{x} = \mathbf{b},$$

where A is the coefficient matrix, $\mathbf{x}$ is the vector of unknowns, and $\mathbf{b}$ is the constants vector.

Depending on the size and structure of the system, as well as available computational tools, there are several fundamental methods to find solutions:

- **Substitution Method**
- **Elimination Method**
- **Matrix Methods** (including Gaussian elimination, inverse matrices, and Cramer's rule)
- **Graphical Method**

Each method offers unique insights and computational advantages, which we will discuss in detail.

5.5.1 Substitution Method

The substitution method is a straightforward algebraic technique ideal for small systems, especially those with two or three variables.

The main idea is to solve one equation for one variable in terms of the others, then substitute this expression into the remaining equations. Repeating this process reduces the number of variables step-by-step until all variable values are found.

Example 5.19 Solve

$$\begin{cases} x + 2y = 5, \\ 3x - y = 2. \end{cases}$$

Step 1- Solve the first equation for x:

$$x = 5 - 2y.$$

Step 2- Substitute into the second equation:

$$3(5 - 2y) - y = 2 \implies 15 - 6y - y = 2 \implies 15 - 7y = 2.$$

Step 3- Solve for y:

$$-7y = 2 - 15 = -13 \implies y = \frac{13}{7}.$$

Step 4- Back-substitute to find x:

$$x = 5 - 2 \times \frac{13}{7} = 5 - \frac{26}{7} = \frac{35 - 26}{7} = \frac{9}{7}.$$

$$\boxed{\left(\tfrac{9}{7}, \tfrac{13}{7}\right)}$$

This method is simple and intuitive but becomes unwieldy for larger systems.

5.5.2 Elimination Method

The elimination method, also called the addition or subtraction method, focuses on eliminating variables by adding or subtracting equations, simplifying the system stepwise.

The idea is to manipulate equations to cancel out one variable, reducing the system to fewer variables.

Example 5.20 Solve

$$\begin{cases} 2x + 3y = 8, \\ 4x - y = 2. \end{cases}$$

Step 1- Make the coefficients of y opposites: Multiply the second equation by 3:

$$4x - y = 2 \implies 12x - 3y = 6.$$

Step 2- Add to the first equation:

$$(2x + 3y) + (12x - 3y) = 8 + 6 \implies 14x = 14.$$

Step 3- Solve for x:

$$x = 1.$$

Step 4- Substitute back to find y:

$$2(1) + 3y = 8 \implies 3y = 6 \implies y = 2.$$

$$\boxed{(1, 2)}$$

Elimination is often more efficient than substitution for larger systems but still less suitable than matrix methods for high-dimensional problems.

5.5.3   Matrix Methods

Matrix methods are powerful, systematic techniques that use matrix algebra to solve linear systems efficiently, especially for large or complex systems.

Elementary Row Operations

Solving a linear system often involves transforming the system into a simpler equivalent system using elementary row operations that preserve the solution set:

1. **Row Interchange**: Swap two rows.
2. **Row Scaling**: Multiply a row by a nonzero scalar.
3. **Row Addition**: Replace a row by the sum of itself and a multiple of another row.

These operations allow us to simplify the system without changing its solutions.

Augmented Matrices

A convenient representation of a system is the **augmented matrix**, which combines the coefficient matrix A and the constants vector $\mathbf{b}$ into one matrix:

$$[A \mid \mathbf{b}] = \begin{bmatrix} a_{11} & a_{12} & \cdots & a_{1n} & | & b_1 \\ a_{21} & a_{22} & \cdots & a_{2n} & | & b_2 \\ \vdots & \vdots & \ddots & \vdots & | & \vdots \\ a_{m1} & a_{m2} & \cdots & a_{mn} & | & b_m \end{bmatrix}.$$

We then apply elementary row operations to simplify this matrix and find solutions.

Example 5.21 Solve

$$\begin{cases} x + 2y = 5, \\ 3x - y = 2. \end{cases}$$

via augmented matrix and row reduction.

$$[A \mid \mathbf{b}] = \begin{bmatrix} 1 & 2 & \mid & 5 \\ 3 & -1 & \mid & 2 \end{bmatrix}.$$

Perform the following steps:

1. $R_2 \leftarrow R_2 - 3R_1$:

$$\begin{bmatrix} 1 & 2 & \mid & 5 \\ 0 & -7 & \mid & -13 \end{bmatrix}.$$

2. $R_2 \leftarrow -\frac{1}{7} R_2$:

$$\begin{bmatrix} 1 & 2 & \mid & 5 \\ 0 & 1 & \mid & \frac{13}{7} \end{bmatrix}.$$

3. $R_1 \leftarrow R_1 - 2R_2$:

$$\begin{bmatrix} 1 & 0 & \mid & \frac{9}{7} \\ 0 & 1 & \mid & \frac{13}{7} \end{bmatrix}.$$

As shown in Fig. 5.12, the solution is

$$\boxed{x = \frac{9}{7}, \, y = \frac{13}{7}}$$

Gaussian Elimination

Gaussian elimination uses elementary row operations to convert the augmented matrix into an upper-triangular or REF, which makes back-substitution straightforward.

REF Requirements:

- All nonzero rows are above any zero rows.
- The first nonzero element (pivot) of each row is 1.
- Each pivot is to the right of the pivot in the row above.
- Entries below each pivot are zero.

RREF further requires zeros above each pivot as well, giving a matrix in its simplest form.

Example 5.22 Solve

$$\begin{cases} 2x + 4y - 2z = 2, \\ 4x + 9y - 3z = 8, \\ -2x - 3y + 7z = 10. \end{cases}$$

$$\Rightarrow \begin{bmatrix} 2 & 4 & -2 & | & 2 \\ 4 & 9 & -3 & | & 8 \\ -2 & -3 & 7 & | & 10 \end{bmatrix}.$$

Gaussian elimination proceeds:

1. $R_1 \leftarrow \frac{1}{2} R_1$
2. $R_2 \leftarrow R_2 - 4R_1$, $R_3 \leftarrow R_3 + 2R_1$
3. $R_3 \leftarrow R_3 - R_2$
4. Back-substitute to get RREF:

$$\begin{bmatrix} 1 & 0 & 0 & | & -1 \\ 0 & 1 & 0 & | & 2 \\ 0 & 0 & 1 & | & 2 \end{bmatrix}.$$

Fig. 5.12 Graphical illustration of the solution to the system

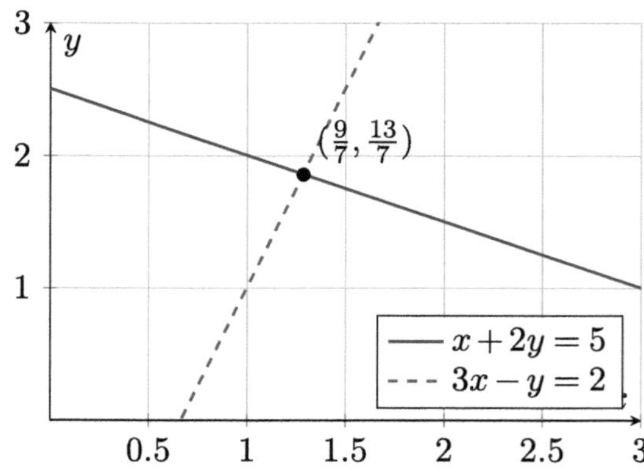

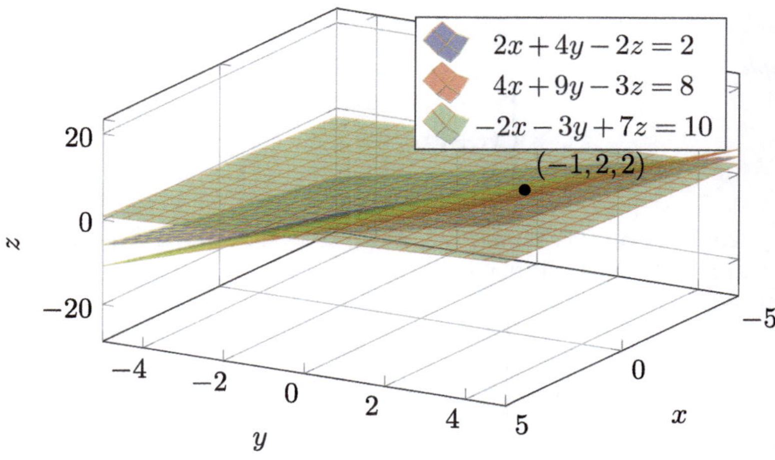

Fig. 5.13 Three planes intersect at the unique solution point

As illustrated in Fig. 5.13, the solution is:

$$\boxed{\mathbf{x} = (-1, 2, 2)}$$

Special Cases: Infinite and No Solutions

In solving systems of linear equations, the outcome may result in either infinitely many solutions or no solution at all. Infinitely many solutions occur when the equations are linearly dependent, leading to free variables and a solution space that forms a line or plane. In contrast, no solution arises when the equations are inconsistent, typically representing parallel hyperplanes that never intersect. These cases highlight the importance of analyzing the rank and consistency of the system to determine its solvability.

(a) **Infinite Solutions**: occur when there are free variables, columns without pivots in RREF, allowing infinitely many parameterized solutions.

Example 5.23

$$\begin{cases} x_1 - 3x_2 + 2x_4 = 7, \\ x_3 + 6x_4 = 9, \\ x_5 = 2. \end{cases}$$

This system has pivots in columns 1, 3, and 5, so x_2 and x_4 are free parameters, e.g. $x_2 = s$, $x_4 = t$. The general solution is

$$\mathbf{x} = \begin{bmatrix} 7 + 3s - 2t \\ s \\ 9 - 6t \\ t \\ 2 \end{bmatrix}, \quad s, t \in \mathbb{R}.$$

(b) **No Solution**: arises when a contradictory equation appears after row reduction, such as

$$0 = c, \quad c \neq 0,$$

implying inconsistency.

Example 5.24

$$\begin{cases} x_1 + x_2 = 1, \\ x_1 + x_2 = 2. \end{cases}$$

Row reduction yields

$$\begin{bmatrix} 1 & 1 & | & 1 \\ 0 & 0 & | & 1 \end{bmatrix},$$

which translates to $0 = 1$, a contradiction.

Figure 5.14 depicts two parallel lines in the $x_1 x_2$-plane corresponding to the equations $x_1 + x_2 = 1$ and $x_1 + x_2 = 2$. Because these lines never intersect, the system has no solution, exemplifying an inconsistent system of linear equations.

5.5.4 Graphical Method

The graphical method provides an intuitive visual approach to solving systems, primarily useful for two-variable systems.

The core idea behind the graphical methods is that each linear equation corresponds to a line in $\mathbb{R}^2$. The solution to the system is the point(s) where these lines intersect.

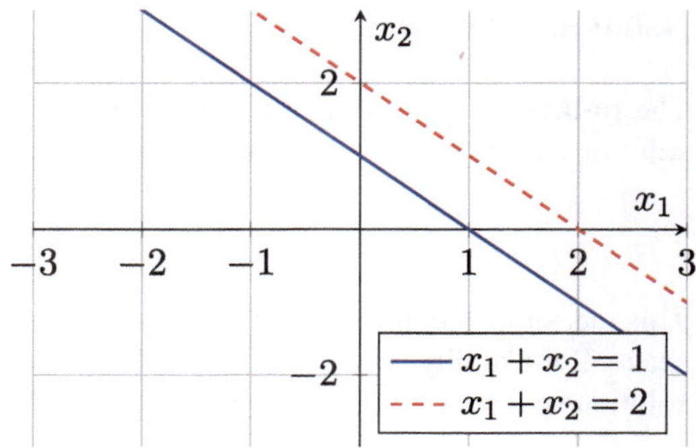

Fig. 5.14 Parallel lines with no intersection imply no solution

Example 5.25 For the system

$$\begin{cases} x + 2y = 5, \\ 3x - y = 2, \end{cases}$$

we plot both lines on the Cartesian plane. Their intersection point corresponds to the solution $\left(\frac{9}{7}, \frac{13}{7}\right)$.

This method gives a geometric understanding but is limited to small systems and cannot provide exact solutions for larger or higher-dimensional problems.

5.6 Matrix Rank and Solutions

Understanding the structure of a linear system hinges on two fundamental properties of matrices: **rank** and **nullity**. These concepts help determine whether a system has no solution, a unique solution, or infinitely many solutions.

Rank of a Matrix

The **rank** of a matrix $A \in \mathbb{R}^{m \times n}$ is the maximum number of linearly independent rows or columns. It reflects the amount of non-redundant information contained in the matrix. More precisely, the rank equals the number of **pivot columns** (or leading 1s) when the matrix is reduced to its RREF:

$$\text{rank}(A) = \# \text{ of pivot columns in RREF}(A).$$

This tells us how many of the input variables (columns) contribute new, non-overlapping information about the system. High rank implies more constraints and potentially a more well-defined solution.

Nullity and the Null Space

The **nullity** of a matrix is the dimension of its **null space**, which is the set of all solutions to the homogeneous equation:

$$Ax = 0.$$

This measures the number of **free variables**, or degrees of freedom, in the solution space. If the nullity is zero, the only solution is the trivial one: $x = 0$. Otherwise, the solution space forms a subspace of $\mathbb{R}^n$ with infinitely many vectors.

The Rank-Nullity Theorem

A key result connecting rank and nullity is the **Rank-Nullity Theorem**, which states:

$$\text{rank}(A) + \text{nullity}(A) = n,$$

where, n is the number of columns of matrix A. This theorem encapsulates a fundamental trade-off: the more constraints (higher rank), the fewer degrees of freedom (lower nullity), and vice versa. We will learn more about it in Chap. 6.

Rank and Solution Structure of Linear Systems

Given a system $Ax = b$, where $A \in \mathbb{R}^{m \times n}$ and $b \in \mathbb{R}^m$, the rank helps classify the nature of solutions:

- **Consistent System**: If $\text{rank}(A) = \text{rank}([A \mid b])$, the system has at least one solution.
- **Unique Solution**: If the system is consistent and $\text{rank}(A) = n$, then all variables are pivot variables, and the solution is unique.
- **Infinitely Many Solutions**: If the system is consistent and $\text{rank}(A) < n$, then there are free variables, leading to an infinite number of solutions.
- **Inconsistent System**: If $\text{rank}(A) < \text{rank}([A \mid b])$, then no solution exists. The vector b lies outside the column space of A.

Remark

Summary of Key Terms:

- **Rank**: Number of linearly independent columns (or rows); determines the dimension of the column space.
- **Nullity**: Dimension of the null space; number of free variables in $Ax = 0$.

- **Column Space**: Span of the columns of A; the set of all possible vectors $\mathbf{b}$ for which $A\mathbf{x} = \mathbf{b}$ has a solution.
- **Null Space**: Set of all vectors $\mathbf{x} \in \mathbb{R}^n$ such that $A\mathbf{x} = \mathbf{0}$; relates to the homogeneous solution structure.

These concepts are not only critical in understanding solution spaces but are also foundational in applications such as dimensionality reduction, feature selection in ML, and understanding over- and under-determined systems.

Example 5.26 For:

$$A = \begin{bmatrix} 1 & 0 & 0 & 0 & 0 \\ 0 & 1 & 0 & 0 & 0 \\ 0 & 0 & 1 & 0 & 0 \\ 0 & 0 & 0 & 1 & 0 \\ 0 & 0 & 0 & 0 & 0 \end{bmatrix}.$$

The matrix is in RREF with 4 pivots. Thus, $\text{rank}(A) = 4$, $\text{nullity}(A) = 5 - 4 = 1$.

The null space is spanned by $\begin{bmatrix} 0 \\ 0 \\ 0 \\ 0 \\ 1 \end{bmatrix}$.

Python Example: Rank and Nullity

```python
import numpy as np

A = np.array([[1, 0, 0, 0, 0],
              [0, 1, 0, 0, 0],
              [0, 0, 1, 0, 0],
              [0, 0, 0, 1, 0],
              [0, 0, 0, 0, 0]])

rank = np.linalg.matrix_rank(A)
nullity = A.shape[1] - rank
print("Rank:", rank, "Nullity:", nullity)
```

Output:

```
Rank: 4 Nullity: 1
```

Example 5.27 Consider the matrix:

$$A = \begin{bmatrix} 1 & 1 & 2 & 3 \\ 3 & 4 & -1 & 2 \\ 3 & 4 & -1 & 2 \\ -1 & -2 & 5 & 4 \end{bmatrix}.$$

We aim to find the rank and nullity of this matrix by row reducing it to its RREF. The rank will be the number of non-zero rows (or pivot positions), and the nullity will be the number of free variables, calculated as nullity $= n -$ rank, where n is the number of columns.

Step 1- Use the first row as the pivot row: Eliminate entries below the leading 1 in column 1.

$$R_2 \leftarrow R_2 - 3R_1, \quad R_3 \leftarrow R_3 - 3R_1, \quad R_4 \leftarrow R_4 + R_1$$

$$\begin{bmatrix} 1 & 1 & 2 & 3 \\ 0 & 1 & -7 & -7 \\ 0 & 1 & -7 & -7 \\ 0 & -1 & 7 & 7 \end{bmatrix}$$

We subtract appropriate multiples of the first row to create zeros below the pivot in the first column. Notice that after this step, rows 2 and 3 become identical, which signals potential redundancy.

Step 2- Use the second row as the next pivot row: Eliminate entries below it.

$$R_3 \leftarrow R_3 - R_2, \quad R_4 \leftarrow R_4 + R_2$$

$$\begin{bmatrix} 1 & 1 & 2 & 3 \\ 0 & 1 & -7 & -7 \\ 0 & 0 & 0 & 0 \\ 0 & 0 & 0 & 0 \end{bmatrix}.$$

Row 3 becomes a zero row, showing that it was linearly dependent on the first two rows. Similarly, row 4 also reduces to a zero row. This confirms that the information in these rows is redundant and does not contribute to increasing the rank. The matrix in RREF has two non-zero rows with leading 1s (pivot positions). Therefore:

$$\text{rank}(A) = 2.$$

Since $A \in \mathbb{R}^{4 \times 4}$, the number of columns is $n = 4$. Applying the Rank-Nullity Theorem:

$$\text{nullity}(A) = 4 - 2 = 2.$$

The null space of A has dimension 2, which means the homogeneous system $A\mathbf{x} = \mathbf{0}$ has infinitely many solutions forming a two-dimensional subspace of $\mathbb{R}^4$. There are two free variables in the general solution, and the solution set can be expressed as a linear combination of two basis vectors in the null space.

5.7 Linear Systems in ML

Linear systems are foundational to ML, modeling relationships between features and outputs or constraints in optimization problems.

5.7.1 Linear Regression (LR)

LR solves $X\mathbf{w} \approx \mathbf{y}$, where $X \in \mathbb{R}^{m \times n}$ is the feature matrix and $\mathbf{y} \in \mathbb{R}^m$ is the target vector. For overdetermined systems ($m > n$), the normal equations are:

$$X^T X \mathbf{w} = X^T \mathbf{y}.$$

5.7.2 Feature Engineering

Linear independence ensures features provide unique information. Redundant (linearly dependent) features cause multicollinearity, destabilizing models. PCA identifies linearly independent directions to reduce dimensionality.

5.7.3 Model Identifiability

Underdetermined systems (more features than samples) have infinite solutions, requiring regularization (e.g., LASSO) to select a unique solution. Overdetermined systems may be inconsistent due to noisy data, solved via least squares.

5.7.4 Other Applications

- **Sparse Coding**: Solves underdetermined systems to find sparse representations.
- **Neural Networks**: Weight updates involve linear systems, with hidden layers expanding the feature span.
- **Kernel Methods**: Decision boundaries are linear combinations of kernel evaluations.

Chapter Summary

This chapter explored the fundamentals of linear systems, establishing them as central to linear algebra and their pivotal role in modeling relationships and constraints in ML. It began with the definition of a system of linear equations in matrix form $A\mathbf{x} = \mathbf{b}$, where A is the coefficient matrix, $\mathbf{x}$ the unknown vector, and $\mathbf{b}$ the constant vector, as detailed in Sect. 5.1. Geometric interpretations were provided, with equations as hyperplanes whose intersections determine solution types: unique (intersecting at a point, Fig. 5.1), none (parallel hyperplanes, Fig. 5.3), or infinitely many (intersecting along a subspace, Fig. 5.5). Examples illustrated substitution for unique solutions like $\left(\frac{15}{4}, \frac{5}{4}\right)$, and parameterization for infinite cases.

Section 5.2 introduced linear combinations $c_1\mathbf{v}_1 + \cdots + c_n\mathbf{v}_n$ as building blocks for solution spaces, with matrix form $A\mathbf{x}$ linking to consistency when $\mathbf{b}$ lies in the column space, as visualized in Fig. 5.6 and inconsistency examples in Fig. 5.7.

The span in Sect. 5.3 was defined as all linear combinations of vectors, forming subspaces like lines (single vector, Fig. 5.9) or planes (two independent vectors, Fig. 5.10), always closed under addition and scalar multiplication, equating to the column space $\mathrm{Col}(A)$.

Linear dependence and independence in Sect. 5.4 were examined via the equation $c_1\mathbf{v}_1 + \cdots + c_n\mathbf{v}_n = \mathbf{0}$, with dependence implying redundancy (e.g., one vector as a multiple of another) and independence requiring only the trivial solution, checked through row reduction to RREF. Examples included dependent sets in $\mathbb{R}^2$ and $\mathbb{R}^3$, connecting to basis and dimension where independent vectors form a basis of dimension equal to their count.

Solving methods in Sect. 5.5 included substitution and elimination for small systems, augmented matrices $[A \mid \mathbf{b}]$ for row operations, and Gaussian elimination to REF/RREF for back-substitution, as in the 3D intersection at $(-1, 2, 2)$ (Fig. 5.13). Graphical methods visualized 2D cases (Fig. 5.12), with special cases of infinite solutions via free variables and none via contradictions (Fig. 5.14).

Matrix rank and solutions in Sect. 5.6 were analyzed through rank (number of pivots in RREF), nullity (dimension of null space, solutions to $A\mathbf{x} = \mathbf{0}$), and the Rank-Nullity Theorem $\mathrm{rank}(A) + \mathrm{nullity}(A) = n$. Solution classification depended on $\mathrm{rank}(A)$ vs. $\mathrm{rank}([A \mid \mathbf{b}])$: equal for consistency, full column rank for uniqueness, lower for infinite solutions, and mismatch for inconsistency.

Section 5.7 connected these to ML: LR via normal equations $X^T X \mathbf{w} = X^T \mathbf{y}$, feature engineering detecting dependence to avoid multicollinearity, identifiability in under/overdetermined systems with regularization, and applications in sparse coding, neural networks, and kernel methods.

Takeaways
Key concepts from the chapter were encapsulated as follows:

- Linear systems $A\mathbf{x} = \mathbf{b}$ as intersections of hyperplanes, classified by consistency and solution multiplicity, with geometric visualizations aiding intuition.
- Linear combinations and spans as subspaces generated by vectors, forming the column space where solutions reside, essential for understanding reachability in $\mathbb{R}^n$.
- Linear dependence (nontrivial null solution) vs. independence (trivial only), determining redundancy, basis formation, and dimension via RREF pivot counts.
- Solution techniques including substitution, elimination, augmented matrices, and Gaussian elimination to RREF, handling unique, parametric infinite, or inconsistent cases.
- Rank (independent columns/rows) and nullity (free variables) via Rank-Nullity Theorem, dictating solution structure: full rank for uniqueness, reduced for multiplicity or inconsistency.
- Applications in ML, from regression solving overdetermined systems to dependence analysis for robust features and regularization for identifiability.

Through these topics, readers developed skills to analyze, solve, and interpret linear systems, bridging algebraic methods with geometric insights and preparing for ML implementations using tools like NumPy.

Exercises

1. Solve the following system using the substitution method and verify the solution:

$$\begin{cases} x_1 + 2x_2 = 5, \\ 3x_1 - x_2 = 2. \end{cases}$$

   Plot the lines to confirm the intersection point.
2. Determine if the system is consistent and find its solution type (unique, infinite, or none):

$$\begin{cases} 2x_1 + 3x_2 = 6, \\ 4x_1 + 6x_2 = 12. \end{cases}$$

3. Compute the linear combination $3\mathbf{v}_1 - 2\mathbf{v}_2$ where $\mathbf{v}_1 = \begin{bmatrix} 1 \\ 2 \end{bmatrix}$ and $\mathbf{v}_2 = \begin{bmatrix} 3 \\ 4 \end{bmatrix}$. Verify using NumPy.

4. Is the vector $\begin{bmatrix} 5 \\ 10 \end{bmatrix}$ in the span of $\begin{bmatrix} 1 \\ 2 \end{bmatrix}$ and $\begin{bmatrix} 2 \\ 4 \end{bmatrix}$? Explain geometrically.

5. Check if the vectors $\begin{bmatrix} 1 \\ 0 \\ 0 \end{bmatrix}, \begin{bmatrix} 0 \\ 1 \\ 0 \end{bmatrix}, \begin{bmatrix} 0 \\ 0 \\ 1 \end{bmatrix}$ are linearly independent. What is their span?

6. Use Gaussian elimination to solve:

$$\begin{cases} x_1 + x_2 + x_3 = 6, \\ 2x_1 - x_2 + 3x_3 = 5, \\ -x_1 + 2x_2 - x_3 = -1. \end{cases}$$

Find the RREF of the augmented matrix.

7. Compute the rank and nullity of the matrix:

$$\begin{bmatrix} 1 & 2 & 3 \\ 2 & 4 & 6 \\ 3 & 6 & 9 \end{bmatrix}.$$

Using Python, confirm your result.

8. Determine if the system has a unique solution, infinite solutions, or no solution:

$$\begin{cases} x_1 - 2x_2 = 3, \\ 2x_1 - 4x_2 = 6, \\ 3x_1 - 6x_2 = 9. \end{cases}$$

9. Express the vector $\begin{bmatrix} 4 \\ 5 \\ 6 \end{bmatrix}$ as a linear combination of $\begin{bmatrix} 1 \\ 1 \\ 1 \end{bmatrix}$ and $\begin{bmatrix} 2 \\ 3 \\ 4 \end{bmatrix}$, or show that it is not in their span.

10. Find the span of the set $\left\{ \begin{bmatrix} 1 \\ 0 \end{bmatrix}, \begin{bmatrix} 0 \\ 1 \end{bmatrix}, \begin{bmatrix} 1 \\ 1 \end{bmatrix} \right\}$. Is this set linearly independent?

11. Solve using the elimination method:

$$\begin{cases} 4x_1 + 2x_2 = 10, \\ 2x_1 + 3x_2 = 9. \end{cases}$$

Verify with a graphical plot.

12. Check linear dependence for $\begin{bmatrix} 1 \\ 2 \\ 3 \end{bmatrix}$, $\begin{bmatrix} 4 \\ 5 \\ 6 \end{bmatrix}$, $\begin{bmatrix} 7 \\ 8 \\ 9 \end{bmatrix}$. Compute the null space dimension.

13. Use row operations to find the solution to:

$$\begin{cases} x_1 + 3x_2 - x_3 = 4, \\ 2x_1 + 6x_2 - 2x_3 = 8. \end{cases}$$

Parameterize if infinite solutions exist.

14. Determine the consistency of:

$$\begin{cases} x_1 + x_2 + x_3 = 1, \\ x_1 + x_2 + x_3 = 2. \end{cases}$$

Explain using rank comparison.

15. Compute the linear combination resulting in the zero vector for the dependent set $\begin{bmatrix} 1 \\ 1 \end{bmatrix}$, $\begin{bmatrix} 2 \\ 2 \end{bmatrix}$, and find nontrivial coefficients.

16. Find the RREF and rank of:

$$\begin{bmatrix} 1 & 1 & 1 \\ 1 & 2 & 3 \\ 1 & 3 & 5 \end{bmatrix}.$$

17. In a ML context, explain how linear dependence among features affects model training (e.g., singular $X^T X$). Provide a small example matrix.

18. Solve the underdetermined system:

$$x_1 + 2x_2 + 3x_3 = 6.$$

Express the infinite solutions parametrically.

19. Verify if $\begin{bmatrix} 0 \\ 0 \\ 0 \end{bmatrix}$ is always in the span of any set of vectors. Prove using the definition.

20. Use NumPy to solve and plot the system:

$$\begin{cases} 3x_1 - x_2 = 10, \\ x_1 - 3x_2 = 0. \end{cases}$$

21. Determine the dimension of the span of $\begin{bmatrix} 1 \\ 0 \\ 0 \end{bmatrix}$, $\begin{bmatrix} 0 \\ 1 \\ 0 \end{bmatrix}$, $\begin{bmatrix} 1 \\ 1 \\ 0 \end{bmatrix}$.

22. For the inconsistent system in Exercise 14, modify **b** to make it consistent and solve.
23. Check if the standard basis vectors in $\mathbb{R}^3$ are linearly independent. What is the nullity of the identity matrix?
24. Apply the Rank-Nullity Theorem to a 3×5 matrix with rank 2. What is the nullity?
25. In LR, explain how the normal equations form a linear system. Provide a small example with 2 features and 3 samples.
26. Solve using Gaussian elimination and discuss the solution type:

$$\begin{cases} x_1 + x_2 = 1, \\ 2x_1 + 2x_2 = 2, \\ 3x_1 + 3x_2 = 3. \end{cases}$$

27. Determine if $\begin{bmatrix} 2 \\ 4 \\ 6 \end{bmatrix}$ is in the span of $\begin{bmatrix} 1 \\ 2 \\ 3 \end{bmatrix}$. If yes, find the scalar.
28. Compute the rank of a zero matrix and discuss its implications for linear systems.
29. For the system with infinite solutions in Exercise 13, find two specific solutions and plot them in 3D to illustrate the solution set.
30. Explain how PCA addresses linear dependence among features by projecting onto a lower-dimensional subspace. Provide a conceptual example.

Chapter 6
Linear Transformations

Introduction

Linear transformations serve as the mechanism by which vectors are mapped from one vector space to another while preserving the operations of addition and scalar multiplication. They formalize operations such as rotation, scaling, projection, and shearing of vector spaces, and they underpin countless applications in ML, computer graphics, physics, and engineering. By representing these transformations as matrices, we gain a powerful tool to analyze and compute how inputs are transformed into outputs, whether it is rotating a 3D model in a video game, compressing data in ML, or solving systems of equations.

This chapter explores linear transformations from both theoretical and practical perspectives, designed to be accessible to undergraduate and early graduate students while providing depth for those with a stronger mathematical background. We begin by defining linear transformations and their properties, then examine common transformations like rotations, projections, and shears, with visualizations to build geometric intuition. We connect these ideas to matrix inverses, exploring how invertibility relates to bijective transformations. A dedicated section on ML applications illustrates how linear transformations are used in algorithms like LR, PCA, and neural networks. Python implementations with NumPy and Matplotlib provide hands-on examples, and a comprehensive set of exercises reinforces the concepts.

We will use several visualizations to illustrate transformations in $\mathbb{R}^2$ and $\mathbb{R}^3$, such as the rotation in Fig. 6.6 or the projection in Fig. 6.7. These figures help bridge the gap between algebraic definitions and geometric intuition, showing how vectors and spaces are altered by transformations. By the end of this chapter, you will understand how to define, compute, and apply linear transformations, and appreciate their critical role in ML and beyond.

© The Author(s), under exclusive license to Springer Nature Singapore Pte Ltd. 2026 169
Md. Jalil Piran, *Linear Algebra with Applications in Machine Learning*,
https://doi.org/10.1007/978-981-95-5167-5_6

Topics Covered

This chapter is organized as follows:

- **Section** 6.1 **Linear Transformations; Definition and Basics**: Defines linear transformations, their matrix representation, kernel, image, and invertibility, with a focus on one-to-one and onto properties.
- **Section** 6.2 **Function Inverses**: Explores the concept of inverse functions, their relation to bijective linear transformations, and the conditions under which a matrix inverse exists, with methods to compute it and applications for solving linear systems.
- **Section** 6.3 **Common Transformations**: Explores specific transformations like rotations, reflections, projections, and shears, with geometric visualizations.
- **Section** 6.4 **Transformations in ML**: Discusses applications in LR, PCA, normalizing flows, autoencoders, and optimization.
- **Summary and Exercises**: Recaps key concepts and provides a set of problems to reinforce theoretical and computational skills.

All code accompanying the chapter is publicly available at https://github.com/jalil-piran/Linear-Algebra-with-Applications-in-Machine-Learning/blob/main/Chapter_06_Linear_Transformations.ipynb.

Acronyms

ML	Machine Learning
LR	Linear Regression
PCA	Principal Component Analysis
RREF	Reduced Row Echelon Form
MSE	Mean Squared Error
3D	Three-Dimensional

6.1   Linear Transformations: Definition and Basics

A **linear transformation**, $T : V \to W$, between vector spaces V and W satisfies:

$$T(\mathbf{u} + \mathbf{v}) = T(\mathbf{u}) + T(\mathbf{v}),$$

$$T(c\mathbf{u}) = cT(\mathbf{u}),$$

for all vectors $\mathbf{u}, \mathbf{v} \in V$ and scalars c. In finite-dimensional spaces, if $V = \mathbb{R}^n$ and $W = \mathbb{R}^m$, T can be represented by a matrix $A \in \mathbb{R}^{m \times n}$ such that:

$$T(\mathbf{x}) = A\mathbf{x}.$$

Fig. 6.1 A function or linear transformation showing the **Domain, Co-domain**, and the **Range**. The domain consists of input vectors v_1, v_2, v_3, the co-domain is the full target space, and the range (shown in red) is the actual image of the transformation

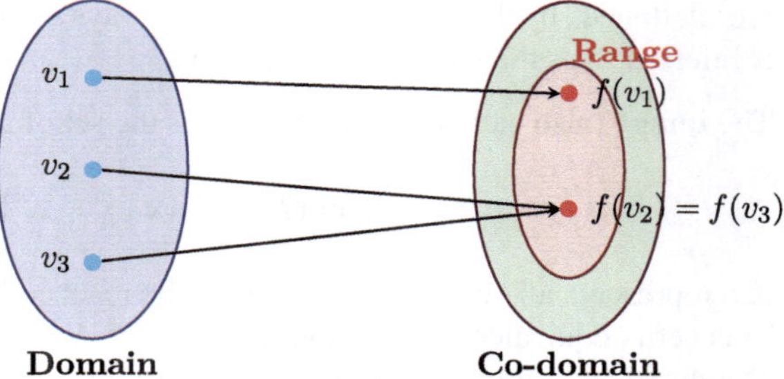

The matrix A encodes how T maps vectors, with its columns representing the images of the standard basis vectors of $\mathbb{R}^n$.

6.1.1   Domain, Codomain, and Range

When describing a linear transformation $T : \mathbb{R}^n \to \mathbb{R}^m$, it is important to distinguish between the domain, codomain, and range. These concepts clarify where the transformation starts, where it maps to, and which outputs are actually attainable, providing a deeper understanding of how the matrix A acts on vectors.

- **Domain**: The input space, $\mathbb{R}^n$, represents all possible input vectors $\mathbf{x}$ that can be fed into the transformation $T(\mathbf{x}) = A\mathbf{x}$. This is where the function T draws its input values from.
- **Codomain**: The output space, $\mathbb{R}^m$, defines the set into which all outputs of the transformation are expected to fall. It is determined by the number of rows in the matrix A, regardless of whether all these outputs are actually attained.
- **Range** (or Image): The set of all actual outputs of the transformation, denoted by $\text{im}(T) = \{A\mathbf{x} \mid \mathbf{x} \in \mathbb{R}^n\}$. This is equivalent to the column space of A, written $\text{Col}(A)$, and consists of all linear combinations of the columns of A. The range is always a subspace of the codomain (Fig. 6.1).

6.1.2   Kernel and Image

The **kernel** of a linear transformation $T : \mathbb{R}^n \to \mathbb{R}^m$, represented by a matrix A, is defined as:

$$\ker(T) = \{\mathbf{x} \in \mathbb{R}^n \mid A\mathbf{x} = \mathbf{0}\},$$

which consists of all input vectors $\mathbf{x}$ that are mapped to the zero vector in the codomain. Geometrically, the kernel describes the directions in the input space that are "flattened" by the transformation. It is always a subspace of $\mathbb{R}^n$, and its dimension is referred to as the *nullity* of T (or of A).

The **image** (also called the *range*) of T is the set of all possible output vectors:

$$\text{im}(T) = \{A\mathbf{x} \mid \mathbf{x} \in \mathbb{R}^n\},$$

and represents all vectors in $\mathbb{R}^m$ that can be reached by applying T to some $\mathbf{x} \in \mathbb{R}^n$. This set is equivalent to the column space of A, $\text{Col}(A)$, and is a subspace of $\mathbb{R}^m$. The dimension of the image is called the *rank* of T.

Example 6.1 Consider $T : \mathbb{R}^2 \to \mathbb{R}^2$, $A = \begin{bmatrix} 1 & 2 \\ 3 & 6 \end{bmatrix}$, we want to find kernel, image, and rank-nullity.

- **Kernel**:
  Solve $A\mathbf{x} = \mathbf{0}$:
  $$\begin{bmatrix} 1 & 2 \\ 3 & 6 \end{bmatrix} \xrightarrow{R_2 \leftarrow R_2 - 3R_1} \begin{bmatrix} 1 & 2 \\ 0 & 0 \end{bmatrix}.$$

  $x_1 + 2x_2 = 0 \implies x_1 = -2x_2.$
  Let $x_2 = t$, so:
  $$\ker(T) = \left\{ t \begin{bmatrix} -2 \\ 1 \end{bmatrix} \mid t \in \mathbb{R} \right\}.$$

- **Image**: The column space is spanned by $\begin{bmatrix} 1 \\ 3 \end{bmatrix}$, as the second column is $2 \times \begin{bmatrix} 1 \\ 3 \end{bmatrix}$.
  Thus:
  $$\text{im}(T) = \text{span}\left(\begin{bmatrix} 1 \\ 3 \end{bmatrix} \right).$$

- **Rank-Nullity**: Figure 6.2 shows kernel and image of this example. The rank is equal to 1 and the nullity is $2 - 1 = 1$.

6.1.3 One-to-One (Injective) and Onto (Surjective)

A transformation $T : \mathbb{R}^n \to \mathbb{R}^m$, represented by the matrix $A \in \mathbb{R}^{m \times n}$, can be classified based on how it maps input vectors to output vectors:

- **One-to-One (Injective)**: The transformation is injective if the kernel of T is trivial, that is,
  $$\ker(T) = \{\mathbf{0}\},$$

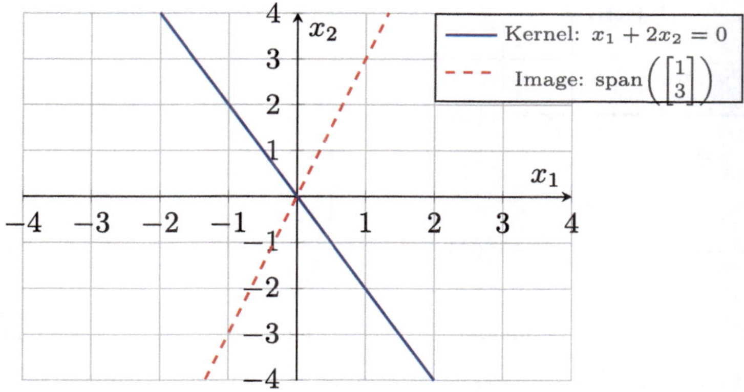

Fig. 6.2 Kernel (blue) and image (red) of $T(\mathbf{x}) = \begin{bmatrix} 1 & 2 \\ 3 & 6 \end{bmatrix} \mathbf{x}$. The kernel and image are 1D subspaces (lines) within $\mathbb{R}^2$

meaning the only solution to $A\mathbf{x} = \mathbf{0}$ is $\mathbf{x} = \mathbf{0}$. This implies that no two distinct input vectors map to the same output. Algebraically, this is equivalent to $\text{rank}(A) = n$, which means the columns of A are linearly independent and span an n-dimensional space.

- **Onto (Surjective)**: The transformation is surjective if its image equals the entire codomain:

$$\text{im}(T) = \mathbb{R}^m,$$

meaning every vector in $\mathbb{R}^m$ is the output of some input vector from $\mathbb{R}^n$. This requires the columns of A to span $\mathbb{R}^m$, which occurs when $\text{rank}(A) = m$. Geometrically, the transformation covers the full output space.

- **Bijective**: A transformation is bijective if it is both injective and surjective. This means each input maps to a unique output, and every possible output is achieved. For this to occur, we must have $m = n$ and $\text{rank}(A) = n$, so A is a square, full-rank matrix and thus invertible. The inverse A^{-1} maps output vectors uniquely back to their corresponding inputs.

Figure 6.3 illustrates these properties, showing how an injective transformation avoids collapsing vectors, while a surjective transformation covers the entire codomain.

Example 6.2 Let $A = \begin{bmatrix} 1 & 2 \\ 3 & 6 \end{bmatrix}$.

We want to check whether this matrix is injective or surjective.

- **Injective?** The rank of A is 1, which is less than the number of columns (2), so the columns are linearly dependent. This means the kernel contains more than just the zero vector, and thus the transformation is not injective.

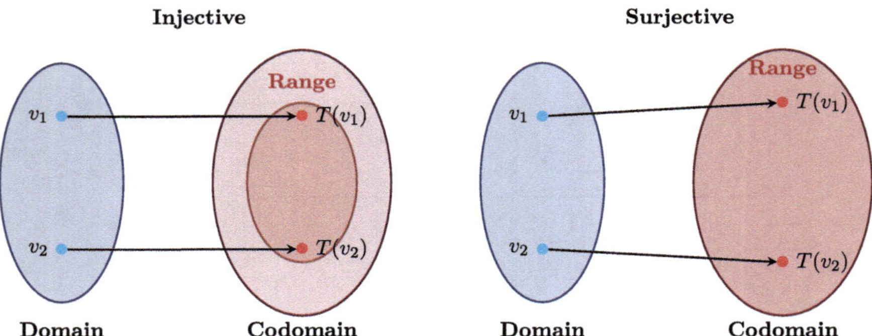

Fig. 6.3 Comparison of injective and surjective linear transformations. An **injective** transformation maps distinct elements in the domain to distinct images in the codomain (left). A **surjective** transformation covers the entire codomain, meaning the range equals the codomain (right)

- **Surjective?** The rank is also less than the number of rows (2), so the image does not span all of $\mathbb{R}^2$. Therefore, the transformation does not reach every possible output, and it is not surjective.

Python Example: Functions

```python
import numpy as np

A = np.array([[1, 2], [3, 6]])
rank = np.linalg.matrix_rank(A)
print("Rank:", rank)
print("Injective:", rank == A.shape[1])
print("Surjective:", rank == A.shape[0])
```

Output:

```
Rank: 1
Injective: False
Surjective: False
```

6.2 Function Inverses

A function $f : X \to Y$ has an **inverse** $f^{-1} : Y \to X$ if it satisfies:

$$f(f^{-1}(y)) = y \quad \forall y \in Y,$$

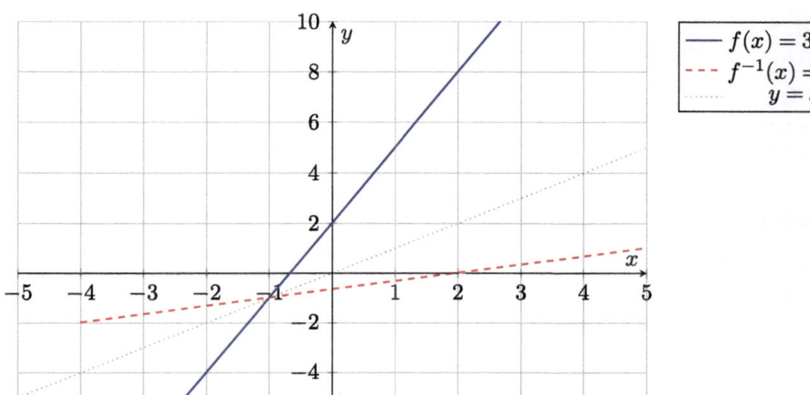

Fig. 6.4 Function $f(x) = 3x + 2$ and its inverse $f^{-1}(x) = \frac{x-2}{3}$

$$f^{-1}(f(x)) = x \quad \forall x \in X.$$

This means the inverse "undoes" the function, effectively reversing the action of the original function. For a function to be invertible, it must be **bijective**, meaning **one-to-one** (injective: distinct inputs produce distinct outputs) and **onto** (surjective: every output in Y is reached). Geometrically, an invertible function reshapes its domain without collapsing or missing parts of the codomain (Fig. 6.4).

Example 6.3 Consider $f(x) = 3x + 2$. To find $f^{-1}(x)$:

- Set $y = f(x)$: $y = 3x + 2$.
- Solve for x: $y - 2 = 3x \implies x = \frac{y-2}{3}$.
- Thus, $f^{-1}(y) = \frac{y-2}{3}$, or $f^{-1}(x) = \frac{x-2}{3}$.

We can verify as follows:

$$f(f^{-1}(x)) = 3\left(\frac{x-2}{3}\right) + 2 = (x - 2) + 2 = x,$$

$$f^{-1}(f(x)) = \frac{(3x + 2) - 2}{3} = \frac{3x}{3} = x.$$

Geometrically, $f(x)$ scales x by 3 and shifts by 2; $f^{-1}(x)$ reverses this by subtracting 2 and scaling by $\frac{1}{3}$.

Python Example: Function Inverse

```python
import numpy as np

def f(x): return 3*x + 2
def f_inv(x): return (x - 2) / 3
```

```
5
6  x = 5
7  print(f"f(f^-1({x})) =", f(f_inv(x)))
8  print(f"f^-1(f({x})) =", f_inv(f(x)))
```

Output:

```
f(f^-1(5)) = 5.0
f^-1(f(5)) = 5.0
```

6.2.1   Non-invertible Functions

Not all functions have inverses. Consider $f(x) = x^2$. Since $f(-2) = f(2) = 4$, it's not one-to-one (multiple inputs map to the same output). Restricting the domain to $x \geq 0$ makes it invertible, with $f^{-1}(x) = \sqrt{x}$.

6.2.2   Matrix Inverse and Bijectivity

A linear transformation $T : \mathbb{R}^n \to \mathbb{R}^n$ with matrix A is bijective if A is invertible, satisfying:

$$AA^{-1} = A^{-1}A = I_n.$$

Invertibility is equivalent to any of the following: $\det(A) \neq 0$ (see Chap. 7 for more details), $\text{rank}(A) = n$, or RREF of A equal to I_n.

For a matrix like

$$A = \begin{bmatrix} a & b \\ c & d \end{bmatrix},$$

The determinant of A is:

$$\det(A) = ad - bc$$

The inverse of matrix $A = \begin{bmatrix} a & b \\ c & d \end{bmatrix}$ is given by $A^{-1} = \frac{1}{ad-bc} \begin{bmatrix} d & -b \\ -c & a \end{bmatrix}$, provided that $ad - bc \neq 0$.

> **Theorem**
> Invertible Matrix Theorem:
> A square matrix $A \in \mathbb{R}^{n \times n}$ is invertible if and only if
>
> $$\det(A) \neq 0.$$
>
> Equivalently, A is invertible if it has full rank, i.e., $\text{rank}(A) = n$.

Conditions for Invertibility

A matrix $A \in \mathbb{R}^{n \times n}$ is invertible if and only if the following equivalent conditions hold:

- $\det(A) \neq 0$: A nonzero determinant means the matrix has full rank and is not singular, ensuring that an inverse exists.
- $\text{rank}(A) = n$: Full rank implies all rows (or columns) are linearly independent, so the transformation represented by A spans the entire space.
- A is row-equivalent to the identity matrix: Using row operations, if A can be transformed into the identity matrix, it confirms the existence of an inverse obtained by applying the same operations to I.
- $A\mathbf{x} = \mathbf{b}$ has a unique solution for every $\mathbf{b} \in \mathbb{R}^n$: This ensures the mapping defined by A is both injective and surjective (bijective), so A^{-1} exists and is well-defined.

Example 6.4 For $A = \begin{bmatrix} 1 & 2 \\ 3 & 5 \end{bmatrix}$:

- **Determinant**: $\det(A) = 1 \cdot 5 - 2 \cdot 3 = -1 \neq 0$. A nonzero determinant indicates that A is invertible and has full rank.
- **Inverse**:

$$A^{-1} = \frac{1}{\det(A)} \begin{bmatrix} 5 & -2 \\ -3 & 1 \end{bmatrix} = \begin{bmatrix} -5 & 2 \\ 3 & -1 \end{bmatrix}.$$

The inverse is computed using the formula for 2×2 matrices, showing that A^{-1} exists because $\det(A) \neq 0$.

- **Verification**:

$$AA^{-1} = \begin{bmatrix} 1 & 2 \\ 3 & 5 \end{bmatrix} \begin{bmatrix} -5 & 2 \\ 3 & -1 \end{bmatrix} = \begin{bmatrix} 1 & 0 \\ 0 & 1 \end{bmatrix}.$$

Multiplying A by its inverse yields the identity matrix, confirming the correctness of the inverse.

Python Example: 2D Matrix Inverse

```python
import numpy as np

A = np.array([[1, 2], [3, 5]])
A_inv = np.linalg.inv(A)
print("Inverse:\n", A_inv)
```

Output:

```
Inverse:
 [[-5.  2.]
 [ 3. -1.]]
```

The following script checks key matrix properties: rank, determinant, and bijectivity. A square matrix is bijective (i.e., invertible) if it has full rank and a non-zero determinant. These properties ensure that a linear transformation is one-to-one and onto.

Python Example: Matrix Properties

```python
import numpy as np

A = np.array([[1, 2], [3, 5]])
rank = np.linalg.matrix_rank(A)
det = np.linalg.det(A)

print("Rank:", rank)
print(f"Determinant: {det:.2f}")
print("Bijective:", rank == A.shape[0] ==
    A.shape[1] and det != 0)
```

Output:

```
Rank: 2
Determinant: -1.00
Bijective: True
```

Methods to Compute A^{-1}

There are several methods to compute as follows.

1. **Using the Definition:** $A^{-1}A = I$

   To confirm or find A^{-1}, multiply it with A to check whether the result is the identity matrix:

$$AA^{-1} = I \quad \text{and} \quad A^{-1}A = I.$$

This property can also be used to verify a given inverse.

2. **Formula for** 2×2 **Matrices**

For a 2×2 matrix:

$$A = \begin{bmatrix} a & b \\ c & d \end{bmatrix}, \quad \det(A) = ad - bc \neq 0,$$

its inverse is given by:

$$A^{-1} = \frac{1}{\det(A)} \begin{bmatrix} d & -b \\ -c & a \end{bmatrix}.$$

6.2.3 Using the Adjugate Formula

The inverse of a square matrix $A \in \mathbb{R}^{n \times n}$, when $\det(A) \neq 0$, can be computed using the **adjugate formula**:

$$A^{-1} = \frac{1}{\det(A)} \cdot \mathrm{adj}(A).$$

This method expresses the inverse in terms of the determinant and the adjugate matrix, and is mainly of theoretical interest for $n > 3$ due to its computational cost.

The adjugate matrix $\mathrm{adj}(A)$ is defined as the transpose of the cofactor matrix:

$$\mathrm{adj}(A) = [C_{ij}]^{\top},$$

where $C_{ij} = (-1)^{i+j} \det(M_{ij})$ is the cofactor of element a_{ij}, and M_{ij} is the minor obtained by deleting the i-th row and j-th column of A.

Example 6.5 Given the matrix:

$$A = \begin{bmatrix} 1 & 2 \\ 3 & 4 \end{bmatrix},$$

we compute the determinant:

$$\det(A) = (1)(4) - (2)(3) = -2.$$

We calculate the cofactor matrix:

$$\mathrm{Cof}(A) = \begin{bmatrix} +4 & -3 \\ -2 & +1 \end{bmatrix} \Rightarrow \mathrm{adj}(A) = \begin{bmatrix} 4 & -2 \\ -3 & 1 \end{bmatrix}.$$

We apply the adjugate formula:

$$A^{-1} = \frac{1}{-2}\begin{bmatrix} 4 & -2 \\ -3 & 1 \end{bmatrix} = \begin{bmatrix} -2 & 1 \\ 1.5 & -0.5 \end{bmatrix}$$

Moreover, we can find the inverse of a matrix using elementary row transformations.

Transform the augmented matrix $[A \mid I]$ into $[I \mid A^{-1}]$ using row operations:

$$[A \mid I] \xrightarrow{\text{RREF}} [I \mid A^{-1}].$$

This algorithmic method is practical for hand and numerical computations.

Example 6.6 Given:

$$A = \begin{bmatrix} 2 & 1 \\ 7 & 4 \end{bmatrix}, \quad \det(A) = 2 \cdot 4 - 7 \cdot 1 = 8 - 7 = 1.$$

Then:

$$A^{-1} = \frac{1}{1}\begin{bmatrix} 4 & -1 \\ -7 & 2 \end{bmatrix} = \begin{bmatrix} 4 & -1 \\ -7 & 2 \end{bmatrix}.$$

We verify as follows:

$$AA^{-1} = \begin{bmatrix} 2 & 1 \\ 7 & 4 \end{bmatrix}\begin{bmatrix} 4 & -1 \\ -7 & 2 \end{bmatrix} = \begin{bmatrix} 1 & 0 \\ 0 & 1 \end{bmatrix} = I.$$

So, in summary:

- The inverse of a matrix allows solving $A\mathbf{x} = \mathbf{b}$ via $\mathbf{x} = A^{-1}\mathbf{b}$.
- A matrix must be square and have a non-zero determinant to be invertible.
- There are several techniques available depending on matrix size and context.

6.2.4   RREF and Invertibility

A square matrix $A \in \mathbb{R}^{n \times n}$ is invertible if its RREF is I_n. The RREF of a matrix is a unique canonical form obtained via elementary row operations, in which each leading entry is 1, each pivot column has zeros elsewhere, and pivot positions move strictly to the right as you go down the rows. RREF properties:

- Leading entry in each nonzero row is 1.
- Each leading 1 is the only nonzero entry in its column.
- Leading 1s form a staircase pattern.
- Zero rows are at the bottom.

If RREF of A is I_n, A has full rank (rank$(A) = n$), and its columns are linearly independent.

Example 6.7 Let $A = \begin{bmatrix} 1 & 2 & 3 \\ 2 & 5 & 6 \\ 3 & 4 & 8 \end{bmatrix}$, we want to check whether A is invertible?

- **Row reduce**:

$$\begin{bmatrix} 1 & 2 & 3 \\ 2 & 5 & 6 \\ 3 & 4 & 8 \end{bmatrix} \xrightarrow{R_2 \leftarrow R_2 - 2R_1} \begin{bmatrix} 1 & 2 & 3 \\ 0 & 1 & 0 \\ 3 & 4 & 8 \end{bmatrix} \xrightarrow{R_3 \leftarrow R_3 - 3R_1} \begin{bmatrix} 1 & 2 & 3 \\ 0 & 1 & 0 \\ 0 & -2 & -1 \end{bmatrix}.$$

$$\xrightarrow{R_3 \leftarrow R_3 + 2R_2} \begin{bmatrix} 1 & 2 & 3 \\ 0 & 1 & 0 \\ 0 & 0 & -1 \end{bmatrix} \xrightarrow{R_3 \leftarrow -R_3} \begin{bmatrix} 1 & 2 & 3 \\ 0 & 1 & 0 \\ 0 & 0 & 1 \end{bmatrix}.$$

$$\xrightarrow{R_1 \leftarrow R_1 - 3R_3} \begin{bmatrix} 1 & 2 & 0 \\ 0 & 1 & 0 \\ 0 & 0 & 1 \end{bmatrix} \xrightarrow{R_1 \leftarrow R_1 - 2R_2} \begin{bmatrix} 1 & 0 & 0 \\ 0 & 1 & 0 \\ 0 & 0 & 1 \end{bmatrix}.$$

- **Result**: RREF is I_3, so A is invertible.

Python Example: Matrix Invertibility

```python
import numpy as np

A = np.array([[1, 2, 3], [2, 5, 6], [3, 4, 8]])
try:
    A_inv = np.linalg.inv(A)
    print("The matrix is invertible.")
except np.linalg.LinAlgError:
    print("The matrix is singular.")
```

Output:

```
The matrix is invertible.
```

Solving Linear Systems with Inverses

For a system $A\mathbf{x} = \mathbf{b}$, if A is invertible, the solution is:

$$\mathbf{x} = A^{-1}\mathbf{b}.$$

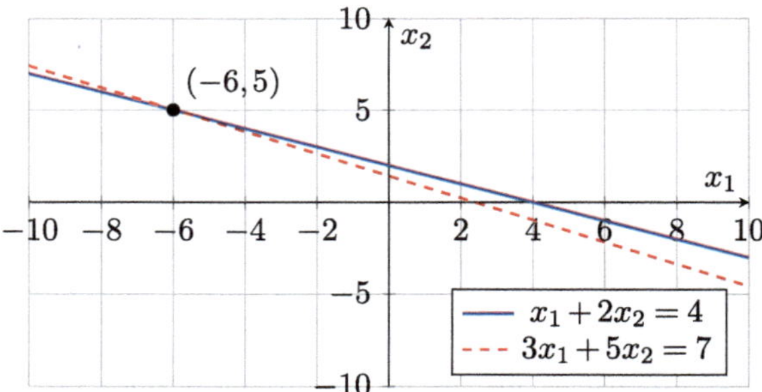

Fig. 6.5 Solution as the intersection of two lines

This method is elegant but computationally expensive for large matrices; Gaussian elimination or matrix factorizations such as LU decomposition are preferred in practice.

Example 6.8 Solve:

$$\begin{cases} x_1 + 2x_2 = 4, \\ 3x_1 + 5x_2 = 7. \end{cases}$$

Matrix form: $A = \begin{bmatrix} 1 & 2 \\ 3 & 5 \end{bmatrix}$, $\mathbf{b} = \begin{bmatrix} 4 \\ 7 \end{bmatrix}$, $A^{-1} = \begin{bmatrix} -5 & 2 \\ 3 & -1 \end{bmatrix}$.

$$\mathbf{x} = A^{-1}\mathbf{b} = \begin{bmatrix} -5 & 2 \\ 3 & -1 \end{bmatrix}\begin{bmatrix} 4 \\ 7 \end{bmatrix} = \begin{bmatrix} -20 + 14 \\ 12 - 7 \end{bmatrix} = \begin{bmatrix} -6 \\ 5 \end{bmatrix}.$$

Now, we can simply do the verification by considering $x_1 = -6$ and $x_2 = 5$: As shown in Fig. 6.5.

$$\begin{bmatrix} 1 & 2 \\ 3 & 5 \end{bmatrix}\begin{bmatrix} -6 \\ 5 \end{bmatrix} = \begin{bmatrix} -6 + 10 \\ -18 + 25 \end{bmatrix} = \begin{bmatrix} 4 \\ 7 \end{bmatrix}.$$

Python Example: Solve Linear System Using Matrix Inverse

```python
import numpy as np

A = np.array([[1, 2], [3, 5]])
b = np.array([4, 7])
x = np.linalg.inv(A) @ b
print("Solution:", x)
```

Output:

```
Solution: [-6.  5.]
```

Properties of Inverses

Now, let's check some properties of inverses.

Inverse of a Product

If A and B are invertible, then:

$$(AB)^{-1} = B^{-1}A^{-1}.$$

This reverses the order, like undoing transformations in reverse sequence.

Example 6.9 Let $A = \begin{bmatrix} 1 & 0 \\ 0 & 2 \end{bmatrix}$, $B = \begin{bmatrix} 2 & 0 \\ 0 & 1 \end{bmatrix}$.

- $A^{-1} = \begin{bmatrix} 1 & 0 \\ 0 & \frac{1}{2} \end{bmatrix}$, $B^{-1} = \begin{bmatrix} \frac{1}{2} & 0 \\ 0 & 1 \end{bmatrix}$.
- $AB = \begin{bmatrix} 2 & 0 \\ 0 & 2 \end{bmatrix}$, $(AB)^{-1} = \begin{bmatrix} \frac{1}{2} & 0 \\ 0 & \frac{1}{2} \end{bmatrix}$.
- $B^{-1}A^{-1} = \begin{bmatrix} \frac{1}{2} & 0 \\ 0 & 1 \end{bmatrix} \begin{bmatrix} 1 & 0 \\ 0 & \frac{1}{2} \end{bmatrix} = \begin{bmatrix} \frac{1}{2} & 0 \\ 0 & \frac{1}{2} \end{bmatrix}$.

Verification confirms $(AB)^{-1} = B^{-1}A^{-1}$.

Inverse of a Transpose

For any invertible matrix A, the inverse of its transpose is the transpose of its inverse:

$$(A^{\mathrm{T}})^{-1} = (A^{-1})^{\mathrm{T}}.$$

The transpose operation reverses the order of multiplication, just like inversion. So applying both still preserves this symmetry.

Example 6.10 Let:

$$A = \begin{bmatrix} 1 & 2 \\ 3 & 5 \end{bmatrix}, \quad A^{-1} = \begin{bmatrix} -5 & 2 \\ 3 & -1 \end{bmatrix}.$$

Then:

$$A^{\mathrm{T}} = \begin{bmatrix} 1 & 3 \\ 2 & 5 \end{bmatrix}, \quad (A^{\mathrm{T}})^{-1} = \begin{bmatrix} -5 & 3 \\ 2 & -1 \end{bmatrix},$$

and

$$(A^{-1})^{\mathrm{T}} = \begin{bmatrix} -5 & 3 \\ 2 & -1 \end{bmatrix}.$$

Thus, Let the matrix be:

$$A = \begin{bmatrix} 1 & 2 \\ 3 & 5 \end{bmatrix}.$$

We are given its inverse:

$$A^{-1} = \begin{bmatrix} -5 & 2 \\ 3 & -1 \end{bmatrix}.$$

Step 1- Compute the transpose of A. The transpose of a matrix is obtained by switching its rows and columns:

$$A^{\mathrm{T}} = \begin{bmatrix} 1 & 3 \\ 2 & 5 \end{bmatrix}.$$

Step 2- Compute the inverse of A^{T}. By directly calculating or using known identities, we obtain:

$$(A^{\mathrm{T}})^{-1} = \begin{bmatrix} -5 & 3 \\ 2 & -1 \end{bmatrix}.$$

Step 3- Compute the transpose of A^{-1}.

$$(A^{-1})^{\mathrm{T}} = \left(\begin{bmatrix} -5 & 2 \\ 3 & -1 \end{bmatrix} \right)^{\mathrm{T}} = \begin{bmatrix} -5 & 3 \\ 2 & -1 \end{bmatrix}.$$

We see that:

$$\boxed{(A^{\mathrm{T}})^{-1} = (A^{-1})^{\mathrm{T}}}$$

This example confirms the important identity in linear algebra:

$$(A^{\mathrm{T}})^{-1} = (A^{-1})^{\mathrm{T}}.$$

This result tells us that taking the inverse of a transpose is equivalent to taking the transpose of the inverse. This property is essential in many applications such as optimization, matrix factorization, and theoretical derivations in ML and numerical linear algebra.

Python Example: Matrix Inverse Properties: Product and Transpose

```python
import numpy as np

A = np.array([[1, 0], [0, 2]])
B = np.array([[2, 0], [0, 1]])

AB_inv = np.linalg.inv(A @ B)
```

```python
B_inv_A_inv = np.linalg.inv(B) @ np.linalg.inv(A)

print("Inverse of product:\n", AB_inv)
print("Product of inverses:\n", B_inv_A_inv)

# Inverse of transpose of
A_T_inv = np.linalg.inv(A.T)
A_inv_T = np.linalg.inv(A).T

print("Inverse of transpose:\n", A_T_inv)
print("Transpose of inverse:\n", A_inv_T)
```

Output:

```
Inverse of product:
 [[0.5 0. ]
 [0.  0.5]]
Product of inverses:
 [[0.5 0. ]
 [0.  0.5]]
Inverse of transpose:
 [[1.  0. ]
 [0.  0.5]]
Transpose of inverse:
 [[1.  0. ]
 [0.  0.5]]
```

Algorithm for Matrix Inversion

The *Gauss-Jordan elimination* method computes A^{-1} by transforming $[A|I_n] \to [I_n|A^{-1}]$.

Example 6.11 For $A = \begin{bmatrix} 1 & 0 & 4 \\ 0 & 2 & 3 \\ 0 & 0 & 1 \end{bmatrix}$:

- **Augmented matrix**:
$$\begin{bmatrix} 1 & 0 & 4 & 1 & 0 & 0 \\ 0 & 2 & 3 & 0 & 1 & 0 \\ 0 & 0 & 1 & 0 & 0 & 1 \end{bmatrix}.$$

- **Normalize row 2**: $R_2 \leftarrow R_2/2$:

$$\left[\begin{array}{ccc|ccc} 1 & 0 & 4 & 1 & 0 & 0 \\ 0 & 1 & \frac{3}{2} & 0 & \frac{1}{2} & 0 \\ 0 & 0 & 1 & 0 & 0 & 1 \end{array}\right].$$

- **Eliminate above pivot in column 3**: $R_1 \leftarrow R_1 - 4R_3$, $R_2 \leftarrow R_2 - \frac{3}{2}R_3$:

$$\left[\begin{array}{ccc|ccc} 1 & 0 & 0 & 1 & 0 & -4 \\ 0 & 1 & 0 & 0 & \frac{1}{2} & -\frac{3}{2} \\ 0 & 0 & 1 & 0 & 0 & 1 \end{array}\right].$$

- $A^{-1} = \begin{bmatrix} 1 & 0 & -4 \\ 0 & \frac{1}{2} & -\frac{3}{2} \\ 0 & 0 & 1 \end{bmatrix}.$

Verification:

$$AA^{-1} = \begin{bmatrix} 1 & 0 & 4 \\ 0 & 2 & 3 \\ 0 & 0 & 1 \end{bmatrix} \begin{bmatrix} 1 & 0 & -4 \\ 0 & \frac{1}{2} & -\frac{3}{2} \\ 0 & 0 & 1 \end{bmatrix} = \begin{bmatrix} 1 & 0 & 0 \\ 0 & 1 & 0 \\ 0 & 0 & 1 \end{bmatrix}.$$

Python Example: Inverse of a 3D Matrix

```python
import numpy as np

A = np.array([[1, 0, 4], [0, 2, 3], [0, 0, 1]])
A_inv = np.linalg.inv(A)
print("Inverse:\n", A_inv)
```

Output:

```
Inverse:
 [[ 1.    0.   -4. ]
 [ 0.    0.5  -1.5]
 [ 0.    0.    1. ]]
```

Conditions for Invertibility

We now consolidate the equivalent conditions for invertibility of a square matrix. A square matrix $A \in \mathbb{R}^{n \times n}$ is invertible if and only if:

- Columns span $\mathbb{R}^n$ (onto).
- $Ax = 0$ has only the trivial solution (one-to-one).
- Columns are linearly independent.
- $\mathrm{rank}(A) = n$.

- $\det(A) \neq 0$.
- RREF of A is I_n.
- A is a product of elementary matrices (see Sect. 6.2.6).

These conditions reflect that an invertible matrix preserves all information in a transformation.

Example 6.12 To find the inverse of a 4×4 matrix, we apply the Gauss-Jordan elimination method by augmenting the matrix with the 4×4 identity matrix and reducing the left side to the identity.
Let:

$$A = \begin{bmatrix} 1 & 2 & 1 & -1 \\ 3 & 8 & 1 & -3 \\ 0 & 4 & 1 & -1 \\ 2 & 6 & 2 & -2 \end{bmatrix}.$$

We want to perform row operations to transform the augmented matrix:

$$[A \mid I] = \left[\begin{array}{cccc|cccc} 1 & 2 & 1 & -1 & 1 & 0 & 0 & 0 \\ 3 & 8 & 1 & -3 & 0 & 1 & 0 & 0 \\ 0 & 4 & 1 & -1 & 0 & 0 & 1 & 0 \\ 2 & 6 & 2 & -2 & 0 & 0 & 0 & 1 \end{array}\right].$$

Step 1- Make pivot in row 1, column 1 (already 1). Eliminate below.

$$R_2 \leftarrow R_2 - 3R_1, \quad R_4 \leftarrow R_4 - 2R_1,$$

$$\Rightarrow \left[\begin{array}{cccc|cccc} 1 & 2 & 1 & -1 & 1 & 0 & 0 & 0 \\ 0 & 2 & -2 & 0 & -3 & 1 & 0 & 0 \\ 0 & 4 & 1 & -1 & 0 & 0 & 1 & 0 \\ 0 & 2 & 0 & 0 & -2 & 0 & 0 & 1 \end{array}\right].$$

Step 2- Make pivot in row 2, column 2 by dividing R_2 by 2.

$$R_2 \leftarrow \frac{1}{2} R_2,$$

$$\Rightarrow \left[\begin{array}{cccc|cccc} 1 & 2 & 1 & -1 & 1 & 0 & 0 & 0 \\ 0 & 1 & -1 & 0 & -1.5 & 0.5 & 0 & 0 \\ 0 & 4 & 1 & -1 & 0 & 0 & 1 & 0 \\ 0 & 2 & 0 & 0 & -2 & 0 & 0 & 1 \end{array}\right].$$

Step 3- Eliminate column 2 using pivot.

$$R_1 \leftarrow R_1 - 2R_2, \quad R_3 \leftarrow R_3 - 4R_2, \quad R_4 \leftarrow R_4 - 2R_2$$

$$\Rightarrow \begin{bmatrix} 1 & 0 & 3 & -1 & 4 & -1 & 0 & 0 \\ 0 & 1 & -1 & 0 & -1.5 & 0.5 & 0 & 0 \\ 0 & 0 & 5 & -1 & 6 & -2 & 1 & 0 \\ 0 & 0 & 2 & 0 & 1 & -1 & 0 & 1 \end{bmatrix}$$

Step 4- Make pivot in row 3, column 3.

$$R_3 \leftarrow \frac{1}{5} R_3$$

$$\Rightarrow \begin{bmatrix} 1 & 0 & 3 & -1 & 4 & -1 & 0 & 0 \\ 0 & 1 & -1 & 0 & -1.5 & 0.5 & 0 & 0 \\ 0 & 0 & 1 & -\frac{1}{5} & \frac{6}{5} & -\frac{2}{5} & \frac{1}{5} & 0 \\ 0 & 0 & 2 & 0 & 1 & -1 & 0 & 1 \end{bmatrix}$$

Step 5- Eliminate column 3.

$$R_1 \leftarrow R_1 - 3R_3, \quad R_2 \leftarrow R_2 + R_3, \quad R_4 \leftarrow R_4 - 2R_3,$$

$$\Rightarrow \begin{bmatrix} 1 & 0 & 0 & -\frac{2}{5} & \frac{2}{5} & \frac{1}{5} & -\frac{3}{5} & 0 \\ 0 & 1 & 0 & -\frac{1}{5} & -\frac{3}{5} & \frac{1}{10} & \frac{1}{5} & 0 \\ 0 & 0 & 1 & -\frac{1}{5} & \frac{6}{5} & -\frac{2}{5} & \frac{1}{5} & 0 \\ 0 & 0 & 0 & \frac{2}{5} & -\frac{7}{5} & -\frac{1}{5} & -\frac{2}{5} & 1 \end{bmatrix}.$$

Step 6- Make pivot in row 4 by dividing by $\frac{2}{5}$.

$$R_4 \leftarrow \frac{5}{2} R_4,$$

$$\Rightarrow \begin{bmatrix} 1 & 0 & 0 & -\frac{2}{5} & \frac{2}{5} & \frac{1}{5} & -\frac{3}{5} & 0 \\ 0 & 1 & 0 & -\frac{1}{5} & -\frac{3}{5} & \frac{1}{10} & \frac{1}{5} & 0 \\ 0 & 0 & 1 & -\frac{1}{5} & \frac{6}{5} & -\frac{2}{5} & \frac{1}{5} & 0 \\ 0 & 0 & 0 & 1 & -\frac{7}{2} & -\frac{1}{2} & -1 & \frac{5}{2} \end{bmatrix}.$$

Step 7- Eliminate above row 4 to get identity on the left.

$$R_1 \leftarrow R_1 + \frac{2}{5} R_4, \quad R_2 \leftarrow R_2 + \frac{1}{5} R_4, \quad R_3 \leftarrow R_3 + \frac{1}{5} R_4,$$

$$\Rightarrow \begin{bmatrix} 1 & 0 & 0 & 0 & -1 & 0 & -1 & 1 \\ 0 & 1 & 0 & 0 & -1 & 0 & 0 & 0.5 \\ 0 & 0 & 1 & 0 & 0.5 & -\frac{1}{2} & 0 & 0.5 \\ 0 & 0 & 0 & 1 & -\frac{7}{2} & -\frac{1}{2} & -1 & \frac{5}{2} \end{bmatrix}.$$

Therefore, the inverse of A is:

$$A^{-1} = \begin{bmatrix} -1 & 0 & -1 & 1 \\ -1 & 0 & 0 & 0.5 \\ 0.5 & -\frac{1}{2} & 0 & 0.5 \\ -\frac{7}{2} & -\frac{1}{2} & -1 & \frac{5}{2} \end{bmatrix}.$$

Note: This process generalizes to any size square matrix using row reduction.

The inverse is:

$$A^{-1} = \frac{1}{\det(A)} \begin{bmatrix} d & -b \\ -c & a \end{bmatrix}.$$

Example 6.13 We are given two invertible matrices:

$$A = \begin{bmatrix} 1 & 2 \\ 3 & 5 \end{bmatrix}, \quad B = \begin{bmatrix} 2 & 1 \\ 7 & 4 \end{bmatrix}.$$

Step 1- Confirm Invertibility

$$\det(A) = (1)(5) - (2)(3) = 5 - 6 = -1 \neq 0$$
$$\det(B) = (2)(4) - (1)(7) = 8 - 7 = 1 \neq 0$$

Both A and B are invertible.

Step 2- Compute AB

$$AB = \begin{bmatrix} 1 & 2 \\ 3 & 5 \end{bmatrix} \begin{bmatrix} 2 & 1 \\ 7 & 4 \end{bmatrix} = \begin{bmatrix} 1 \cdot 2 + 2 \cdot 7 & 1 \cdot 1 + 2 \cdot 4 \\ 3 \cdot 2 + 5 \cdot 7 & 3 \cdot 1 + 5 \cdot 4 \end{bmatrix} = \begin{bmatrix} 16 & 9 \\ 41 & 23 \end{bmatrix}.$$

Step 3- Compute $\det(AB)$

$$\det(AB) = (16)(23) - (9)(41) = 368 - 369 = -1 \neq 0.$$

So AB is invertible.

Step 4- Compute A^{-1} and B^{-1}

For a 2×2 matrix:

$$M = \begin{bmatrix} a & b \\ c & d \end{bmatrix}, \quad M^{-1} = \frac{1}{ad - bc} \begin{bmatrix} d & -b \\ -c & a \end{bmatrix}.$$

Inverse of A:

$$\det(A) = -1, \quad A^{-1} = \frac{1}{-1}\begin{bmatrix} 5 & -2 \\ -3 & 1 \end{bmatrix} = \begin{bmatrix} -5 & 2 \\ 3 & -1 \end{bmatrix}.$$

Inverse of B:

$$\det(B) = 1, \quad B^{-1} = \frac{1}{1}\begin{bmatrix} 4 & -1 \\ -7 & 2 \end{bmatrix} = \begin{bmatrix} 4 & -1 \\ -7 & 2 \end{bmatrix}$$

Step 5- Compute $(AB)^{-1} = B^{-1}A^{-1}$

$$\begin{aligned}
(AB)^{-1} &= \begin{bmatrix} 4 & -1 \\ -7 & 2 \end{bmatrix}\begin{bmatrix} -5 & 2 \\ 3 & -1 \end{bmatrix} \\
&= \begin{bmatrix} 4(-5) + (-1)(3) & 4(2) + (-1)(-1) \\ -7(-5) + 2(3) & -7(2) + 2(-1) \end{bmatrix} \\
&= \begin{bmatrix} -20 - 3 & 8 + 1 \\ 35 + 6 & -14 - 2 \end{bmatrix}, \\
&= \begin{bmatrix} -23 & 9 \\ 41 & -16 \end{bmatrix}.
\end{aligned}$$

Step 6- Verify $AB \cdot (AB)^{-1} = I$

$$AB = \begin{bmatrix} 16 & 9 \\ 41 & 23 \end{bmatrix}, \quad (AB)^{-1} = \begin{bmatrix} -23 & 9 \\ 41 & -16 \end{bmatrix}.$$

Multiply:

$$\begin{aligned}
AB \cdot (AB)^{-1} &= \begin{bmatrix} 16(-23) + 9(41) & 16(9) + 9(-16) \\ 41(-23) + 23(41) & 41(9) + 23(-16) \end{bmatrix} \\
&= \begin{bmatrix} -368 + 369 & 144 - 144 \\ -943 + 943 & 369 - 368 \end{bmatrix} = \begin{bmatrix} 1 & 0 \\ 0 & 1 \end{bmatrix}.
\end{aligned}$$

$$\boxed{(AB)^{-1} = \begin{bmatrix} -23 & 9 \\ 41 & -16 \end{bmatrix}, \quad \text{and} \quad AB \cdot (AB)^{-1} = I.}$$

6.2.5   General Case ($n \times n$)

For an $n \times n$ matrix A, the inverse is:

$$A^{-1} = \frac{1}{\det(A)}\text{adj}(A), \quad \det(A) \neq 0,$$

where $\mathrm{adj}(A)$ is the adjugate (transpose of the cofactor matrix).
To do that we need to follow these steps:

- Compute $\det(A)$.
- Find the matrix of minors (determinants of $(n-1) \times (n-1)$ submatrices).
- Apply signs to get the cofactor matrix: $C_{ij} = (-1)^{i+j} \cdot \mathrm{minor}_{ij}$.
- Transpose to get $\mathrm{adj}(A) = C^{\mathrm{T}}$.
- Scale by $\frac{1}{\det(A)}$.

Example 6.14 Let $A = \begin{bmatrix} 1 & 2 & 3 \\ 0 & 1 & 4 \\ 5 & 6 & 0 \end{bmatrix}$.

- **Determinant**:

$$
\begin{aligned}
\det(A) &= 1(1 \cdot 0 - 4 \cdot 6) - 2(0 \cdot 0 - 4 \cdot 5) + 3(0 \cdot 6 - 1 \cdot 5) \\
&= 1(-24) - 2(-20) + 3(-5) \\
&= -24 + 40 - 15 = 1.
\end{aligned}
$$

- **Matrix of Minors**: For each a_{ij}, compute the determinant of the submatrix excluding row i and column j. For a_{11}:

$$
\begin{bmatrix} 1 & 4 \\ 6 & 0 \end{bmatrix}, \quad \det = 1 \cdot 0 - 4 \cdot 6 = -24.
$$

Full matrix of minors:

$$
\begin{bmatrix} -24 & -20 & -5 \\ -18 & -15 & -4 \\ 5 & 4 & 1 \end{bmatrix}.
$$

- **Cofactor Matrix**: Apply $(-1)^{i+j}$:

$$
\begin{bmatrix} (-1)^2(-24) & (-1)^3(20) & (-1)^4(-5) \\ (-1)^3(12) & (-1)^4(-15) & (-1)^5(4) \\ (-1)^4(-8) & (-1)^5(-4) & (-1)^6(1) \end{bmatrix} = \begin{bmatrix} -24 & 20 & -5 \\ -12 & -15 & 4 \\ -8 & -4 & 1 \end{bmatrix}.
$$

- **Adjugate**: Transpose:

$$
\mathrm{adj}(A) = \begin{bmatrix} -24 & 18 & 5 \\ 20 & -15 & -4 \\ -5 & 4 & 1 \end{bmatrix}.
$$

- **Inverse**: Since $\det(A) = 1$, $A^{-1} = \mathrm{adj}(A)$.
- **Verification**:

$$AA^{-1} = \begin{bmatrix} 1 & 2 & 3 \\ 0 & 1 & 4 \\ 5 & 6 & 0 \end{bmatrix} \begin{bmatrix} -24 & 18 & 5 \\ 20 & -15 & -4 \\ -5 & 4 & 1 \end{bmatrix} = \begin{bmatrix} 1 & 0 & 0 \\ 0 & 1 & 0 \\ 0 & 0 & 1 \end{bmatrix}.$$

Not All Square Matrices Are Invertible

It is a common misconception that every square matrix is invertible. In reality, a square matrix is invertible if and only if it is *non-singular*, meaning its determinant is non-zero. To illustrate this, consider the following matrix:

$$A = \begin{bmatrix} 1 & 2 \\ 0 & 0 \end{bmatrix}.$$

We do not yet know if A is invertible. Suppose we attempt to find its inverse by introducing a general 2×2 matrix:

$$A^{-1} = \begin{bmatrix} a & b \\ c & d \end{bmatrix}.$$

Then the matrix multiplication $A \cdot A^{-1}$ becomes:

$$\begin{bmatrix} 1 & 2 \\ 0 & 0 \end{bmatrix} \begin{bmatrix} a & b \\ c & d \end{bmatrix} = \begin{bmatrix} a + 2c & b + 2d \\ 0 & 0 \end{bmatrix}.$$

This product is clearly not equal to the identity matrix:

$$\begin{bmatrix} a + 2c & b + 2d \\ 0 & 0 \end{bmatrix} \neq \begin{bmatrix} 1 & 0 \\ 0 & 1 \end{bmatrix}.$$

No matter how we choose the values of a, b, c, and d, the second row of the product will always be zero. Therefore, A does not have an inverse. This confirms that A is a *singular matrix*.

A square matrix is invertible if and only if it has full rank (i.e., its rows or columns are linearly independent). In the example above, the second row of A is a zero row, which immediately implies the matrix is not full-rank and hence non-invertible.

6.2.6   *Elementary Matrices*

An **elementary matrix** is obtained by applying a single row operation to the identity matrix. These matrices formalize row operations used in Gaussian elimination, acting as single-step transformations. Types:

- **Swap**: Interchange rows i and j.
- **Scale**: Multiply row i by a nonzero scalar k.

- **Add**: Add k times row i to row j.

- **Swap**: $E_1 = \begin{bmatrix} 0 & 1 \\ 1 & 0 \end{bmatrix}$ for rows 1 and 2.

- **Scale**: $E_2 = \begin{bmatrix} 1 & 0 \\ 0 & k \end{bmatrix}$ for scaling row 2 by $k \neq 0$.

- **Add**: $E_3 = \begin{bmatrix} 1 & 0 \\ k & 1 \end{bmatrix}$ for adding k times row 1 to row 2.

The inverse of an elementary matrix reverses the operation:

- Swap: Same matrix (self-inverse).
- Scale by k: Scale by $\frac{1}{k}$.
- Add k: Add $-k$.

Example 6.15 For $E = \begin{bmatrix} 1 & 0 & 0 \\ 0 & 1 & 0 \\ 2 & 0 & 1 \end{bmatrix}$, the inverse is:

$$E^{-1} = \begin{bmatrix} 1 & 0 & 0 \\ 0 & 1 & 0 \\ -2 & 0 & 1 \end{bmatrix}.$$

Verification:

$$EE^{-1} = \begin{bmatrix} 1 & 0 & 0 \\ 0 & 1 & 0 \\ 2 & 0 & 1 \end{bmatrix} \begin{bmatrix} 1 & 0 & 0 \\ 0 & 1 & 0 \\ -2 & 0 & 1 \end{bmatrix} = \begin{bmatrix} 1 & 0 & 0 \\ 0 & 1 & 0 \\ 0 & 0 & 1 \end{bmatrix}.$$

Example 6.16 Consider an elementary matrix of $\begin{bmatrix} 1 & 0 & 0 \\ 0 & 1 & 0 \\ 0 & 0 & 1 \end{bmatrix}$, we can perform elementary row operations on this elementary matrix as follows:

- Swap rows 1 and 2: $E_1 = \begin{bmatrix} 0 & 1 & 0 \\ 1 & 0 & 0 \\ 0 & 0 & 1 \end{bmatrix}$.

- Scale row 2 by -4: $E_2 = \begin{bmatrix} 1 & 0 & 0 \\ 0 & -4 & 0 \\ 0 & 0 & 1 \end{bmatrix}$.

- Add 2 times row 1 to row 3: $E_3 = \begin{bmatrix} 1 & 0 & 0 \\ 0 & 1 & 0 \\ 2 & 0 & 1 \end{bmatrix}$.

Multiplying a matrix A by an elementary matrix on the left applies the corresponding row operation.

6.2.7 Inverse of Elementary Matrices

Elementary matrices are invertible:

- Swap: Same swap (self-inverse).
- Scale by k: Scale by $\frac{1}{k}$.
- Add k times row i to row j: Add $-k$ times row i to row j.

Example 6.17 For E_3, the inverse is given by:

$$
E_3^{-1} = \begin{bmatrix} 1 & 0 & 0 \\ 0 & 1 & 0 \\ -2 & 0 & 1 \end{bmatrix}.
$$

This matrix represents an *elementary row operation*, specifically a **row replacement** on a 3×3 identity matrix. In this case, the operation is:

$$
Row\,3 \to Row\,3 - 2 \times Row\,1.
$$

To understand why, observe the transformation encoded by E_3^{-1}. When this matrix is left-multiplied to another matrix A, it updates the third row of A by subtracting 2 times the first row from it. This effectively *reverses* the elementary operation:

- If the operation encoded by E_3 is: Row 3 $\to$ Row 3 $+2\times$ Row 1,
- then its inverse must undo this, i.e., Row 3 $\to$ Row 3 $-2\times$ Row 1.

Verification

Let us denote:

$$
E_3 = \begin{bmatrix} 1 & 0 & 0 \\ 0 & 1 & 0 \\ 2 & 0 & 1 \end{bmatrix}, \quad
E_3^{-1} = \begin{bmatrix} 1 & 0 & 0 \\ 0 & 1 & 0 \\ -2 & 0 & 1 \end{bmatrix}.
$$

Now compute:

$$
E_3 \cdot E_3^{-1} = \begin{bmatrix} 1 & 0 & 0 \\ 0 & 1 & 0 \\ 2 & 0 & 1 \end{bmatrix} \begin{bmatrix} 1 & 0 & 0 \\ 0 & 1 & 0 \\ -2 & 0 & 1 \end{bmatrix} = \begin{bmatrix} 1 & 0 & 0 \\ 0 & 1 & 0 \\ 0 & 0 & 1 \end{bmatrix} = I_3.
$$

E_3^{-1} correctly reverses the effect of the elementary row operation encoded by E_3, confirming it is indeed the inverse.

Python Example: Swap Matrix Rows Using Permutation Matrix

```python
import numpy as np

A = np.array([[1, 4], [2, 5], [3, 6]])
E1 = np.array([[0, 1, 0], [1, 0, 0], [0, 0, 1]])
A_swapped = E1 @ A
print("After swapping rows 1 and 2:\n",
    A_swapped)
```

Output:

```
After swapping rows 1 and 2:
 [[2 5]
 [1 4]
 [3 6]]
```

6.3 Common Transformations

This section explores specific linear transformations, their matrix representations, and geometric effects, with visualizations to enhance understanding.

6.3.1 Rotation

A rotation in $\mathbb{R}^2$ by angle θ is:

$$A = \begin{bmatrix} \cos\theta & -\sin\theta \\ \sin\theta & \cos\theta \end{bmatrix}.$$

It preserves lengths and angles. The determinant is $\det(A) = 1$, confirming the transformation is orientation-preserving.

Example 6.18 Rotate $\mathbf{v} = \begin{bmatrix} 1 \\ 0 \end{bmatrix}$ by $90°$:

$$A = \begin{bmatrix} 0 & -1 \\ 1 & 0 \end{bmatrix}, \quad T(\mathbf{v}) = \begin{bmatrix} 0 \\ 1 \end{bmatrix}.$$

Figure 6.6 shows the vector rotating from the x-axis to the y-axis.

Fig. 6.6 Rotation by 90° in $\mathbb{R}^2$

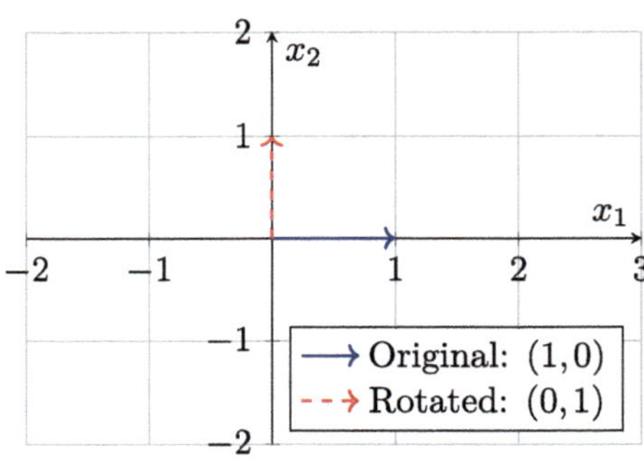

6.3.2   Projection

A projection onto a line through the origin in $\mathbb{R}^2$ along a vector $\mathbf{u} = \begin{bmatrix} u_1 \\ u_2 \end{bmatrix}$ is:

$$A = \frac{1}{u_1^2 + u_2^2} \begin{bmatrix} u_1^2 & u_1 u_2 \\ u_1 u_2 & u_2^2 \end{bmatrix}.$$

It maps vectors onto the line span($\mathbf{u}$).

Example 6.19 Project onto $\mathbf{u} = \begin{bmatrix} 1 \\ 1 \end{bmatrix}$:

$$A = \frac{1}{2} \begin{bmatrix} 1 & 1 \\ 1 & 1 \end{bmatrix}.$$

For $\mathbf{v} = \begin{bmatrix} 2 \\ 1 \end{bmatrix}$:

$$T(\mathbf{v}) = \frac{1}{2} \begin{bmatrix} 1 & 1 \\ 1 & 1 \end{bmatrix} \begin{bmatrix} 2 \\ 1 \end{bmatrix} = \begin{bmatrix} 1.5 \\ 1.5 \end{bmatrix}.$$

Figure 6.7 shows the projection onto the line $x_1 = x_2$.

6.3.3   Reflection

A reflection across the hyperplane with normal vector $\mathbf{u}$ in $\mathbb{R}^2$ uses a Householder matrix

$$A = I - \frac{2\mathbf{u}\mathbf{u}^\mathsf{T}}{\mathbf{u}^\mathsf{T}\mathbf{u}}.$$

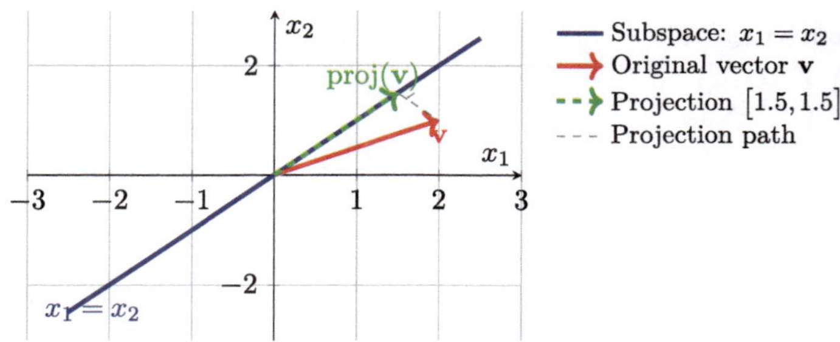

Fig. 6.7 Orthogonal projection of vector $\mathbf{v} = \begin{bmatrix} 2 \\ 1 \end{bmatrix}$ onto the subspace $x_1 = x_2$. The dashed gray line shows the perpendicular projection path, with the right angle indicating orthogonality

Fig. 6.8 Reflection over the y-axis

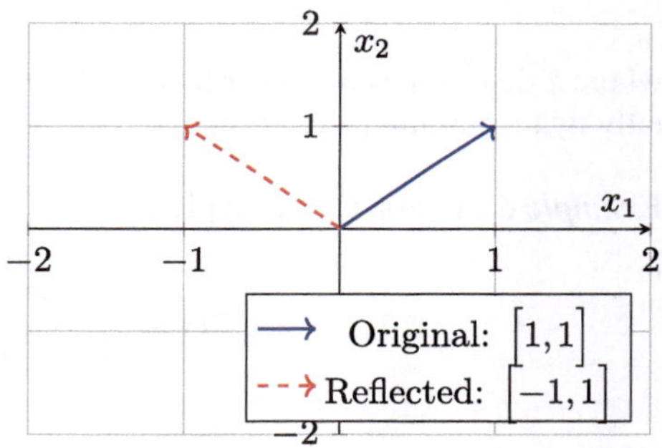

For $\mathbf{u} = \begin{bmatrix} 1 \\ 0 \end{bmatrix}$:

$$A = \begin{bmatrix} -1 & 0 \\ 0 & 1 \end{bmatrix}.$$

Example 6.20 Reflect $\mathbf{v} = \begin{bmatrix} 1 \\ 1 \end{bmatrix}$ over the y-axis:

$$T(\mathbf{v}) = \begin{bmatrix} -1 & 0 \\ 0 & 1 \end{bmatrix} \begin{bmatrix} 1 \\ 1 \end{bmatrix} = \begin{bmatrix} -1 \\ 1 \end{bmatrix}.$$

Figure 6.8 shows the reflection.

6.3.4 Shear

A horizontal shear in $\mathbb{R}^2$ is a linear transformation that shifts points parallel to the x-axis while keeping the y-coordinate unchanged. It is represented by the matrix:

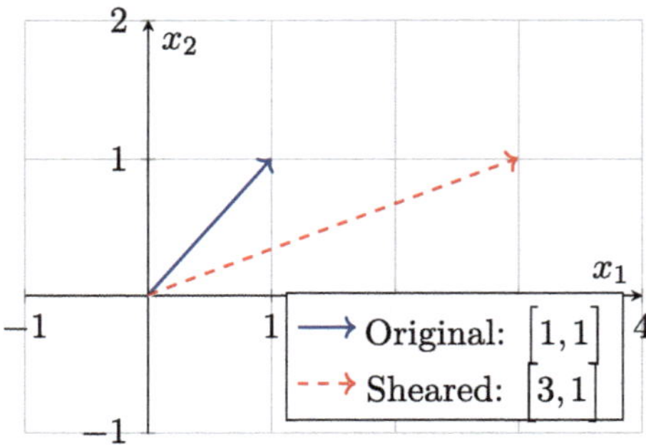

Fig. 6.9 Horizontal shear with $k = 2$

$$A = \begin{bmatrix} 1 & k \\ 0 & 1 \end{bmatrix},$$

where k determines the shear factor. This transformation moves each point horizontally by an amount proportional to its vertical position.

Example 6.21 For $k = 2$, applying the shear matrix to the vector $\mathbf{v} = \begin{bmatrix} 1 \\ 1 \end{bmatrix}$ results in:

$$T(\mathbf{v}) = \begin{bmatrix} 1 & 2 \\ 0 & 1 \end{bmatrix} \begin{bmatrix} 1 \\ 1 \end{bmatrix} = \begin{bmatrix} 3 \\ 1 \end{bmatrix}.$$

Figure 6.9 illustrates the effect of a horizontal shear on the vector $\begin{bmatrix} 1 \\ 1 \end{bmatrix}$, showing how it is displaced to $\begin{bmatrix} 3 \\ 1 \end{bmatrix}$ along the x-axis, while the y-component remains unchanged. The transformation effectively "slides" points horizontally based on their vertical position.

6.4   Transformations in ML

Linear transformations are fundamental operations in ML. They support various tasks such as data preprocessing, dimensionality reduction, optimization, and model design. Mathematically, they are represented by matrix operations that map input vectors to transformed outputs in structured and predictable ways.

6.4.1   *Linear Regression (LR)*

In LR, we aim to find a set of weights $\mathbf{w}$ that best map input features X to output targets $\mathbf{y}$. The *normal equation*:

$$\mathbf{w} = (X^\mathsf{T}X)^{-1}X^\mathsf{T}\mathbf{y},$$

is a closed-form solution derived from minimizing the MSE. It involves taking the inverse of the Gram matrix $X^\mathsf{T}X$, assuming it is invertible.

Python Example: Least Squares Solution Using Normal Equation

```python
import numpy as np

X = np.array([[1, 2], [3, 4], [5, 6]])
y = np.array([1, 2, 3])
w = np.linalg.pinv(X) @ y
print("Weights:", w)
```

Output:

```
Weights: [0.  0.5]
```

This code creates a simple dataset X with 3 samples and 2 features, along with target values y. It computes the weight vector $\mathbf{w}$ using the normal equation. The result shows how linear algebra directly provides optimal coefficients for regression.

6.4.2 PCA

PCA is a dimensionality reduction technique that transforms data into a new coordinate system aligned with directions of maximum variance (principal components). These directions are obtained by computing the eigenvectors of the data's covariance matrix. The resulting transformation matrix reduces redundancy while preserving structure in high-dimensional data.

6.4.3 Normalizing Flows

Normalizing flows use invertible linear and nonlinear transformations to convert complex data distributions into simpler ones (e.g., Gaussian). These transformations are bijective, allowing exact computation of both the forward and inverse operations, which is essential in generative modeling.

6.4.4 Autoencoders

Autoencoders learn to compress data into a lower-dimensional latent space using an *encoder*, and then reconstruct the original input through a **decoder**. The encoder typically applies a linear transformation followed by a non-linearity. The decoder approximates the inverse transformation to recover the input, thereby learning meaningful representations.

6.4.5 Optimization

In second-order optimization methods such as Newton's method, the **Hessian matrix** (second derivative of the loss function) is used to refine parameter updates (see Chap. 12 for more details). The update rule involves the inverse of the Hessian, making it a critical linear transformation for efficient convergence in non-convex landscapes.

6.4.6 Other Applications

- **Sparse Coding**: Uses linear transformations to represent signals with few non-zero coefficients, promoting interpretability and efficiency.
- **Neural Networks**: Each layer performs a linear transformation using weight matrices before applying a non-linear activation function.
- **Kernel Methods**: Implicitly apply linear transformations in high-dimensional spaces to enable complex decision boundaries via the kernel trick.

Python Example: Projection of a Vector onto a Line Using Outer Product

```python
import numpy as np
import matplotlib.pyplot as plt

u = np.array([1, 1])
A = np.outer(u, u) / np.dot(u, u)
v = np.array([2, 1])
v_projected = A @ v

plt.quiver(0, 0, v[0], v[1], color='blue',
    angles='xy', scale_units='xy', scale=1,
    label='Original')
plt.quiver(0, 0, v_projected[0], v_projected[1],
    color='red', angles='xy', scale_units='xy',
    scale=1, label='Projected')
plt.plot([-3, 3], [-3, 3], 'k--', label='Line')
```

```
12  plt.axis('equal')
13  plt.legend()
14  plt.show()
```

Output:

        See Figure 6.10.

This code calculates the orthogonal projection of vector $\mathbf{v} = \begin{bmatrix} 2 \\ 1 \end{bmatrix}$ onto the line spanned by $\mathbf{u} = \begin{bmatrix} 1 \\ 1 \end{bmatrix}$. The projection matrix A is formed using the outer product formula. The plot visually contrasts the original and projected vectors.

Chapter Summary

This chapter delved into linear transformations as mappings between vector spaces that preserve addition and scalar multiplication, represented by matrices A such that $T(\mathbf{x}) = A\mathbf{x}$, with a focus on their geometric and algebraic properties relevant to ML. Section 6.1 defined key concepts: domain as the input space $\mathbb{R}^n$, codomain as the output space $\mathbb{R}^m$, and range (image) as the actual outputs $\mathrm{im}(T) = \mathrm{Col}(A)$, illustrated in Fig. 6.1. The kernel $\ker(T) = \{\mathbf{x} \mid A\mathbf{x} = \mathbf{0}\}$ captured directions flattened to zero, and the Rank-Nullity Theorem $\dim(\ker(T)) + \dim(\mathrm{im}(T)) = n$ linked nullity and rank, as exemplified with $A = \begin{bmatrix} 1 & 2 \\ 3 & 6 \end{bmatrix}$ where both dimensions were 1 (Fig. 6.2).

Injectivity (one-to-one) required trivial kernel ($\mathrm{rank}(A) = n$), surjectivity (onto) required full row span ($\mathrm{rank}(A) = m$), and bijectivity both, necessitating square full-rank matrices, as visualized in Fig. 6.3. Section 6.2 examined function inverses, where f^{-1} undid f for bijective functions, like $f(x) = 3x + 2$ with $f^{-1}(x) = \frac{x-2}{3}$. For matrices, invertibility ($\det(A) \neq 0$, $\mathrm{rank}(A) = n$) enabled $A^{-1}A = I$, with $2{\times}2$ formula $\frac{1}{\det(A)} \begin{bmatrix} d & -b \\ -c & a \end{bmatrix}$, adjugate method $A^{-1} = \frac{1}{\det(A)} \mathrm{adj}(A)$, and Gauss-Jordan on $[A \mid I]$ to $[I \mid A^{-1}]$, verified for examples like $\begin{bmatrix} 1 & 2 \\ 3 & 5 \end{bmatrix}$. Properties included $(AB)^{-1} = B^{-1}A^{-1}$ and $(A^{\mathrm{T}})^{-1} = (A^{-1})^{\mathrm{T}}$, with elementary matrices encoding row operations and their inverses reversing them.

Fig. 6.10 Orthogonal projection of a vector onto a line in $\mathbb{R}^2$. The original vector $\mathbf{v} = \begin{bmatrix} 2 \\ 1 \end{bmatrix}$ (in blue) is projected onto the line $\mathbf{u} = \begin{bmatrix} 1 \\ 1 \end{bmatrix}$ defined by (dashed black). The red vector represents the projection $\mathbf{v}_{\text{proj}} = A\mathbf{v}$, where $A = \frac{\mathbf{u}\mathbf{u}^{\text{T}}}{\mathbf{u}^{\text{T}}\mathbf{u}}$

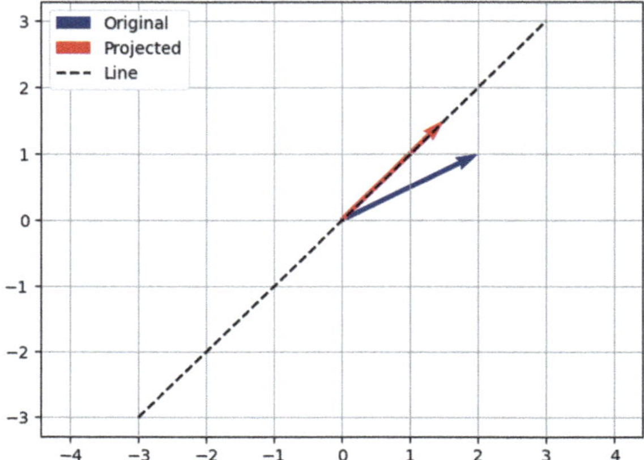

Common transformations in Sect. 6.3 included rotations $\begin{bmatrix} \cos\theta & -\sin\theta \\ \sin\theta & \cos\theta \end{bmatrix}$ (Fig. 6.6), projections $A = \frac{\mathbf{u}\mathbf{u}^{\text{T}}}{\mathbf{u}^{\text{T}}\mathbf{u}}$ onto lines (Fig. 6.7), reflections like over y-axis $\begin{bmatrix} -1 & 0 \\ 0 & 1 \end{bmatrix}$ (Fig. 6.8), and shears $\begin{bmatrix} 1 & k \\ 0 & 1 \end{bmatrix}$ (Fig. 6.9), all linear and often invertible.

Applications in ML (Sect. 6.4) highlighted LR via normal equations $(X^{\text{T}}X)^{-1}X^{\text{T}}\mathbf{y}$, PCA for variance-aligned coordinates, normalizing flows for bijective density transformations, autoencoders for compression-reconstruction, and optimization using Hessian inverses, with examples in sparse coding and kernel methods. Python snippets using NumPy for inverses, properties, and projections (Fig. 6.10) bridged theory to implementation.

Takeaways

Key elements from the chapter were distilled into:

- Linear transformations $T(\mathbf{x}) = A\mathbf{x}$ preserving operations, with domain/codomain/range, kernel (null space), image (column space), and Rank-Nullity Theorem governing dimensions.
- Injectivity (trivial kernel, full column rank), surjectivity (full row rank), and bijectivity (square invertible matrices) determining mapping completeness.
- Inverses for bijective transformations, computed via 2×2 formula, adjugate, or Gauss-Jordan, with properties like product reversal $(AB)^{-1} = B^{-1}A^{-1}$ and transpose symmetry.
- Elementary matrices encoding row swaps or scales or adds, invertible with reversing operations, facilitating RREF for invertibility checks.

- Common transformations, rotations (angle-preserving), projections (onto subspaces), reflections (over lines or planes), shears (parallel shifts), visualized geometrically.
- ML integrations: regression solving via inverses, PCA eigenvector transformations, flows or autoencoders for bijective mappings, and optimization with Hessians.

Through these explorations, readers mastered defining, computing, and applying linear transformations, gaining insights into their role in reshaping data and spaces for ML algorithms.

Exercises

1. Verify if $T(\mathbf{x}) = \begin{bmatrix} 2 & 1 \\ 1 & 3 \end{bmatrix} \mathbf{x}$ is linear.

2. Find the matrix of $T : \mathbb{R}^2 \to \mathbb{R}^2$ where $T(\begin{bmatrix} 1 \\ 0 \end{bmatrix}) = \begin{bmatrix} 2 \\ 1 \end{bmatrix}, T(\begin{bmatrix} 0 \\ 1 \end{bmatrix}) = \begin{bmatrix} -1 \\ 3 \end{bmatrix}$.

3. Compute the kernel of $T(\mathbf{x}) = \begin{bmatrix} 1 & 2 \\ 2 & 4 \end{bmatrix} \mathbf{x}$.

4. Determine the image of $T(\mathbf{x}) = \begin{bmatrix} 1 & 0 \\ 0 & 1 \\ 1 & 1 \end{bmatrix} \mathbf{x}$.

5. Is $T(\mathbf{x}) = \begin{bmatrix} 1 & 2 \\ 3 & 4 \end{bmatrix} \mathbf{x}$ injective? Surjective?

6. Find the inverse of $\begin{bmatrix} 4 & 7 \\ 2 & 6 \end{bmatrix}$ using the 2x2 formula.

7. Use Gauss-Jordan to find the inverse of $\begin{bmatrix} 1 & 0 & 4 \\ 0 & 2 & 3 \\ 0 & 0 & 1 \end{bmatrix}$.

8. Verify that $\begin{bmatrix} 0 & 1 \\ 1 & 0 \end{bmatrix}$ is its own inverse.

9. For $A = \begin{bmatrix} 1 & 2 \\ 3 & 5 \end{bmatrix}$, compute $(A^{\mathrm{T}})^{-1}$ and $(A^{-1})^{\mathrm{T}}$.

10. Show that the shear matrix $\begin{bmatrix} 1 & 2 \\ 0 & 1 \end{bmatrix}$ is invertible and find its inverse.

11. Compute the image of $\mathbf{v} = \begin{bmatrix} 1 \\ 1 \end{bmatrix}$ under the rotation matrix

$$\mathbf{Q} = \begin{bmatrix} \cos(\pi/4) & -\sin(\pi/4) \\ \sin(\pi/4) & \cos(\pi/4) \end{bmatrix}.$$

12. Project $\mathbf{v} = \begin{bmatrix} 3 \\ 2 \end{bmatrix}$ onto the line spanned by $\begin{bmatrix} 1 \\ 1 \end{bmatrix}$.

13. Reflect $\mathbf{v} = \begin{bmatrix} 2 \\ 1 \end{bmatrix}$ over the x-axis using the matrix $\begin{bmatrix} 1 & 0 \\ 0 & -1 \end{bmatrix}$ and plot the result.

14. Apply the horizontal shear matrix $\begin{bmatrix} 1 & k \\ 0 & 1 \end{bmatrix}$, $k = 1.5$, to $\mathbf{v} = \begin{bmatrix} 1 \\ 2 \end{bmatrix}$.

15. Check if $\begin{bmatrix} 1 & 2 & 1 \\ 2 & 4 & 2 \\ 0 & 0 & 1 \end{bmatrix}$ is invertible using RREF.

16. Solve $\begin{bmatrix} 1 & 2 \\ 3 & 5 \end{bmatrix} \mathbf{x} = \begin{bmatrix} 4 \\ 7 \end{bmatrix}$ using the inverse.

17. For $A = \begin{bmatrix} 1 & 2 \\ 3 & 4 \end{bmatrix}$, $B = \begin{bmatrix} 2 & 1 \\ 0 & 1 \end{bmatrix}$, verify $(AB)^{-1} = B^{-1}A^{-1}$.

18. Compute the kernel and image of $T(\mathbf{x}) = \begin{bmatrix} 1 & 1 & 1 \\ 0 & 1 & 1 \end{bmatrix} \mathbf{x}$.

19. Is $T(\mathbf{x}) = \begin{bmatrix} 1 & 0 \\ 0 & 0 \end{bmatrix} \mathbf{x}$ bijective? Explain.

20. Write a Python script to visualize the reflection of $\begin{bmatrix} 1 \\ 1 \end{bmatrix}$ over the line $x_1 = x_2$ using the matrix $\begin{bmatrix} 0 & 1 \\ 1 & 0 \end{bmatrix}$.

21. Find the matrix for a 30° rotation in $\mathbb{R}^2$ and apply it to $\begin{bmatrix} 2 \\ 0 \end{bmatrix}$.

22. Prove that the inverse of an elementary matrix is an elementary matrix.

23. Compute the inverse of $\begin{bmatrix} 1 & 2 & 1 \\ 0 & 1 & 2 \\ 1 & 0 & 1 \end{bmatrix}$ using the adjugate method.

24. Determine if the transformation $T : \mathbb{R}^3 \to \mathbb{R}^2$ defined by $T(\mathbf{x}) = \begin{bmatrix} 1 & 2 & 3 \\ 0 & 1 & 2 \end{bmatrix} \mathbf{x}$ is surjective.

25. Write a Python script to plot the projection of $\begin{bmatrix} 2 \\ 3 \end{bmatrix}$ onto the x-axis.

Chapter 7
Determinants

Introduction

The determinant of a square matrix is a powerful scalar that distills critical information about a linear transformation, acting as a mathematical lens through which we can understand invertibility, volume scaling, and orientation changes. In the context of linear algebra for ML, determinants are indispensable, providing insights into the behavior of matrices in algorithms like LR, PCA, and normalizing flows. Geometrically, the determinant quantifies the scaling factor of areas in 2D or volumes in 3D, while algebraically, the determinant determines whether a matrix is invertible and how systems of linear equations can be solved.

In this chapter, we begin with the definition and geometric interpretation of determinants, using visualizations to clarify concepts like the signed area of a parallelogram or the volume of a parallelepiped. We then explore computational methods, including cofactor expansion, upper triangular matrices, and Cramer's rule, with detailed examples to build computational fluency. A dedicated section on ML applications highlights the determinant's role in real-world algorithms, supported by Python implementations using NumPy and Matplotlib. Comprehensive exercises at the end reinforce both theoretical and computational skills, ensuring a deep understanding of determinants in the context of linear algebra and ML.

Topics Covered

This chapter is organized as follows:

- **Section 7.1 Determinants; Definition and Basics**: Defines determinants for 2×2 and 3×3 matrices, explores their geometric meaning as area or volume, and discusses orientation and invertibility.

- **Section** 7.2 **Computing Determinants**: Covers methods like cofactor expansion, upper triangular matrices, properties (e.g., row operations, matrix products), and Cramer's rule for solving linear systems.
- **Section** 7.3 **Properties of Determinants**: Outlines key properties of determinants, including their behavior under matrix operations like scalar multiplication, matrix products, inverses, and transposes, highlighting their multiplicative nature and applications in solving linear systems with Cramer's rule.
- **Section** 7.4 **Determinants in ML**: Examines applications in LR, PCA, normalizing flows, covariance matrices, optimization, and numerical stability.
- **Summary and Exercises**: Recaps key concepts and provides a comprehensive set of problems to reinforce theoretical and computational understanding.

All code accompanying the chapter is publicly available at https://github.com/jalil-piran/Linear-Algebra-with-Applications-in-Machine-Learning/blob/main/Chapter_07_Determinants.ipynb.

Acronyms

ML	Machine Learning
LR	Linear Regression
PCA	Principal Component Analysis
2D	Two-Dimensional
3D	Three-Dimensional
LU	Lower-Upper Decomposition
QR	QR Decomposition

The visualizations, such as Fig. 7.1 for 2D area scaling and Fig. 7.3 for Cramer's rule, bridge algebraic computations with geometric intuition, making abstract concepts tangible. By the end of this chapter, you will be equipped to compute determinants, interpret their significance, and apply them in ML contexts.

7.1 Determinants: Definition and Basics

The determinant is a fundamental scalar quantity associated with a square matrix that encodes important properties of the matrix, including invertibility, scaling of volume, and the behavior of linear transformations. For a matrix $A \in \mathbb{R}^{n \times n}$, the determinant, denoted $\det(A)$ or $|A|$, provides a measure of how the matrix transforms space: a nonzero determinant indicates that the matrix is invertible and preserves the dimension of the space, while a zero determinant signifies linear dependence among rows or columns and a loss of dimensionality. Determinants also play a crucial role in solving systems of linear equations, evaluating eigenvalues, and understanding the geometric interpretation of matrices, such as areas and volumes in two- and three-dimensional spaces. Various methods exist to compute determinants, including cofactor expansion, row reduction, and leveraging properties of triangular matrices, making them an indispensable tool in linear algebra and its applications.

> **Definition**
> The **determinant** of a square matrix A, denoted $\det(A)$, is a scalar that encapsulates key properties of the linear transformation $T(\mathbf{x}) = A\mathbf{x}$. It provides insights into invertibility, volume scaling, and orientation preservation, making it a cornerstone of linear algebra.

7.1.1 Key Properties

- **Invertibility**: Matrix A is invertible if and only if $\det(A) \neq 0$. A zero determinant indicates a singular matrix, collapsing the space to a lower dimension.
- **Volume Scaling**: The absolute value $|\det(A)|$ measures the factor by which areas (in 2D) or volumes (in 3D) are scaled under the transformation.
- **Orientation**: A positive $\det(A)$ preserves orientation, while a negative $\det(A)$ reverses it (e.g., via reflection).

7.1.2 Determinant of 2×2 Matrices

For a 2×2 matrix:

$$A = \begin{bmatrix} a & b \\ c & d \end{bmatrix}, \quad \det(A) = ad - bc.$$

Geometrically, $|\det(A)|$ is the area of the parallelogram formed by the column vectors $\begin{bmatrix} a \\ c \end{bmatrix}$ and $\begin{bmatrix} b \\ d \end{bmatrix}$. The sign indicates whether the transformation preserves or reverses orientation.

Example 7.1 Consider:

$$A = \begin{bmatrix} 1 & 2 \\ 3 & 4 \end{bmatrix}.$$

$$\det(A) = (1)(4) - (2)(3) = 4 - 6 = -2.$$

The area of the parallelogram is $|-2| = 2$, and the negative sign indicates an orientation flip. Figure 7.1 visualizes this, showing the parallelogram formed by $\begin{bmatrix} 1 \\ 3 \end{bmatrix}$ and $\begin{bmatrix} 2 \\ 4 \end{bmatrix}$.

Fig. 7.1 The shaded region's area equals $|det(A)|$, which gives the unsigned area of the parallelogram in 2D

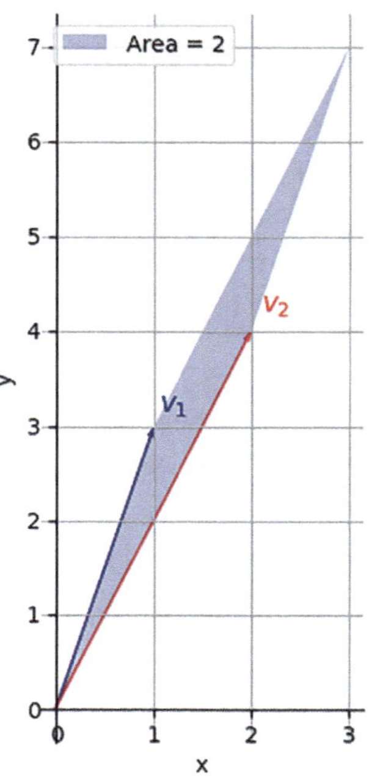

Python Example: Parallelogram Area

```python
import numpy as np
import matplotlib.pyplot as plt

A = np.array([[1, 2],
              [3, 4]])

v1, v2 = A[:, 0], A[:, 1]

det_A = np.abs(np.linalg.det(A))

fig, ax = plt.subplots(figsize=(6, 6))

ax.spines['left'].set_position('zero')
ax.spines['bottom'].set_position('zero')
ax.spines['right'].set_color('none')
ax.spines['top'].set_color('none')

# Plot vectors
ax.quiver([0, 0], [0, 0],
          [v1[0], v2[0]],
          [v1[1], v2[1]],
          angles='xy',
```

```
23              scale_units='xy',
24              scale=1,
25              color=['blue', 'red'])
26
27 # Add vector labels
28 ax.text(v1[0]*1.05, v1[1]*1.05, r'$v_1$',
       fontsize=12, color='blue')
29 ax.text(v2[0]*1.05, v2[1]*1.05, r'$v_2$',
       fontsize=12, color='red')
30
31 # Parallelogram points
32 parallelogram = np.array([
33     [0, 0],
34     v1,
35     v1 + v2,
36     v2,
37     [0, 0]
38 ])
39
40 # Fill parallelogram
41 ax.fill(parallelogram[:, 0],
42         parallelogram[:, 1],
43         'b',
44         alpha=0.2,
45         label=f'Area = {det_A:.0f}')
46
47 # Formatting
48 ax.set_aspect('equal')
49 ax.grid(True)
50 ax.legend()
51 ax.set_xlabel('x')
52 ax.set_ylabel('y')
53 plt.show()
```

Output:

    See Figure 7.1.

Python Example: Determinant of a 2x2 Matrix

```python
import numpy as np

A = np.array([[1, 2], [3, 4]])
det_A = np.linalg.det(A)
print("2x2 Determinant:
    {det_A:.2f}".format(det_A=det_A))
```

Output:

```
2x2 Determinant: -2.00
```

Example 7.2 Given the matrix:

$$A = \begin{bmatrix} 4 & 1 \\ 0 & 2 \end{bmatrix}.$$

To compute the determinant of a 2×2 matrix, we use the formula:

$$\det(A) = ad - bc,$$

where the matrix A is:

$$A = \begin{bmatrix} a & b \\ c & d \end{bmatrix}.$$

Step 1- Identify the entries.
From the matrix A, we have:

$$a = 4, \quad b = 1, \quad c = 0, \quad d = 2.$$

Step 2- Plug into the determinant formula.

$$\det(A) = (4)(2) - (1)(0) = 8 - 0 = 8.$$

$$\boxed{\det(A) = 8}$$

This result tells us that the matrix A is **invertible**, since its determinant is nonzero.

7.1.3 *Determinant of* 3 × 3 *Matrices*

The determinant of a 3 × 3 matrix captures important geometric and algebraic properties. It plays a central role in understanding invertibility, volume scaling, and orientation in three-dimensional space.

Let $A \in \mathbb{R}^{3\times 3}$ be a square matrix:

$$A = \begin{bmatrix} a_{11} & a_{12} & a_{13} \\ a_{21} & a_{22} & a_{23} \\ a_{31} & a_{32} & a_{33} \end{bmatrix}.$$

Laplace Expansion: The determinant of A can be computed using cofactor expansion along the first row:

$$\det(A) = a_{11}(a_{22}a_{33} - a_{23}a_{32}) - a_{12}(a_{21}a_{33} - a_{23}a_{31}) + a_{13}(a_{21}a_{32} - a_{22}a_{31}).$$

Each term consists of an entry from the first row of A multiplied by the determinant of the corresponding 2 × 2 minor, with alternating signs. This expansion provides an intuitive decomposition of the total volume into contributions from each coordinate direction.

The determinant of a 3 × 3 matrix also represents the **signed volume** of the parallelepiped defined by the column vectors (or row vectors) of the matrix. If the determinant is zero, the vectors are linearly dependent and lie in the same plane, meaning the volume collapses to zero. A positive determinant indicates a right-handed orientation, while a negative value indicates a left-handed orientation.

Mnemonic Formula (Rule of Sarrus): An alternative, memory-friendly formula for computing the determinant of a 3 × 3 matrix is:

$$\det(A) = a_{11}a_{22}a_{33} + a_{12}a_{23}a_{31} + a_{13}a_{21}a_{32} - a_{13}a_{22}a_{31} - a_{12}a_{21}a_{33} - a_{11}a_{23}a_{32}.$$

where the first three terms represent the "forward diagonals" and the last three terms the "backward diagonals" when the matrix is visualized with its first two columns repeated to the right.

In summary, the determinant of a 3 × 3 matrix:

- The matrix is invertible if and only if $\det(A) \neq 0$.
- Gives the signed volume of the parallelepiped spanned by the column vectors.
- Acts as the scaling factor of volume under the linear transformation defined by A.

Example 7.3 Compute the determinant of matrix A:

$$A = \begin{bmatrix} 1 & 2 & 3 \\ 4 & 5 & 6 \\ 7 & 8 & 0 \end{bmatrix}.$$

Using the formula:

$$\det(A) = 1(5 \cdot 0 - 6 \cdot 8) - 2(4 \cdot 0 - 6 \cdot 7) + 3(4 \cdot 8 - 5 \cdot 7) = -48 + 84 - 9 = 27.$$

The volume of the parallelepiped is 27, with a positive sign indicating orientation preservation. Figure 7.2 illustrates a 3D parallelepiped for a simpler matrix.

Example 7.4 Given the matrix:

$$A = \begin{bmatrix} 1 & 2 & 1 \\ 0 & 3 & 4 \\ 3 & 1 & 4 \end{bmatrix}.$$

We compute the determinant of a 3×3 matrix using cofactor expansion along the first row:

$$\det(A) = a_{11} \begin{vmatrix} 3 & 4 \\ 1 & 4 \end{vmatrix} - a_{12} \begin{vmatrix} 0 & 4 \\ 3 & 4 \end{vmatrix} + a_{13} \begin{vmatrix} 0 & 3 \\ 3 & 1 \end{vmatrix}.$$

Step 1- Compute the 2×2 minors.

$$\begin{vmatrix} 3 & 4 \\ 1 & 4 \end{vmatrix} = (3)(4) - (1)(4) = 12 - 4 = 8,$$

$$\begin{vmatrix} 0 & 4 \\ 3 & 4 \end{vmatrix} = (0)(4) - (3)(4) = 0 - 12 = -12,$$

$$\begin{vmatrix} 0 & 3 \\ 3 & 1 \end{vmatrix} = (0)(1) - (3)(3) = 0 - 9 = -9.$$

Step 2- Plug into the cofactor expansion.

$$\begin{aligned} \det(A) &= 1(8) - 2(-12) + 1(-9) \\ &= 8 + 24 - 9 \\ &= \boxed{23} \end{aligned}$$

The determinant of matrix A is $\boxed{23}$, which is nonzero. Therefore, A is invertible.

Geometric Insights

- **2D**: The determinant is the signed area of the parallelogram spanned by the columns. If $\det(A) = 0$, the vectors are linearly dependent, collapsing the area to zero.
- **3D**: The determinant is the signed volume of the parallelepiped. A zero determinant indicates coplanar vectors, collapsing the volume to zero (see Fig. 7.2).
- **Intuition**: A zero determinant signifies a transformation that "squashes" space, aligning with non-invertibility.

Fig. 7.2 The determinant of
$$A = \begin{bmatrix} 1 & 0 & 0 \\ 0 & 2 & 0 \\ 0 & 0 & 3 \end{bmatrix}$$
matrix gives the volume of the parallelepiped spanned by its column vectors. Here, $\det(A) = 6$, which equals the volume of the box

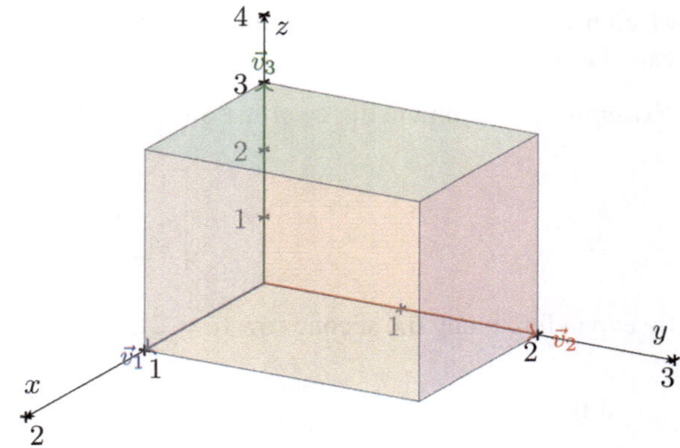

Example 7.5 For:
$$A = \begin{bmatrix} 1 & 2 \\ 2 & 4 \end{bmatrix}.$$

$$\det(A) = (1)(4) - (2)(2) = 4 - 4 = 0.$$

The zero determinant indicates the columns are linearly dependent (the second is twice the first), collapsing the parallelogram to a line, and the matrix is not invertible.

7.2 Computing Determinants

This section covers methods to compute determinants, including cofactor expansion, upper triangular matrices, properties of determinants, and Cramer's rule for solving linear systems.

7.2.1 Cofactor Expansion

The **cofactor expansion** computes the determinant by breaking it into smaller submatrices. For an $n \times n$ matrix $A = [a_{ij}]$, the cofactor of element a_{ij} is:

$$C_{ij} = (-1)^{i+j} \det(A_{ij}),$$

where A_{ij} is the submatrix obtained by deleting row i and column j. The determinant is:

$$\det(A) = \sum_{j=1}^{n} a_{ij} C_{ij} \quad \text{(along row } i\text{)},$$

or equivalently along any column. Choosing a row/column with zeros simplifies calculations.

Example 7.6 Compute the determinant for matrix A:

$$A = \begin{bmatrix} 1 & 2 & 3 \\ 4 & 5 & 6 \\ 7 & 8 & 9 \end{bmatrix},$$

by expanding along the second row ($i = 2$):

$$\det(A) = 4 \cdot (-1)^{2+1} \cdot \det \begin{bmatrix} 2 & 3 \\ 8 & 9 \end{bmatrix} + 5 \cdot (-1)^{2+2} \cdot \det \begin{bmatrix} 1 & 3 \\ 7 & 9 \end{bmatrix} + 6 \cdot (-1)^{2+3} \cdot \det \begin{bmatrix} 1 & 2 \\ 7 & 8 \end{bmatrix}$$

$$= 4 \cdot (-1) \cdot (2 \cdot 9 - 3 \cdot 8) + 5 \cdot (+1) \cdot (1 \cdot 9 - 3 \cdot 7) + 6 \cdot (-1) \cdot (1 \cdot 8 - 2 \cdot 7)$$

$$= 24 - 60 + 36 = \boxed{0}$$

Since $\det(A) = 0$, the matrix A is *singular*, it is not invertible. The columns of A are linearly dependent, and the volume of the 3D parallelepiped they form is zero. The transformation collapses 3D space into a plane.

7.2.2   Upper Triangular Matrices

For an upper triangular matrix:

$$A = \begin{bmatrix} a & * & * \\ 0 & b & * \\ 0 & 0 & c \end{bmatrix}, \quad \det(A) = a \cdot b \cdot c.$$

The determinant is the product of diagonal entries, since row operations that add multiples of one row to another preserve the determinant, while row swaps negate it and row scalings multiply it by the scaling factor.

Example 7.7 For:

$$A = \begin{bmatrix} 2 & 1 & 3 \\ 0 & 4 & 5 \\ 0 & 0 & 6 \end{bmatrix}, \quad \det(A) = 2 \cdot 4 \cdot 6 = 48.$$

We can do verification by expanding row 3:

$$\det(A) = 6 \cdot \det \begin{bmatrix} 2 & 1 \\ 0 & 4 \end{bmatrix} = 6 \cdot 8 = 48.$$

7.3 Properties of Determinants

Determinants connect matrix operations to their geometric effects, simplifying computations and providing intuition.

1. Identity Matrix:
$$\det(I_n) = 1.$$

The identity matrix preserves volume and orientation.

Example 7.8
$$I_2 = \begin{bmatrix} 1 & 0 \\ 0 & 1 \end{bmatrix}, \quad \det(I_2) = 1.$$

2. Row Exchange: Swapping two rows negates the determinant, reflecting an orientation flip.
$$A = \begin{bmatrix} 1 & 2 \\ 3 & 4 \end{bmatrix}, \quad \det(A) = -2,$$

$$A' = \begin{bmatrix} 3 & 4 \\ 1 & 2 \end{bmatrix}, \quad \det(A') = 2 = -\det(A).$$

If two rows are identical, $\det(A) = 0$.

3. Linearity in Rows:

- Scaling a Row: Scaling a row by k scales the determinant by k.

$$A = \begin{bmatrix} 1 & 2 \\ 3 & 4 \end{bmatrix}, \quad A'' = \begin{bmatrix} 2 & 4 \\ 3 & 4 \end{bmatrix}, \quad \det(A'') = -4 = 2 \cdot (-2).$$

- Adding a Multiple: Adding a multiple of one row to another preserves the determinant.
- Row as Sum: The determinant is linear in each row:

$$\det \begin{bmatrix} \mathbf{u} + \mathbf{v} & \mathbf{w} \\ \mathbf{x} & \mathbf{y} \end{bmatrix} = \det \begin{bmatrix} \mathbf{u} & \mathbf{w} \\ \mathbf{x} & \mathbf{y} \end{bmatrix} + \det \begin{bmatrix} \mathbf{v} & \mathbf{w} \\ \mathbf{x} & \mathbf{y} \end{bmatrix}.$$

Python Example: Determinant

```python
import numpy as np

A = np.array([[1, 2], [3, 4]])
A_swapped = np.array([[3, 4], [1, 2]])
A_scaled = np.array([[2, 4], [3, 4]])
```

```
6
7  print(f"Original det: {np.linalg.det(A):.2f}")
8  print(f"Swapped rows det:
       {np.linalg.det(A_swapped):.2f}")
9  print(f"Scaled row det:
       {np.linalg.det(A_scaled):.2f}")
```

Output:

```
Original det: -2.00
Swapped rows det: 2.00
Scaled row det: -4.00
```

4. **Scalar Multiplication**: Multiplying an entire $n \times n$ matrix A by a scalar k scales its determinant by k^n, not just by k. This reflects how volume scales in n-dimensional space.

$$\det(kA) = k^n \det(A).$$

Example 7.9 Consider matrix A and scalar k as follows:

$$A = \begin{bmatrix} 1 & 2 \\ 3 & 4 \end{bmatrix},$$

Then, the determinant of kA is:

$$k = 4, \quad \det(4A) = 4^2 \cdot (-2) = -32.$$

5. **Sum of Matrices**: The determinant of a sum is *not* equal to the sum of determinants. Determinants are nonlinear with respect to matrix addition.

$$\det(A + B) \neq \det(A) + \det(B).$$

Example 7.10 Consider the following matrices:

$$A = \begin{bmatrix} 1 & 2 \\ 3 & 4 \end{bmatrix}, \quad B = \begin{bmatrix} 5 & 6 \\ 7 & 8 \end{bmatrix}.$$

Then, determinant of $A + B$ is:

$$\det(A + B) = \det\left(\begin{bmatrix} 6 & 8 \\ 10 & 12 \end{bmatrix}\right) = -8 \neq -2 + (-2).$$

6. **Product of Matrices**: The determinant of a product is the product of the determinants. This is one of the most important multiplicative properties of the determinant.

$$\det(AB) = \det(A) \cdot \det(B).$$

Example 7.11

$$A = \begin{bmatrix} 1 & 2 \\ 3 & 4 \end{bmatrix}, \quad \det(A) = -2,$$

$$B = \begin{bmatrix} 2 & 0 \\ 1 & 2 \end{bmatrix}, \quad \det(B) = 4,$$

$$\det(AB) = (-2)(4) = -8.$$

7. **Inverse**: The determinant of the inverse matrix is the reciprocal of the original determinant, as long as the matrix is invertible.

$$\det(A^{-1}) = \frac{1}{\det(A)} \quad \text{if } \det(A) \neq 0.$$

Example 7.12

$$A = \begin{bmatrix} 1 & 2 \\ 3 & 4 \end{bmatrix}, \quad \det(A) = -2$$

$$\Rightarrow \det(A^{-1}) = \frac{1}{-2} = -0.5$$

8. **Power of a Matrix**: Raising a matrix to a power raises the determinant to the same power. This stems from repeated multiplication.

$$\det(A^k) = (\det(A))^k.$$

Example 7.13

$$A = \begin{bmatrix} 1 & 2 \\ 3 & 4 \end{bmatrix}, \quad \det(A) = -2$$

$$A^2 = A \cdot A = \begin{bmatrix} 7 & 10 \\ 15 & 22 \end{bmatrix}, \quad \det(A^2) = 4 = (-2)^2$$

9. **Transpose**: Taking the transpose of a matrix does not change its determinant. The orientation of row and column vectors does not affect volume.

$$\det(A^T) = \det(A).$$

Example 7.14

$$A = \begin{bmatrix} 1 & 2 \\ 3 & 4 \end{bmatrix}, \quad A^T = \begin{bmatrix} 1 & 3 \\ 2 & 4 \end{bmatrix},$$

$$\det(A^T) = -2 = \det(A)$$

Python Example: Determinant Properties

```python
import numpy as np

A = np.array([[1, 2], [3, 4]])
B = np.array([[2, 0], [1, 2]])
k = 4

print(f"det(kA): {np.linalg.det(k * A):.2f}")
print(f"k^2 * det(A): {k**2 *
    np.linalg.det(A):.2f}")
print(f"det(AB): {np.linalg.det(A @ B):.2f}")
print(f"det(A) * det(B): {np.linalg.det(A) *
    np.linalg.det(B):.2f}")
print(f"det(A^T): {np.linalg.det(A.T):.2f}")
print(f"det(A): {np.linalg.det(A):.2f}")
```

Output:

```
det(kA): -32.00
k^2 * det(A): -32.00
det(AB): -8.00
det(A) * det(B): -8.00
det(A^T): -2.00
det(A): -2.00
```

7.3.1 Cramer's Rule

Cramer's Rule provides a direct method to solve a system of linear equations of the form:

$$A\mathbf{x} = \mathbf{b},$$

where $A \in \mathbb{R}^{n \times n}$ is an invertible square matrix, $\mathbf{x} \in \mathbb{R}^n$ is the vector of unknowns, and $\mathbf{b} \in \mathbb{R}^n$ is the right-hand side vector.

The rule expresses each component x_i of the solution vector $\mathbf{x}$ as a ratio of two determinants:

$$x_i = \frac{\det(A_i)}{\det(A)}, \quad i = 1, 2, \ldots, n$$

where A_i is the matrix obtained by replacing the ith column of A with the vector $\mathbf{b}$, while keeping the other columns unchanged.

Key Ideas:

- Cramer's Rule only applies when A is a square and invertible matrix (i.e., $\det(A) \neq 0$).
- The solution relies on computing $n + 1$ determinants: one for A and one for each A_i.
- Geometrically, this rule can be interpreted in terms of volume scaling and orientation in $\mathbb{R}^n$.

Example 7.15 Solve the system:

$$\begin{cases} x_1 + 2x_2 = 5, \\ 3x_1 + 4x_2 = 11. \end{cases}$$

First, write the coefficient matrix and the right-hand side:

$$A = \begin{bmatrix} 1 & 2 \\ 3 & 4 \end{bmatrix}, \quad \mathbf{b} = \begin{bmatrix} 5 \\ 11 \end{bmatrix}.$$

We compute the determinant of A:

$$\det(A) = (1)(4) - (2)(3) = 4 - 6 = -2.$$

Now construct matrices A_1 and A_2 by replacing the respective columns of A with $\mathbf{b}$:

$$A_1 = \begin{bmatrix} 5 & 2 \\ 11 & 4 \end{bmatrix}, \quad \det(A_1) = (5)(4) - (2)(11) = 20 - 22 = -2 \Rightarrow x_1 = \frac{-2}{-2} = 1$$

$$A_2 = \begin{bmatrix} 1 & 5 \\ 3 & 11 \end{bmatrix}, \quad \det(A_2) = (1)(11) - (5)(3) = 11 - 15 = -4 \Rightarrow x_2 = \frac{-4}{-2} = 2$$

As shown in Fig. 7.3, the intersection of the two lines is $(1, 2)$.

$$\boxed{\mathbf{x} = \begin{bmatrix} 1 \\ 2 \end{bmatrix}}$$

Verification:

$$A\mathbf{x} = \begin{bmatrix} 1 & 2 \\ 3 & 4 \end{bmatrix} \begin{bmatrix} 1 \\ 2 \end{bmatrix} = \begin{bmatrix} 5 \\ 11 \end{bmatrix}.$$

Example 7.16 Solve:

$$\begin{cases} 3x_1 + x_2 - 2x_3 = 4, \\ -x_1 + 2x_2 + 3x_3 = 1, \\ 2x_1 + x_2 + 4x_3 = -2. \end{cases}$$

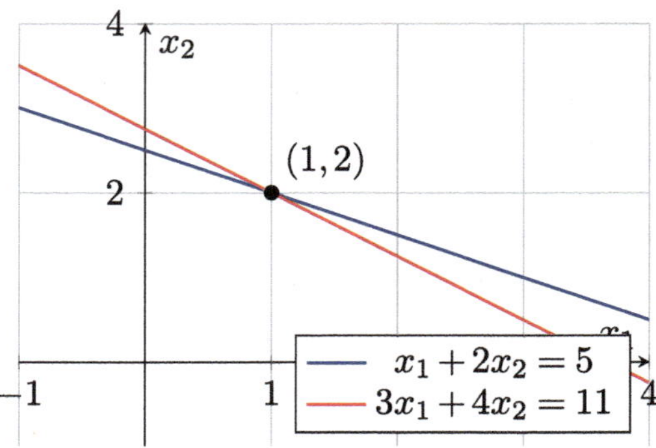

Fig. 7.3 Cramer's rule finds the intersection point $(1, 2)$, which solves the system by evaluating determinant ratios

Coefficient matrix and right-hand side:

$$A = \begin{bmatrix} 3 & 1 & -2 \\ -1 & 2 & 3 \\ 2 & 1 & 4 \end{bmatrix}, \quad \mathbf{b} = \begin{bmatrix} 4 \\ 1 \\ -2 \end{bmatrix}.$$

Compute $\det(A)$ using cofactor expansion:

$$\begin{aligned} \det(A) &= 3(2 \cdot 4 - 3 \cdot 1) - 1(-1 \cdot 4 - 3 \cdot 2) + (-2)(-1 \cdot 1 - 2 \cdot 2) \\ &= 3(8 - 3) - 1(-4 - 6) - 2(-1 - 4) \\ &= 15 + 10 + 10 = 35. \end{aligned}$$

Compute $\det(A_1)$, $\det(A_2)$, and $\det(A_3)$:

$$A_1 = \begin{bmatrix} 4 & 1 & -2 \\ 1 & 2 & 3 \\ -2 & 1 & 4 \end{bmatrix} \Rightarrow \det(A_1) = 0 \Rightarrow x_1 = 0.$$

$$A_2 = \begin{bmatrix} 3 & 4 & -2 \\ -1 & 1 & 3 \\ 2 & -2 & 4 \end{bmatrix} \Rightarrow \det(A_2) = 70 \Rightarrow x_2 = \frac{70}{35} = 2.$$

$$A_3 = \begin{bmatrix} 3 & 1 & 4 \\ -1 & 2 & 1 \\ 2 & 1 & -2 \end{bmatrix} \Rightarrow \det(A_3) = -35 \Rightarrow x_3 = \frac{-35}{35} = -1$$

$$\mathbf{x} = \begin{bmatrix} 0 \\ 2 \\ -1 \end{bmatrix}$$

Verification:

$$A\mathbf{x} = \begin{bmatrix} 3 & 1 & -2 \\ -1 & 2 & 3 \\ 2 & 1 & 4 \end{bmatrix} \begin{bmatrix} 0 \\ 2 \\ -1 \end{bmatrix} = \begin{bmatrix} 4 \\ 1 \\ -2 \end{bmatrix}$$

Python Example: Cramer's Rule (2x2)

```python
import numpy as np

A = np.array([[1, 2], [3, 4]])
b = np.array([5, 11])
det_A = np.linalg.det(A)
A1 = A.copy(); A1[:, 0] = b
A2 = A.copy(); A2[:, 1] = b
x1 = np.linalg.det(A1) / det_A
x2 = np.linalg.det(A2) / det_A
print("Solution:", [x1, x2])
```

Solution: [1.0, 2.0]

7.4 Determinants in ML

Determinants play a critical role in ML influencing model design, optimization, and data analysis.

1. **LR**: The matrix $A^T A$ must have $\det(A^T A) \neq 0$ for a unique solution:

$$\mathbf{w} = (A^T A)^{-1} A^T \mathbf{b}.$$

   A zero determinant indicates multicollinearity, making the solution unstable.
2. **PCA**: The determinant of the covariance matrix measures the volume of the data's confidence ellipsoid, guiding dimensionality reduction.
3. **Normalizing Flows**: For an invertible linear transformation $x = Az$ with base density $p_z(z)$, the transformed density satisfies $\log p(x) = \log p_{z(z)} - \log|\det(A)|$
4. **Optimization**: The Hessian's determinant in Newton's method informs curvature and convergence properties.
5. **Numerical Stability**: Log-determinant penalties in algorithms like Gaussian processes ensure robust training.
6. **LU and QR**: Determinants simplify computations in regression and eigenvalue problems.

Example 7.17 For:

$$A = \begin{bmatrix} 1 & 1 \\ 1 & 2 \\ 1 & 3 \end{bmatrix}, \quad \mathbf{y} = \begin{bmatrix} 1 \\ 2 \\ 3 \end{bmatrix},$$

$$A^T A = \begin{bmatrix} 3 & 6 \\ 6 & 14 \end{bmatrix}, \quad \det(A^T A) = 42 - 36 = 6 \neq 0.$$

The system is solvable, yielding weights for LR.

Chapter Summary

This chapter examined determinants as scalar measures encapsulating essential properties of square matrices, particularly their role in assessing invertibility, volume scaling, and orientation in linear transformations critical to ML. It commenced with the definition for 2×2 matrices as $ad - bc$, geometrically interpreted as the signed area of the parallelogram spanned by columns (Fig. 7.1), where the absolute value gave area and the sign indicated orientation preservation or reversal. For 3×3 matrices, cofactor expansion along a row or column, such as $\det(A) = \sum_j a_{1j} C_{1j}$ with cofactors $C_{ij} = (-1)^{i+j} \det(A_{ij})$, yielded the signed volume of the parallelepiped (Fig. 7.2), with zero determinant signaling linear dependence and collapse to lower dimensions, as exemplified by $\det \begin{bmatrix} 1 & 2 & 3 \\ 4 & 5 & 6 \\ 7 & 8 & 9 \end{bmatrix} = 0$ (Example in Sect. 7.2).

Computational methods were detailed, including the Rule of Sarrus for 3×3 and upper triangular forms where $\det(A)$ equaled the product of diagonals, simplifying via row reduction while adjusting for swaps (negation) and scalings. Section 7.3 outlined properties: $\det(I_n) = 1$, row swaps negate $\det(A)$, scaling a row by k multiplies by k, adding multiples preserves, $\det(AB) = \det(A)\det(B)$, $\det(A^{-1}) = 1/\det(A)$, $\det(A^k) = (\det(A))^k$, and $\det(A^T) = \det(A)$, verified through examples and Python code demonstrating $\det(kA) = k^n \det(A)$ and nonlinearity in sums.

Cramer's Rule emerged as a determinant-based solver for $A\mathbf{x} = \mathbf{b}$ ($\det(A) \neq 0$), with $x_i = \det(A_i)/\det(A)$ where A_i replaces column i with $\mathbf{b}$, illustrated for 2D systems like $x_1 + 2x_2 = 5$, $3x_1 + 4x_2 = 11$ yielding $(1, 2)$ (Fig. 7.3) and 3D cases. Section 7.4 connected determinants to ML: nonzero $\det(A^T A)$ ensured unique regression weights, covariance determinants gauged data volume in PCA, Jacobian log-determinants adjusted densities in normalizing flows, and Hessian determinants informed optimization curvature.

Takeaways

Key insights from the chapter were summarized as:

- Determinants as signed volumes (2×2: area, 3×3: parallelepiped), with zero indicating singularity and linear dependence, linking algebra to geometry.

- Computation via cofactor expansion, Sarrus rule, or triangular forms, efficient for small matrices and preserved under row operations with adjustments.
- Properties enabling simplification: multiplicative over products, reciprocal for inverses, unchanged by transpose, scaled by k^n for scalar multiples, and negated by row swaps.
- Cramer's Rule for solving invertible systems through column-replaced determinants, providing explicit solutions tied to volume ratios.
- ML roles in regression (invertibility of Gram matrices), PCA (data ellipsoid volume), flows (density transformations), and optimization (curvature analysis).

Through these elements, readers acquired proficiency in computing and interpreting determinants, equipping them to analyze matrix behavior and apply these tools in ML contexts like model stability and dimensionality reduction.

Exercises

1. Compute $\det \begin{bmatrix} 4 & 3 \\ 1 & 2 \end{bmatrix}$.

2. Compute $\det \begin{bmatrix} 2 & 1 & 0 \\ 3 & 4 & 1 \\ 5 & 6 & 3 \end{bmatrix}$ using cofactor expansion along row 1.

3. Solve using Cramer's rule: $3x_1 + x_2 = 5$, $x_1 + 2x_2 = 4$.

4. Verify $\det(AB) = \det(A)\det(B)$ for $A = \begin{bmatrix} 2 & 1 \\ 1 & 3 \end{bmatrix}$, $B = \begin{bmatrix} 1 & 0 \\ 0 & 2 \end{bmatrix}$.

5. For $A = \begin{bmatrix} 5 & 0 & 0 \\ 0 & 2 & 0 \\ 0 & 0 & 1 \end{bmatrix}$, compute $\det(A)$, $\det(3A)$, and compare with $3^3 \cdot \det(A)$.

6. Verify $\det(A^T) = \det(A)$ for $A = \begin{bmatrix} 1 & 2 & 3 \\ 0 & 4 & 5 \\ 1 & 0 & 6 \end{bmatrix}$.

7. Show $\det \begin{bmatrix} 3 & 1 \\ 3 & 1 \end{bmatrix} = 0$.

8. Compute $\det(A^{-1})$ for $A = \begin{bmatrix} 2 & 1 \\ 1 & 3 \end{bmatrix}$ and verify $\det(A^{-1}) = 1/\det(A)$.

9. Compute $\det \begin{bmatrix} 1 & 2 & 3 \\ 4 & 5 & 6 \\ 7 & 8 & 9 \end{bmatrix}$ using cofactor expansion along column 3.

10. Compute $\det(A^2)$ for $A = \begin{bmatrix} 1 & 2 \\ 3 & 4 \end{bmatrix}$ and compare with $(\det(A))^2$.

11. Compute $\det \begin{bmatrix} 2 & 0 & 1 \\ 0 & 3 & 2 \\ 1 & 0 & 4 \end{bmatrix}$ using row reduction, keeping track of how row operations affect the determinant.

12. Solve using Cramer's rule: $2x_1 - x_2 + x_3 = 1$, $x_1 + 2x_2 - x_3 = 2$, $x_1 - x_2 + 2x_3 = 0$.

13. Show that swapping two rows of $A = \begin{bmatrix} 1 & 2 \\ 3 & 4 \end{bmatrix}$ negates the determinant.

14. For $A = \begin{bmatrix} 1 & 0 \\ 0 & 2 \end{bmatrix}$, compute $\det(5A)$ and verify $\det(5A) = 5^2 \det(A)$.

15. Compute the adjugate of $A = \begin{bmatrix} 1 & 2 \\ 3 & 4 \end{bmatrix}$ and use it to find A^{-1}.

16. Verify $\det(AA^T) = (\det(A))^2$ for $A = \begin{bmatrix} 1 & 2 \\ 3 & 4 \end{bmatrix}$.

17. Compute the determinant of $A = \begin{bmatrix} 1 & 1 & 1 \\ 0 & 2 & 2 \\ 0 & 0 & 3 \end{bmatrix}$ and interpret it as the volume scaling factor of the associated linear transformation.

18. Write a Python script to compute and visualize the area (given by $|\det(A)|$) of the parallelogram formed by the column vectors. $A = \begin{bmatrix} 2 & 1 \\ 1 & 3 \end{bmatrix}$.

19. Use Cramer's rule to solve: $4x_1 + 2x_2 = 6$, $x_1 + 3x_2 = 7$.

20. Show that $\det \begin{bmatrix} 1 & 2 & 3 \\ 4 & 5 & 6 \\ 0 & 0 & 0 \end{bmatrix} = 0$ and explain why geometrically.

21. Compute the determinant of $A = \begin{bmatrix} 1 & 0 & 2 \\ 0 & 2 & 1 \\ 1 & 1 & 3 \end{bmatrix}$ using cofactor expansion along row 2.

22. Verify that scaling a row of $A = \begin{bmatrix} 1 & 2 \\ 3 & 4 \end{bmatrix}$ by 3 scales the determinant by 3.

Chapter 8
Eigenvalues and Eigenvectors

Introduction

Eigenvalues and eigenvectors are central to linear algebra, serving as powerful tools to decode the intrinsic behavior of linear transformations. The term 'eigen' (German for 'own') was introduced by David Hilbert in 1904 in the context of integral equations, though the underlying concepts trace back to earlier work by Euler, Cauchy, and others. This property is akin to finding the "natural axes" of a transformation, simplifying complex matrix operations into scaling along specific directions. In ML, eigenvalues and eigenvectors underpin algorithms like PCA, spectral clustering, and optimization techniques, making them indispensable for data analysis and model design.

This chapter begins by defining these concepts and their geometric interpretations, using figures like Fig. 8.1 to illustrate invariant directions. Computational methods, including the characteristic polynomial and eigenspace calculations, are detailed with step-by-step examples to ensure clarity. We also cover the role of eigenvalues in linear operators and diagonalization, providing a foundation for advanced topics like SVD. Python implementations with NumPy and Matplotlib offer hands-on tools for computation and visualization, while a dedicated section on ML applications connects these concepts to real-world problems. Extensive exercises reinforce both theoretical and computational skills, ensuring a deep understanding of eigenvalues and eigenvectors in linear algebra and ML.

© The Author(s), under exclusive license to Springer Nature Singapore Pte Ltd. 2026   225
Md. Jalil Piran, *Linear Algebra with Applications in Machine Learning*,
https://doi.org/10.1007/978-981-95-5167-5_8

Topics Covered

This chapter is organized as follows:

- **Section** 8.1 **Eigenvalues and Eigenvectors; Definition and Basics**: Defines eigenvalues, eigenvectors, and eigenspaces, with geometric interpretations and examples.
- **Section** 8.2 **Computing Eigenvalues and Eigenvectors**: Details methods for finding eigenvalues via the characteristic polynomial and eigenvectors via linear systems.
- **Section** 8.2.5 **Characteristic Polynomial and Diagonalization**: Explores the characteristic polynomial, eigenspaces, linear independence, and conditions for diagonalizability.
- **Section** 8.3 **Linear Operators and Eigenvalues**: Discusses eigenvalues in the context of linear operators and their matrix representations.
- **Section** 8.4 **Eigenvalues in ML**: Examines applications in PCA, spectral clustering, optimization, and more.
- **Summary and Exercises**: Recaps key concepts and provides a comprehensive set of problems to deepen understanding.

All code accompanying the chapter is publicly available at https://github.com/jalil-piran/Linear-Algebra-with-Applications-in-Machine-Learning/blob/main/Chapter_08_Eigenvalues.ipynb.

Visualizations, such as Fig. 8.1, clarify how eigenvectors remain invariant under transformations, while Python examples bridge theory and practice. By the end, readers will master eigenvalue computations, understand their geometric significance, and appreciate their applications in ML.

List of Acronyms

ML	Machine Learning
PCA	Principal Component Analysis
SVD	Singular Value Decomposition
SVMs	Support Vector Machines
REF	Row Echelon Form
RREF	Reduced Row Echelon Form

8.1 Eigenvalues and Eigenvectors: Definition and Basics

For a square matrix $A \in \mathbb{R}^{n \times n}$, a non-zero vector $\mathbf{v} \in \mathbb{R}^n$ is an **eigenvector** if:

$$A\mathbf{v} = \lambda\mathbf{v},$$

where λ (real or complex) is the **eigenvalue**. This implies A scales $\mathbf{v}$ by λ without altering its direction. The **eigenspace** of λ is the set of all vectors satisfying this equation, including the zero vector, forming the null space of $A - \lambda I$:

$$\text{Eigenspace}(\lambda) = \text{Null}(A - \lambda I).$$

Geometrically, eigenvectors are invariant directions under the transformation A. Imagine A as a machine that stretches, compresses, or rotates vectors. Most vectors change direction, but eigenvectors remain aligned with their original direction, scaled by λ. If $\lambda > 1$, the vector stretches; if $0 < \lambda < 1$, it shrinks; if $\lambda < 0$, it flips; if $\lambda = 0$, it collapses to zero. For example, in a scaling transformation, eigenvectors lie along axes that are stretched or compressed, as shown in Fig. 8.1.

Definition
A vector $\mathbf{v} \neq \mathbf{0}$ is an **eigenvector** of A if $A\mathbf{v} = \lambda\mathbf{v}$, for some scalar λ (eigenvalue). The eigenspace of λ is:

$$\text{Eigenspace}(\lambda) = \{\mathbf{v} \in \mathbb{R}^n \mid (A - \lambda I)\mathbf{v} = \mathbf{0}\}.$$

Example 8.1 Consider:

$$A = \begin{bmatrix} 2 & 0 \\ 0 & 3 \end{bmatrix}, \quad \mathbf{v} = \begin{bmatrix} 1 \\ 0 \end{bmatrix}.$$

$$A\mathbf{v} = \begin{bmatrix} 2 & 0 \\ 0 & 3 \end{bmatrix} \begin{bmatrix} 1 \\ 0 \end{bmatrix} = \begin{bmatrix} 2 \\ 0 \end{bmatrix} = 2 \begin{bmatrix} 1 \\ 0 \end{bmatrix}.$$

Thus, $\mathbf{v} = \begin{bmatrix} 1 \\ 0 \end{bmatrix}$ is an eigenvector with eigenvalue $\lambda = 2$. Similarly, $\begin{bmatrix} 0 \\ 1 \end{bmatrix}$ has eigenvalue $\lambda = 3$.

Example 8.2 To verify an eigenvector, consider:

$$A = \begin{bmatrix} 5 & 2 & 1 \\ -2 & 1 & -1 \\ 2 & 2 & 4 \end{bmatrix}, \quad \mathbf{v} = \begin{bmatrix} 1 \\ -1 \\ 1 \end{bmatrix}.$$

$$A\mathbf{v} = \begin{bmatrix} 5(1) + 2(-1) + 1(1) \\ -2(1) + 1(-1) - 1(1) \\ 2(1) + 2(-1) + 4(1) \end{bmatrix} = \begin{bmatrix} 4 \\ -4 \\ 4 \end{bmatrix} = 4 \begin{bmatrix} 1 \\ -1 \\ 1 \end{bmatrix}.$$

Thus, $\mathbf{v}$ is an eigenvector with $\lambda = 4$.

Fig. 8.1 Eigenvectors of $A = \begin{bmatrix} 2 & 0 \\ 0 & 3 \end{bmatrix}$. The blue vector $\mathbf{v}_1 = \begin{bmatrix} 1 \\ 0 \end{bmatrix}$ is scaled by $\lambda = 2$, and the red vector $\mathbf{v}_2 = \begin{bmatrix} 0 \\ 1 \end{bmatrix}$ by $\lambda = 3$

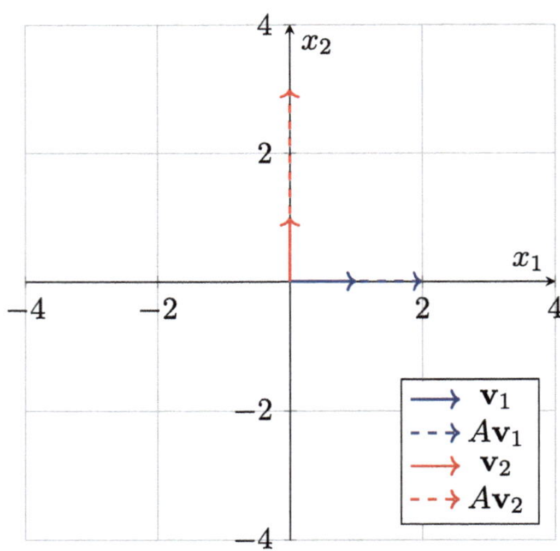

Python Example: Eigenvalue Verification

```python
import numpy as np
A = np.array([[5, 2, 1], [-2, 1, -1], [2, 2, 4]])
v = np.array([1, -1, 1])
Av = A @ v
lambda_v = 4
print("A*v:", Av)
print("lambda*v:", lambda_v * v)
```

Output:

```
A*v: [ 4 -4  4]
lambda*v: [ 4 -4  4]
```

8.2   Computing Eigenvalues and Eigenvectors

To find eigenvalues, solve the **characteristic equation**:

$$\det(A - \lambda I) = 0.$$

The roots of the resulting **characteristic polynomial** are the eigenvalues, which may be real or complex. For each eigenvalue λ, solve:

$$(A - \lambda I)\mathbf{v} = \mathbf{0},$$

to find eigenvectors, which form the eigenspace. Eigenvalues quantify how a transformation scales vectors along eigenvector directions, with large eigenvalues indicating stretching and small or zero eigenvalues suggesting compression or collapse.

Example 8.3 Consider the matrix:

$$A = \begin{bmatrix} -6 & 3 \\ 4 & 5 \end{bmatrix}.$$

Step 1- Compute the characteristic equation.

Subtract λI from A:

$$A - \lambda I = \begin{bmatrix} -6 - \lambda & 3 \\ 4 & 5 - \lambda \end{bmatrix}.$$

Step 2- Find the determinant of $A - \lambda I$.

This gives the characteristic polynomial:

$$\det(A - \lambda I) = (-6 - \lambda)(5 - \lambda) - (3)(4).$$

Simplify:

$$= (-6)(5) - (-6)\lambda - 5\lambda + \lambda^2 - 12 = -30 + 6\lambda - 5\lambda + \lambda^2 - 12.$$

$$= \lambda^2 + \lambda - 42.$$

$$\det(A - \lambda I) = \lambda^2 + \lambda - 42 = 0.$$

Step 3- Solve the quadratic equation.

$$\lambda = \frac{-1 \pm \sqrt{1^2 + 4 \cdot 42}}{2} = \frac{-1 \pm \sqrt{169}}{2} = \frac{-1 \pm 13}{2}.$$

So the eigenvalues are:

$$\lambda_1 = 6, \quad \lambda_2 = -7.$$

Step 4- Find eigenvector for $\lambda = 6$.

Substitute into $A - 6I$:

$$A - 6I = \begin{bmatrix} -12 & 3 \\ 4 & -1 \end{bmatrix}.$$

Perform row reduction:

$$R_2 \rightarrow R_2 + \frac{1}{3}R_1 \Rightarrow \begin{bmatrix} -12 & 3 \\ 0 & 0 \end{bmatrix}.$$

This gives the equation:

$$-12x_1 + 3x_2 = 0 \Rightarrow x_2 = 4x_1.$$

So the eigenvector is:

$$\boxed{\mathbf{v}_1 = t\begin{bmatrix} 1/4 \\ 1 \end{bmatrix}}$$

Verification:

$$A\begin{bmatrix} 1/4 \\ 1 \end{bmatrix} = \begin{bmatrix} (-6/4)+3 \\ 1+5 \end{bmatrix} = \begin{bmatrix} 3/2 \\ 6 \end{bmatrix} = 6\begin{bmatrix} 1/4 \\ 1 \end{bmatrix}.$$

Step 5- Find eigenvector for $\lambda = -7$.
Substitute into $A - (-7)I = A + 7I$:

$$A + 7I = \begin{bmatrix} 1 & 3 \\ 4 & 12 \end{bmatrix}.$$

Row reduction:

$$R_2 \rightarrow R_2 - 4R_1 \Rightarrow \begin{bmatrix} 1 & 3 \\ 0 & 0 \end{bmatrix}.$$

This gives:

$$x_1 + 3x_2 = 0 \Rightarrow x_1 = -3x_2.$$

So the eigenvector is:

$$\boxed{\mathbf{v}_2 = t\begin{bmatrix} -3 \\ 1 \end{bmatrix}}$$

Example 8.4 For:

$$A = \begin{bmatrix} 2 & 0 & 0 \\ 0 & 4 & 5 \\ 0 & 4 & 3 \end{bmatrix}.$$

$$\det(A - \lambda I) = (2 - \lambda)\begin{vmatrix} 4-\lambda & 5 \\ 4 & 3-\lambda \end{vmatrix} = (2 - \lambda)[(4 - \lambda)(3 - \lambda) - 20].$$

$$= (2 - \lambda)(\lambda^2 - 7\lambda - 8) = (2 - \lambda)(\lambda - 8)(\lambda + 1) = 0.$$

Eigenvalues: $\lambda = 2, 8, -1$. For $\lambda = 2$:

$$A - 2I = \begin{bmatrix} 0 & 0 & 0 \\ 0 & 2 & 5 \\ 0 & 4 & 1 \end{bmatrix}.$$

Row reduce:

$$R_3 \to R_3 - 2R_2 : \begin{bmatrix} 0 & 0 & 0 \\ 0 & 2 & 5 \\ 0 & 0 & -9 \end{bmatrix}, \quad x_3 = 0, x_2 = -\frac{5}{2}x_3, x_2 = 0, x_1 \text{ free} \implies \mathbf{v}_1 = t \begin{bmatrix} 1 \\ 0 \\ 0 \end{bmatrix}.$$

For $\lambda = 8$:

$$A - 8I = \begin{bmatrix} -6 & 0 & 0 \\ 0 & -4 & 5 \\ 0 & 4 & -5 \end{bmatrix}.$$

Row reduce:

$$R_3 \to R_3 + R_2 : \begin{bmatrix} -6 & 0 & 0 \\ 0 & -4 & 5 \\ 0 & 0 & 0 \end{bmatrix}, \quad x_2 = \frac{5}{4}x_3 \implies \mathbf{v}_2 = t \begin{bmatrix} 0 \\ 5/4 \\ 1 \end{bmatrix}.$$

For $\lambda = -1$:

$$A + I = \begin{bmatrix} 3 & 0 & 0 \\ 0 & 5 & 5 \\ 0 & 4 & 4 \end{bmatrix}.$$

Row reduce:

$$R_3 \to R_3 - \frac{4}{5}R_2 : \begin{bmatrix} 3 & 0 & 0 \\ 0 & 5 & 5 \\ 0 & 0 & 0 \end{bmatrix}, \quad x_2 = -x_3 \implies \mathbf{v}_3 = t \begin{bmatrix} 0 \\ -1 \\ 1 \end{bmatrix}.$$

Example 8.5 For:

$$A = \begin{bmatrix} 1 & 2 \\ 2 & 4 \end{bmatrix}.$$

$$\det(A - \lambda I) = (1 - \lambda)(4 - \lambda) - 4 = \lambda^2 - 5\lambda = \lambda(\lambda - 5) = 0.$$

Eigenvalues: $\lambda = 0, 5$. For $\lambda = 0$:

$$A = \begin{bmatrix} 1 & 2 \\ 2 & 4 \end{bmatrix}, \quad x_1 + 2x_2 = 0 \implies x_1 = -2x_2, \quad \mathbf{v}_1 = t \begin{bmatrix} -2 \\ 1 \end{bmatrix}.$$

For $\lambda = 5$:

$$A - 5I = \begin{bmatrix} -4 & 2 \\ 2 & -1 \end{bmatrix}, \quad -4x_1 + 2x_2 = 0 \implies x_2 = 2x_1, \quad \mathbf{v}_2 = t \begin{bmatrix} 1 \\ 2 \end{bmatrix}.$$

Example 8.6 For:

$$A = \begin{bmatrix} 1 & 0 & 0 \\ 3 & -2 & 0 \\ 2 & 3 & 4 \end{bmatrix}.$$

Since A is lower triangular, eigenvalues are the diagonal entries: $\lambda = 1, -2, 4$.

For $\lambda = 1$:

$$A - I = \begin{bmatrix} 0 & 0 & 0 \\ 3 & -3 & 0 \\ 2 & 3 & 3 \end{bmatrix}.$$

Row reduce:

$$R_3 \rightarrow R_3 - \tfrac{2}{3}R_2 : \begin{bmatrix} 0 & 0 & 0 \\ 3 & -3 & 0 \\ 0 & 5 & 3 \end{bmatrix},$$

$$\text{swap } R_1 \leftrightarrow R_3, \ R_2 \leftrightarrow R_3 : \begin{bmatrix} 3 & -3 & 0 \\ 0 & 5 & 3 \\ 0 & 0 & 0 \end{bmatrix},$$

$$R_1 \rightarrow \tfrac{1}{3}R_1, \ R_2 \rightarrow \tfrac{1}{5}R_2 : \begin{bmatrix} 1 & -1 & 0 \\ 0 & 1 & \tfrac{3}{5} \\ 0 & 0 & 0 \end{bmatrix},$$

$$R_1 \rightarrow R_1 + R_2 : \begin{bmatrix} 1 & 0 & \tfrac{3}{5} \\ 0 & 1 & \tfrac{3}{5} \\ 0 & 0 & 0 \end{bmatrix}.$$

$$x_1 = -\tfrac{3}{5}x_3, \ x_2 = -\tfrac{3}{5}x_3 \implies \mathbf{v}_1 = t \begin{bmatrix} -\tfrac{3}{5} \\ -\tfrac{3}{5} \\ 1 \end{bmatrix}.$$

For $\lambda = -2$:

$$A + 2I = \begin{bmatrix} 3 & 0 & 0 \\ 3 & 0 & 0 \\ 2 & 3 & 6 \end{bmatrix}.$$

Row reduce:

$$R_2 \rightarrow R_2 - R_1 : \begin{bmatrix} 3 & 0 & 0 \\ 0 & 0 & 0 \\ 2 & 3 & 6 \end{bmatrix}, \quad R_3 \rightarrow R_3 - \tfrac{2}{3}R_1 : \begin{bmatrix} 3 & 0 & 0 \\ 0 & 0 & 0 \\ 0 & 3 & 6 \end{bmatrix}.$$

$$x_1 = 0, \ x_2 = -2x_3 \implies \mathbf{v}_2 = t \begin{bmatrix} 0 \\ -2 \\ 1 \end{bmatrix}.$$

For $\lambda = 4$:

$$A - 4I = \begin{bmatrix} -3 & 0 & 0 \\ 3 & -6 & 0 \\ 2 & 3 & 0 \end{bmatrix}.$$

Row reduce:

$$R_2 \to R_2 + R_1, \ R_3 \to R_3 + \tfrac{2}{3}R_1 : \begin{bmatrix} -3 & 0 & 0 \\ 0 & -6 & 0 \\ 0 & 3 & 0 \end{bmatrix}.$$

$$x_1 = 0, \ x_2 = 0 \implies \mathbf{v}_3 = t \begin{bmatrix} 0 \\ 0 \\ 1 \end{bmatrix}.$$

Python Example: Eigenvalues and Eigenvectors in SymPy

```python
import sympy as sp

A = sp.Matrix([[1, 2], [2, 4]])

eigen_data = A.eigenvects()

for val, multiplicity, vectors in eigen_data:
    print(f"Eigenvalue: {val} (multiplicity
        {multiplicity})")
    for vec in vectors:
        print("Eigenvector:", sp.pretty(vec))
```

Output:

```
 Eigenvalue: 0 (multiplicity 1)
Eigenvector:
[  -2 ]
[   1 ]
Eigenvalue: 5 (multiplicity 1)
Eigenvector:
[ 1/2]
[   1]
```

8.2.1 Finding Eigenvalues

As established above, eigenvalues are roots of

$$\det(A - \lambda I) = 0.$$

The roots of the resulting **characteristic polynomial** are the eigenvalues, which may be real or complex. We now examine this procedure in detail.

Eigenvalues measure how a transformation scales vectors along their corresponding eigenvector directions. Large eigenvalues indicate strong stretching, whereas small or zero eigenvalues signify compression or collapse, highlighting the underlying structure of the transformation.

Example 8.7 For:

$$A = \begin{bmatrix} 1 & 1 \\ 4 & 1 \end{bmatrix},$$

$$A - \lambda I = \begin{bmatrix} 1 - \lambda & 1 \\ 4 & 1 - \lambda \end{bmatrix}.$$

We compute the determinant:

$$\det(A - \lambda I) = (1 - \lambda)(1 - \lambda) - 1 \cdot 4 = (1 - \lambda)^2 - 4 = \lambda^2 - 2\lambda + 1 - 4 = \lambda^2 - 2\lambda - 3.$$

Then, we solve:

$$\lambda^2 - 2\lambda - 3 = 0 \implies (\lambda - 3)(\lambda + 1) = 0 \implies \lambda = 3, -1.$$

Python Example: Eigenvalues with NumPy

```python
import numpy as np
A = np.array([[1, 1], [4, 1]])
eigenvalues, eigenvectors = np.linalg.eig(A)
print("Eigenvalues:", eigenvalues)
```

Output:

```
Eigenvalues: [ 3. -1.]
```

The characteristic equation ensures $A - \lambda I$ is singular, meaning there exist non-zero vectors $\mathbf{v}$ in its null space, which are the eigenvectors. Complex eigenvalues arise for matrices like rotations, indicating no real invariant directions.

8.2.2 Finding Eigenvectors

For each eigenvalue λ, solve:
$$(A - \lambda I)\mathbf{v} = \mathbf{0}.$$

The solutions form the eigenspace, a subspace of $\mathbb{R}^n$.

Example 8.8 For $A = \begin{bmatrix} 1 & 1 \\ 4 & 1 \end{bmatrix}$, with $\lambda = 3, -1$.

For $\lambda = 3$:
$$A - 3I = \begin{bmatrix} -2 & 1 \\ 4 & -2 \end{bmatrix}.$$

Row reduce:

$$R_2 \to R_2 + 2R_1 : \begin{bmatrix} -2 & 1 \\ 0 & 0 \end{bmatrix}, \quad R_1 \to -\frac{1}{2}R_1 : \begin{bmatrix} 1 & -0.5 \\ 0 & 0 \end{bmatrix}.$$

$$x_1 = 0.5x_2 \implies \mathbf{v} = x_2 \begin{bmatrix} 0.5 \\ 1 \end{bmatrix} = t \begin{bmatrix} 1 \\ 2 \end{bmatrix}.$$

For $\lambda = -1$:
$$A + I = \begin{bmatrix} 2 & 1 \\ 4 & 2 \end{bmatrix}.$$

Row reduce:

$$R_2 \to R_2 - 2R_1 : \begin{bmatrix} 2 & 1 \\ 0 & 0 \end{bmatrix}, \quad R_1 \to \frac{1}{2}R_1 : \begin{bmatrix} 1 & 0.5 \\ 0 & 0 \end{bmatrix}.$$

$$x_1 = -0.5x_2 \implies \mathbf{v} = x_2 \begin{bmatrix} -0.5 \\ 1 \end{bmatrix} = t \begin{bmatrix} -1 \\ 2 \end{bmatrix}.$$

Python Example: Eigenvectors

```python
import sympy as sp

A = sp.Matrix([[1, 1],
[4, 1]])

eigen_data = A.eigenvects()

for eigenval, multiplicity, eigvecs in eigen_data:
    print(f"Eigenvalue: {eigenval}
        (multiplicity: {multiplicity})")
```

```python
10        for vec in eigvecs:
11            print("Eigenvector:")
12            print(sp.pretty(vec, use_unicode=False))
```

Output:

```
Eigenvalue: -1 (multiplicity: 1)
Eigenvector:
[ -1/2]
[   1 ]
Eigenvalue: 3 (multiplicity: 1)
Eigenvector:
[ 1/2]
[   1]
```

Geometric Interpretation

Eigenvectors are like fixed axes in a transformation, scaled by eigenvalues. For example, a matrix may stretch space along one eigenvector while compressing along another, simplifying the transformation's behavior.

8.2.3 Eigenspaces

The **eigenspace** associated with an eigenvalue λ of a square matrix $A \in \mathbb{R}^{n \times n}$ is the set of all vectors $\mathbf{v} \in \mathbb{R}^n$ such that:

$$(A - \lambda I_n)\mathbf{v} = \mathbf{0},$$

where I_n is the identity matrix of size $n \times n$. This eigenspace is denoted by:

$$\text{Eigenspace}(\lambda) = \text{Null}(A - \lambda I_n).$$

Definition
The **eigenspace** of A corresponding to an eigenvalue λ is the null space of $A - \lambda I_n$. It contains all eigenvectors for λ along with the zero vector and forms a subspace of $\mathbb{R}^n$.

Geometric vs. Algebraic Multiplicity

- The **geometric multiplicity (geom. mult.)** of λ is the dimension of its eigenspace, i.e., the number of linearly independent eigenvectors associated with λ.
- The **algebraic multiplicity (alg. mult.)** of λ is the number of times it appears as a root of the characteristic polynomial:

$$p(\lambda) = \det(A - \lambda I_n).$$

For any eigenvalue λ of a matrix A, the geometric multiplicity is always less than or equal to the algebraic multiplicity:

$$\text{geom. mult.}(\lambda) \le \text{alg. mult.}(\lambda).$$

Why It Matters?
This distinction is crucial in determining whether a matrix is **diagonalizable**. A matrix A is diagonalizable if and only if, for each eigenvalue λ, the geometric multiplicity equals its algebraic multiplicity, and the sum of all geometric multiplicities equals n, the size of the matrix.

Example 8.9 Suppose $\lambda = 3$ is an eigenvalue of A with:

$$\text{alg. mult.} = 2, \quad \text{geom. mult.} = 1.$$

Then A is *not* diagonalizable, since we lack enough linearly independent eigenvectors (Fig. 8.2).

In summary:

- Eigenspaces are vector subspaces formed by eigenvectors.
- Algebraic multiplicity counts how many times λ is a root of the characteristic polynomial.
- Geometric multiplicity measures how many independent directions (eigenvectors) are associated with λ.
- Always: $\text{geom. mult.}(\lambda) \le \text{alg. mult.}(\lambda)$.

Fig. 8.2 Eigenvectors $\mathbf{v}_1$ and $\mathbf{v}_2$ of matrix $A = \begin{bmatrix} 2 & 3 \\ 4 & 1 \end{bmatrix}$, with their projections on axes

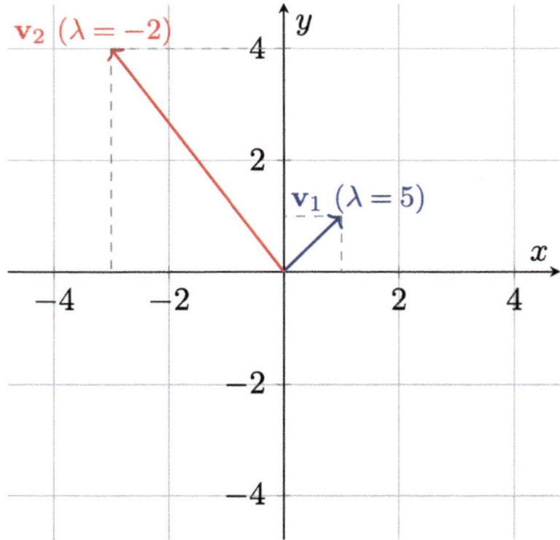

Example 8.10 For $A = \begin{bmatrix} 2 & 3 \\ 4 & 1 \end{bmatrix}$, with $\lambda = 5, -2$.

For $\lambda = 5$:

$$A - 5I = \begin{bmatrix} -3 & 3 \\ 4 & -4 \end{bmatrix}.$$

Row reduce:

$$R_1 \to -\frac{1}{3}R_1 : \begin{bmatrix} 1 & -1 \\ 4 & -4 \end{bmatrix}, \quad R_2 \to R_2 - 4R_1 : \begin{bmatrix} 1 & -1 \\ 0 & 0 \end{bmatrix}.$$

$$x_1 = x_2 \implies \mathbf{v} = x_2 \begin{bmatrix} 1 \\ 1 \end{bmatrix}.$$

$$\text{Eigenspace}(5) = \text{Span}\left\{ \begin{bmatrix} 1 \\ 1 \end{bmatrix} \right\}, \quad \text{dimension} = 1.$$

For $\lambda = -2$:

$$A + 2I = \begin{bmatrix} 4 & 3 \\ 4 & 3 \end{bmatrix}.$$

Row reduce:

$$R_2 \to R_2 - R_1 : \begin{bmatrix} 4 & 3 \\ 0 & 0 \end{bmatrix}, \quad R_1 \to \frac{1}{4}R_1 : \begin{bmatrix} 1 & 0.75 \\ 0 & 0 \end{bmatrix}.$$

$$x_1 = -0.75x_2 \implies \mathbf{v} = x_2 \begin{bmatrix} -0.75 \\ 1 \end{bmatrix} = x_2 \begin{bmatrix} -3 \\ 4 \end{bmatrix}.$$

$$\text{Eigenspace}(-2) = \text{Span}\left\{\begin{bmatrix} -3 \\ 4 \end{bmatrix}\right\}, \quad \text{dimension} = 1.$$

Example 8.11 For:

$$A = \begin{bmatrix} -1 & 0 & 0 \\ 0 & 1 & 2 \\ 0 & 2 & 1 \end{bmatrix}.$$

Compute the characteristic polynomial:

$$\det(A - \lambda I) = \begin{vmatrix} -1-\lambda & 0 & 0 \\ 0 & 1-\lambda & 2 \\ 0 & 2 & 1-\lambda \end{vmatrix} = (-1-\lambda)\begin{vmatrix} 1-\lambda & 2 \\ 2 & 1-\lambda \end{vmatrix}.$$

$$= (-1-\lambda)[(1-\lambda)^2 - 4]$$
$$= (-1-\lambda)(\lambda^2 - 2\lambda + 1 - 4)$$
$$= (-1-\lambda)(\lambda^2 - 2\lambda - 3).$$
$$= (-1-\lambda)(\lambda - 3)(\lambda + 1) = 0 \implies \lambda = -1, 3.$$

For $\lambda = 3$:

$$A - 3I = \begin{bmatrix} -4 & 0 & 0 \\ 0 & -2 & 2 \\ 0 & 2 & -2 \end{bmatrix}.$$

Row reduce:

$$R_2 \to \frac{1}{2}R_2,\, R_3 \to \frac{1}{2}R_3 : \begin{bmatrix} -4 & 0 & 0 \\ 0 & -1 & 1 \\ 0 & 1 & -1 \end{bmatrix},$$

$$R_3 \to R_3 + R_2 : \begin{bmatrix} -4 & 0 & 0 \\ 0 & -1 & 1 \\ 0 & 0 & 0 \end{bmatrix},$$

$$R_1 \to -\frac{1}{4}R_1,\, R_2 \to -R_2 : \begin{bmatrix} 1 & 0 & 0 \\ 0 & 1 & -1 \\ 0 & 0 & 0 \end{bmatrix}.$$

$$x_1 = 0, \quad x_2 = x_3$$

$$\implies \mathbf{v} = x_3 \begin{bmatrix} 0 \\ 1 \\ 1 \end{bmatrix}.$$

$$\text{Eigenspace}(3) = \text{Span} \left\{ \begin{bmatrix} 0 \\ 1 \\ 1 \end{bmatrix} \right\}$$

$$\boxed{\text{dimension} = 1}$$

For $\lambda = -1$:

$$A + I = \begin{bmatrix} 0 & 0 & 0 \\ 0 & 2 & 2 \\ 0 & 2 & 2 \end{bmatrix}.$$

Row reduce:

$$R_2 \to \frac{1}{2} R_2 : \begin{bmatrix} 0 & 0 & 0 \\ 0 & 1 & 1 \\ 0 & 2 & 2 \end{bmatrix},$$

$$R_3 \to R_3 - 2R_2 : \begin{bmatrix} 0 & 0 & 0 \\ 0 & 1 & 1 \\ 0 & 0 & 0 \end{bmatrix}.$$

$$x_2 = -x_3, \quad x_1 \text{ free} \implies \mathbf{v} = x_1 \begin{bmatrix} 1 \\ 0 \\ 0 \end{bmatrix} + x_3 \begin{bmatrix} 0 \\ -1 \\ 1 \end{bmatrix}.$$

$$\text{Eigenspace}(-1) = \text{Span} \left\{ \begin{bmatrix} 1 \\ 0 \\ 0 \end{bmatrix}, \begin{bmatrix} 0 \\ -1 \\ 1 \end{bmatrix} \right\}, \quad \text{dimension} = 2$$

Remark

- Eigenspaces capture all invariant directions for an eigenvalue.
- geom. mult. $\leq$ alg. mult. (root multiplicity in the characteristic polynomial).
- Symmetric matrices have orthogonal eigenvectors, enabling orthogonal bases.

Fig. 8.3 Eigenvectors of $A = \begin{bmatrix} 1 & 2 \\ 2 & 1 \end{bmatrix}$, with $\lambda_1 = 3$, $\lambda_2 = -1$

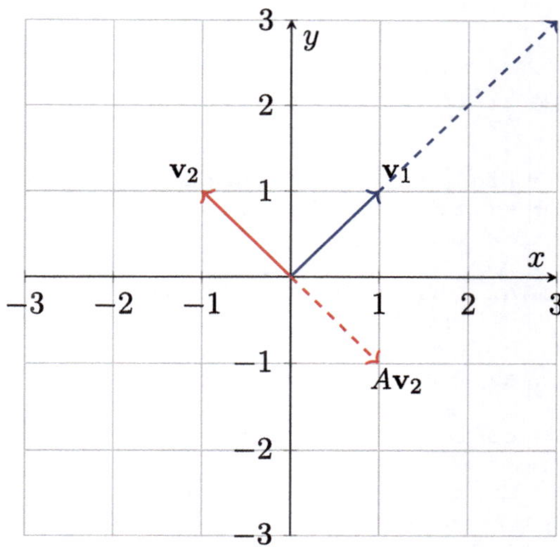

Geometric Interpretation:

Eigenvectors are directions invariant under A, scaled by eigenvalues (Fig. 8.3). For example:

- **Scaling**: $A = \begin{bmatrix} 2 & 0 \\ 0 & 0.5 \end{bmatrix}$ has eigenvectors $\begin{bmatrix} 1 \\ 0 \end{bmatrix}$, $\begin{bmatrix} 0 \\ 1 \end{bmatrix}$, with $\lambda = 2, 0.5$, stretching x-axis and compressing y-axis.
- **Reflection**: $A = \begin{bmatrix} 0 & 1 \\ 1 & 0 \end{bmatrix}$ reflects over $y = x$, with $\lambda = 1$ (along $y = x$) and $\lambda = -1$ (orthogonal).
- **Rotation**: $A = \begin{bmatrix} 0 & -1 \\ 1 & 0 \end{bmatrix}$ (90° rotation) has no real eigenvectors, only complex $\lambda = i, -i$.

Python Example: Eigenvectors

```python
import numpy as np
import matplotlib.pyplot as plt

A = np.array([[1, 2],
[2, 1]])

eigvals, eigvecs = np.linalg.eig(A)

v1 = eigvecs[:, 0] / np.linalg.norm(eigvecs[:,
    0])
v2 = eigvecs[:, 1] / np.linalg.norm(eigvecs[:,
    1])
```

```python
11
12 # Compute transformed vectors A*v1 and A*v2
13 Av1 = A @ v1
14 Av2 = A @ v2
15
16 fig, ax = plt.subplots(figsize=(6, 6))
17 ax.axhline(0, color='gray', linewidth=0.5)
18 ax.axvline(0, color='gray', linewidth=0.5)
19
20 ax.quiver(0, 0, v1[0], v1[1], angles='xy',
       scale_units='xy', scale=1,
21 color='blue', label=r'$\mathbf{v}_1$')
22 ax.quiver(0, 0, v2[0], v2[1], angles='xy',
       scale_units='xy', scale=1,
23 color='red', label=r'$\mathbf{v}_2$')
24
25 ax.quiver(0, 0, Av1[0], Av1[1], angles='xy',
       scale_units='xy', scale=1,
26 color='blue', alpha=0.4,
       label=r'$A\mathbf{v}_1$')
27 ax.quiver(0, 0, Av2[0], Av2[1], angles='xy',
       scale_units='xy', scale=1,
28 color='red', alpha=0.4, label=r'$A\mathbf{v}_2$')
29
30 ax.set_xlim(-3, 3)
31 ax.set_ylim(-3, 3)
32 ax.set_aspect('equal', adjustable='box')
33
34 ax.legend()
35 ax.set_title(r"Eigenvectors $\mathbf{v}_1,
       \mathbf{v}_2$ and their transforms
       $A\mathbf{v}_1, A\mathbf{v}_2$")
36
37 plt.show()
```

Output:

```
See Figure 8.4.
```

As shown in Fig. 8.4, the eigenvectors of a matrix point in directions that remain invariant under the linear transformation defined by the matrix. The solid arrows represent the original eigenvectors $\mathbf{v}_1$ and $\mathbf{v}_2$, while the faded arrows show the result of applying matrix A to each vector. These transformed vectors lie on the same line as the originals, scaled by their respective eigenvalues.

Eigenvectors simplify a transformation by reducing it to scaling along fixed directions, making complex matrices easier to analyze.

Fig. 8.4 Visualization of eigenvectors $\mathbf{v}_1$ and $\mathbf{v}_2$ of matrix $A = \begin{bmatrix} 1 & 2 \\ 2 & 1 \end{bmatrix}$, along with their transformed counterparts $A\mathbf{v}_1$ and $A\mathbf{v}_2$. The transformed vectors are shown with reduced opacity to distinguish them from the original eigenvectors

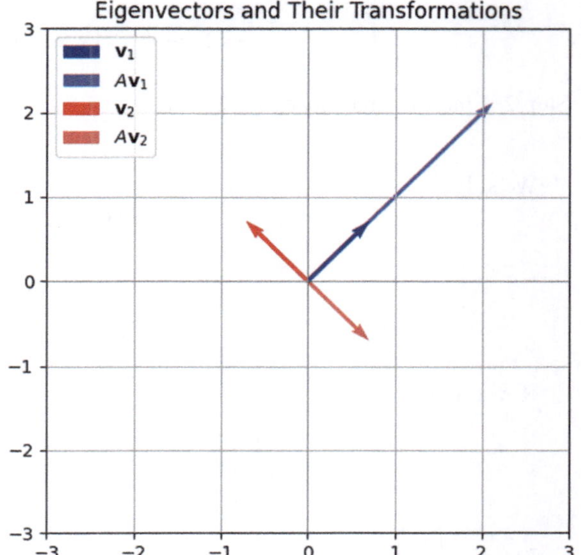

Example 8.12 Finding the Eigenspaces of Matrix $\mathbf{A} = \begin{bmatrix} 2 & 3 \\ 2 & 1 \end{bmatrix}$.

Step 1- Compute the Characteristic Polynomial

To find the eigenvalues, we solve:

$$\det(A - \lambda I) = 0,$$

$$A - \lambda I = \begin{bmatrix} 2 - \lambda & 3 \\ 2 & 1 - \lambda \end{bmatrix},$$

$$\Rightarrow \det(A - \lambda I) = (2 - \lambda)(1 - \lambda) - (2)(3),$$

$$= (2 - \lambda)(1 - \lambda) - 6 = \lambda^2 - 3\lambda - 4$$

Now we solve the quadratic:

$$\lambda^2 - 3\lambda - 4 = 0,$$

$$\lambda = \frac{3 \pm \sqrt{9 + 16}}{2} = \frac{3 \pm \sqrt{25}}{2} = \frac{3 \pm 5}{2},$$

$$\Rightarrow \lambda_1 = 4, \quad \lambda_2 = -1.$$

Step 2- Find the Eigenspace for $\lambda = 4$

We solve:

$$(A - 4I)\mathbf{x} = 0,$$

$$A - 4I = \begin{bmatrix} -2 & 3 \\ 2 & -3 \end{bmatrix}.$$

Row reduce:

$$\begin{bmatrix} -2 & 3 \\ 2 & -3 \end{bmatrix} \xrightarrow{R_2 \leftarrow R_2 + R_1} \begin{bmatrix} -2 & 3 \\ 0 & 0 \end{bmatrix} \Rightarrow \text{one pivot.}$$

Let $\mathbf{x} = \begin{bmatrix} x_1 \\ x_2 \end{bmatrix}$. Then:

$$-2x_1 + 3x_2 = 0 \Rightarrow x_1 = \frac{3}{2}x_2.$$

So general solution is:

$$\mathbf{x} = x_2 \begin{bmatrix} \frac{3}{2} \\ 1 \end{bmatrix} \sim \begin{bmatrix} 3 \\ 2 \end{bmatrix}.$$

$$\boxed{\text{Eigenspace for } \lambda = 4 : \quad \text{Span} \left\{ \begin{bmatrix} 3 \\ 2 \end{bmatrix} \right\}}$$

Step 3- Find the Eigenspace for $\lambda = -1$

We solve:

$$(A + I)\mathbf{x} = 0 \Rightarrow \begin{bmatrix} 3 & 3 \\ 2 & 2 \end{bmatrix}$$

Row reduce:

$$\begin{bmatrix} 3 & 3 \\ 2 & 2 \end{bmatrix} \xrightarrow{R_2 \leftarrow R_2 - \frac{2}{3} R_1} \begin{bmatrix} 3 & 3 \\ 0 & 0 \end{bmatrix} \Rightarrow \text{one pivot}$$

$$3x_1 + 3x_2 = 0 \Rightarrow x_1 = -x_2$$

So:

$$\mathbf{x} = x_2 \begin{bmatrix} -1 \\ 1 \end{bmatrix}$$

$$\text{Eigenspace for } \lambda = -1: \quad \text{Span} \left\{ \begin{bmatrix} -1 \\ 1 \end{bmatrix} \right\}$$

Final Answer

- Eigenvalue $\lambda = 4$, eigenspace: $\text{Span} \left\{ \begin{bmatrix} 3 \\ 2 \end{bmatrix} \right\}$
- Eigenvalue $\lambda = -1$, eigenspace: $\text{Span} \left\{ \begin{bmatrix} -1 \\ 1 \end{bmatrix} \right\}$

8.2.4 *Properties of Eigenvalues and Eigenvectors*

- **Eigenspaces**: The eigenspace of λ is a subspace of $\mathbb{R}^n$.
- **Orthogonality (Symmetric Matrices)**: For $A = A^{\mathrm{T}}$, eigenvectors for distinct eigenvalues are orthogonal, enabling an orthonormal basis.
- **Multiplicities**:

  - **Algebraic multiplicity**: Number of times λ is a root of $\det(A - \lambda I) = 0$.
  - **Geometric multiplicity**: Dimension of $\text{Null}(A - \lambda I)$.
  - Always: $1 \leq$ geom. mult. $\leq$ alg. mult..

- **Scalar Multiples**: If $\mathbf{v}$ is an eigenvector, so is $c\mathbf{v}$ ($c \neq 0$).
- **Zero Vector**: Not an eigenvector, despite being in the eigenspace.

Example 8.13 For $A = \begin{bmatrix} 1 & 2 \\ 2 & 1 \end{bmatrix}$, eigenvalues are $\lambda = 3, -1$, and eigenvectors are:

$$\mathbf{v}_1 = \begin{bmatrix} 1 \\ 1 \end{bmatrix}, \quad \mathbf{v}_2 = \begin{bmatrix} 1 \\ -1 \end{bmatrix}.$$

$$\mathbf{v}_1 \cdot \mathbf{v}_2 = 1 \cdot 1 + 1 \cdot (-1) = 0.$$

Python Example: Eigenvectors and Their Dot Product

```python
import numpy as np
A = np.array([[1, 2], [2, 1]])
eigenvalues, eigenvectors = np.linalg.eig(A)
v1, v2 = eigenvectors[:, 0], eigenvectors[:, 1]
dot_product = np.dot(v1, v2)
print("Dot product of eigenvectors:",
    dot_product)
```

Output:

```
Dot product of eigenvectors: 0.0
```

8.2.5 *Characteristic Polynomial and Diagonalization*

As established at the beginning of this section, the **characteristic polynomial** is used to determine the eigenvalues of a square matrix. These eigenvalues describe how a matrix $A \in \mathbb{R}^{n \times n}$ stretches, compresses, or reflects vectors along specific directions.

A scalar λ is called an **eigenvalue** of the matrix A if there exists a non-zero vector $\mathbf{v} \in \mathbb{R}^n$ such that:

$$A\mathbf{v} = \lambda \mathbf{v}.$$

Rewriting this equation, we bring all terms to one side:

$$(A - \lambda I_n)\mathbf{v} = \mathbf{0},$$

where I_n is the identity matrix of size $n \times n$. For this homogeneous system to have a nontrivial solution $\mathbf{v} \neq \mathbf{0}$, the coefficient matrix $A - \lambda I_n$ must be singular. Therefore, we require:

$$\det(A - \lambda I_n) = 0.$$

This leads to a polynomial equation in λ, called the **characteristic equation**, and the left-hand side expression is known as the **characteristic polynomial**.

Definition

Let $A \in \mathbb{R}^{n \times n}$. The **characteristic polynomial** of A is defined as:

$$p(\lambda) = \det(A - \lambda I_n),$$

where the roots of $p(\lambda) = 0$ are the eigenvalues of A.

Key Insights and Conditions:
The following statements are equivalent ways to identify whether λ is an eigenvalue of A:

- λ is an eigenvalue of A if there exists a non-zero vector $\mathbf{v}$ such that $A\mathbf{v} = \lambda\mathbf{v}$.
- λ is an eigenvalue if and only if the matrix $A - \lambda I_n$ is not invertible.
- The determinant $\det(A - \lambda I_n) = 0$ is the necessary and sufficient condition for λ to be an eigenvalue.
- The columns of $A - \lambda I_n$ are linearly dependent when λ is an eigenvalue.
- The rank of $A - \lambda I_n$ is less than n, implying it has a nontrivial null space (kernel).

The characteristic polynomial reflects how the linear transformation associated with A acts on space. Each eigenvalue λ represents a scaling factor along a specific direction (the corresponding eigenvector). When $\lambda = 0$, this indicates the transformation collapses some directions to zero, revealing the singular nature of A.

In summary, the characteristic polynomial provides a compact and powerful tool to extract the eigenvalues of a matrix. By solving:

$$\det(A - \lambda I_n) = 0,$$

we obtain the spectrum (set of eigenvalues) of A, which in turn informs us about the matrix's behavior in terms of transformation, stability, diagonalizability, and more.

Example 8.14 We want to try a matrix analysis of $A = \begin{bmatrix} 2 & 1 \\ 2 & 3 \end{bmatrix}$.

(i) **Characteristic Polynomial**
To find eigenvalues, we compute the characteristic polynomial:
$$\det(A - \lambda I) = 0,$$

$$A - \lambda I = \begin{bmatrix} 2-\lambda & 1 \\ 2 & 3-\lambda \end{bmatrix} \Rightarrow \det(A - \lambda I) = (2-\lambda)(3-\lambda) - (2)(1),$$

$$= (2-\lambda)(3-\lambda) - 2 = \lambda^2 - 5\lambda + 4.$$

Characteristic Polynomial:
$$p(\lambda) = \lambda^2 - 5\lambda + 4.$$

(ii) **Eigenvalues**
Solve the quadratic equation:
$$\lambda^2 - 5\lambda + 4 = 0 \Rightarrow \lambda = \frac{5 \pm \sqrt{25 - 16}}{2} = \frac{5 \pm 3}{2},$$

$$\Rightarrow \lambda_1 = 4, \quad \lambda_2 = 1.$$

(iii) **Eigenvectors**

(a) **Eigenvector for $\lambda_1 = 4$**
Solve $(A - 4I)\mathbf{x} = 0$:

$$A - 4I = \begin{bmatrix} -2 & 1 \\ 2 & -1 \end{bmatrix} \Rightarrow \text{Row reduce:} \begin{bmatrix} -2 & 1 \\ 2 & -1 \end{bmatrix} \to \begin{bmatrix} 1 & -0.5 \\ 0 & 0 \end{bmatrix} \Rightarrow x_1 = 0.5x_2.$$

So eigenvectors are:

$$\mathbf{x} = x_2 \begin{bmatrix} 0.5 \\ 1 \end{bmatrix} \sim \begin{bmatrix} 1 \\ 2 \end{bmatrix}$$

(b) **Eigenvector for $\lambda_2 = 1$**

$$A - I = \begin{bmatrix} 1 & 1 \\ 2 & 2 \end{bmatrix} \to \begin{bmatrix} 1 & 1 \\ 0 & 0 \end{bmatrix} \Rightarrow x_1 = -x_2$$

Eigenvectors:

$$\mathbf{x} = x_2 \begin{bmatrix} -1 \\ 1 \end{bmatrix}$$

(iv) **Null Space and Nullity**
To find the nullity, solve $A\mathbf{x} = 0$:

$$\begin{bmatrix} 2 & 1 \\ 2 & 3 \end{bmatrix} \begin{bmatrix} x_1 \\ x_2 \end{bmatrix} = \begin{bmatrix} 0 \\ 0 \end{bmatrix} \Rightarrow \text{Augmented matrix:} \begin{bmatrix} 2 & 1 & | & 0 \\ 2 & 3 & | & 0 \end{bmatrix} \to \text{Row reduce}$$

$$R_2 \leftarrow R_2 - R_1 : \begin{bmatrix} 2 & 1 \\ 0 & 2 \end{bmatrix} \Rightarrow \text{Pivot in each column} \Rightarrow \text{Nullity} = 0$$

Nullity = 0,   Null space = {zero vector}

(v) **Rank**
By rank-nullity theorem:

$$\text{Rank} + \text{Nullity} = n = 2 \Rightarrow \text{Rank} = 2.$$

(vi) **Trace and Determinant**

- **Trace:** $\text{tr}(A) = 2 + 3 = 5.$
- **Determinant:** $\det(A) = 2 \cdot 3 - 2 \cdot 1 = 6 - 2 = 4.$

(vii) **Column Space**

Columns of A:

$$\begin{bmatrix} 2 \\ 2 \end{bmatrix}, \quad \begin{bmatrix} 1 \\ 3 \end{bmatrix}.$$

These are linearly independent $(rank = 2)$ $\implies$ Span is all of $\mathbb{R}^2$

(viii) **Eigenspace Representation as Linear Combinations**

For eigenvalue $\lambda = 4$, we had eigenvector:

$$\mathbf{x} = \begin{bmatrix} 1 \\ 2 \end{bmatrix} \Rightarrow x_1 = x, \quad x_2 = 2x \Rightarrow x_2 = 2x_1.$$

We can represent it in parametric vector form as:

$$\mathbf{x} = x_1 \begin{bmatrix} 1 \\ 2 \end{bmatrix}, \quad \text{so } E_1 = \begin{bmatrix} 1 \\ 2 \end{bmatrix}.$$

This kind of representation is useful in describing nullspaces or eigenspaces via free variables.

In summary:

- **Eigenvalues**: $\lambda = 4, 1$
- **Eigenvectors**: $\begin{bmatrix} 1 \\ 2 \end{bmatrix}, \begin{bmatrix} -1 \\ 1 \end{bmatrix}$
- **Characteristic Polynomial**: $\lambda^2 - 5\lambda + 4$
- **Nullity**: 0
- **Rank**: 2
- **Trace**: 5
- **Determinant**: 4
- **Column space**: All of $\mathbb{R}^2$
- **Invertible**: Yes $(\det \neq 0)$

Example 8.15 For:

$$A = \begin{bmatrix} 2 & 3 \\ 4 & 1 \end{bmatrix},$$

compute:

$$A - \lambda I = \begin{bmatrix} 2 - \lambda & 3 \\ 4 & 1 - \lambda \end{bmatrix}.$$

The determinant is:

$$\det(A - \lambda I) = (2 - \lambda)(1 - \lambda) - 3 \cdot 4 = (2 - \lambda)(1 - \lambda) - 12.$$

Expand:

$$(2 - \lambda)(1 - \lambda) = 2 - 2\lambda - \lambda + \lambda^2 = \lambda^2 - 3\lambda + 2,$$

$$\det(A - \lambda I) = \lambda^2 - 3\lambda + 2 - 12 = \lambda^2 - 3\lambda - 10.$$

Solve:

$$\lambda^2 - 3\lambda - 10 = 0 \implies (\lambda - 5)(\lambda + 2) = 0 \implies \lambda = 5, -2.$$

The characteristic polynomial is $p(\lambda) = \lambda^2 - 3\lambda - 10$, with eigenvalues $\lambda = 5, -2$.

Python Example: Characteristic Polynomial and Eigenvalues

```python
import numpy as np
A = np.array([[2, 3], [4, 1]])
char_poly = np.poly(A)
eigenvalues = np.roots(char_poly)
print("Characteristic polynomial coefficients:",
    char_poly)
print("Eigenvalues:", eigenvalues)
```

Output:

```
Characteristic polynomial coefficients:
[  1.   -3. -10.]
Eigenvalues: [ 5. -2.]
```

Example 8.16   For:

$$A = \begin{bmatrix} -4 & -3 & 0 \\ 3 & 6 & 0 \\ 0 & 0 & 2 \end{bmatrix}.$$

Compute:

$$A - \lambda I = \begin{bmatrix} -4 - \lambda & -3 & 0 \\ 3 & 6 - \lambda & 0 \\ 0 & 0 & 2 - \lambda \end{bmatrix}.$$

This matrix is block-diagonal because all entries outside the top-left 2×2 block and the bottom-right 1×1 block are zero.

$$\det(A - \lambda I) = (2 - \lambda) \det \begin{bmatrix} -4 - \lambda & -3 \\ 3 & 6 - \lambda \end{bmatrix}.$$

For the 2×2 block:

$$\det \begin{bmatrix} -4 - \lambda & -3 \\ 3 & 6 - \lambda \end{bmatrix} = (-4 - \lambda)(6 - \lambda) - (-3)(3).$$

$$= (-4 - \lambda)(6 - \lambda) + 9 = -24 + 4\lambda - 6\lambda + \lambda^2 + 9$$

$$= \lambda^2 - 2\lambda - 15.$$

$$= (\lambda - 5)(\lambda + 3).$$

Thus:

$$\det(A - \lambda I) = (2 - \lambda)(\lambda - 5)(\lambda + 3).$$

Eigenvalues: $\lambda = 2, 5, -3$.

The block-diagonal structure simplifies the determinant, with $\lambda = 2$ from the third component and $\lambda = 5, -3$ from the 2×2 block.

Example 8.17

For:

$$A = \begin{bmatrix} 2 & 3 \\ 2 & 1 \end{bmatrix}.$$

$$\det(A - \lambda I) = (2 - \lambda)(1 - \lambda) - 6 = \lambda^2 - 3\lambda - 4 = (\lambda - 4)(\lambda + 1) = 0.$$

Eigenvalues: $\lambda = 4, -1$. For $\lambda = 4$:

$$A - 4I = \begin{bmatrix} -2 & 3 \\ 2 & -3 \end{bmatrix}, \quad R_2 \to R_2 + R_1 : \begin{bmatrix} -2 & 3 \\ 0 & 0 \end{bmatrix}, \quad x_1 = \frac{3}{2}x_2 \implies \mathbf{v}_1 = t \begin{bmatrix} 3 \\ 2 \end{bmatrix}.$$

For $\lambda = -1$:

$$A + I = \begin{bmatrix} 3 & 3 \\ 2 & 2 \end{bmatrix}, \quad R_2 \to R_2 - \frac{2}{3}R_1 : \begin{bmatrix} 3 & 3 \\ 0 & 0 \end{bmatrix}, \quad x_1 = -x_2 \implies \mathbf{v}_2 = t \begin{bmatrix} -1 \\ 1 \end{bmatrix}.$$

Diagonalizable with:

$$P = \begin{bmatrix} 3 & -1 \\ 2 & 1 \end{bmatrix}, \quad D = \begin{bmatrix} 4 & 0 \\ 0 & -1 \end{bmatrix}.$$

Example 8.18 Following Example 8.11:

$$\text{Eigenspace}(3) = \text{Span}\left\{\begin{bmatrix} 0 \\ 1 \\ 1 \end{bmatrix}\right\},$$

$$\text{dimension} = 1.$$

For $\lambda = -1$:

$$A + I = \begin{bmatrix} 0 & 0 & 0 \\ 0 & 2 & 2 \\ 0 & 2 & 2 \end{bmatrix}, \quad R_3 \to R_3 - R_2 : \begin{bmatrix} 0 & 0 & 0 \\ 0 & 2 & 2 \\ 0 & 0 & 0 \end{bmatrix}, \quad x_2 = -x_3, x_1 \text{ free.}$$

$$\text{Eigenspace}(-1) = \text{Span}\left\{\begin{bmatrix} 1 \\ 0 \\ 0 \end{bmatrix}, \begin{bmatrix} 0 \\ -1 \\ 1 \end{bmatrix}\right\}, \quad \text{dimension} = 2.$$

8.2.6   Diagonalization

A matrix $A \in \mathbb{R}^{n \times n}$ is **diagonalizable** if:

$$A = PDP^{-1},$$

where P's columns are eigenvectors, and D's diagonal entries are eigenvalues. This requires n linearly independent eigenvectors, i.e., geometric multiplicity equals algebraic multiplicity for all eigenvalues.

Procedure:

1. Solve $\det(A - \lambda I) = 0$ for eigenvalues.
2. Find eigenvectors by solving $(A - \lambda I)\mathbf{v} = \mathbf{0}$.
3. Form P with eigenvectors as columns.
4. Form D with corresponding eigenvalues.
5. Verify $A = PDP^{-1}$.

Python Example: Diagonalization and Matrix Reconstruction

```python
import numpy as np
A = np.array([[-1, 0, 0], [0, 1, 2], [0, 2, 1]])
eigenvalues, eigenvectors = np.linalg.eig(A)
P = eigenvectors
D = np.diag(eigenvalues)
P_inv = np.linalg.inv(P)
```

```
7  A_reconstructed = P @ D @ P_inv
8  print("Original A:\n", A)
9  print("Reconstructed A:\n", A_reconstructed)
```

Output:

```
 Original A:
 [[-1   0   0]
  [ 0   1   2]
  [ 0   2   1]]
 Reconstructed A:
 [[-1.   0.   0.]
  [ 0.   1.   2.]
  [ 0.   2.   1.]]
```

Example 8.19
For:

$$A = \begin{bmatrix} 0 & 1 \\ 0 & 0 \end{bmatrix}.$$

$$\det(A - \lambda I) = \lambda^2 \implies \lambda = 0 \text{ (alg. mult. 2)}.$$

$$A - 0I = A = \begin{bmatrix} 0 & 1 \\ 0 & 0 \end{bmatrix}.$$

The above matrix is already in REF.

$$x_2 = 0 \implies \mathbf{v} = x_1 \begin{bmatrix} 1 \\ 0 \end{bmatrix}.$$

geom. mult.: $1 <$ alg. mult.: 2, so A is not diagonalizable (Fig. 8.5).

Example 8.20
Let:

$$C = \begin{bmatrix} -6 & -3 & 1 \\ 5 & 2 & -1 \\ 2 & 3 & -5 \end{bmatrix}.$$

polynomial $\det(C - \lambda I_3)$ as follows:

Characteristic Polynomial: $-(\lambda + 1)(\lambda + 4)^2 \Rightarrow \lambda_1 = -1, \quad \lambda_2 = \lambda_3 = -4$
- Algebraic multiplicity of $\lambda = -1$: 1
- Algebraic multiplicity of $\lambda = -4$: 2

Fig. 8.5 Non-diagonalizable matrix $A = \begin{bmatrix} 0 & 1 \\ 0 & 0 \end{bmatrix}$

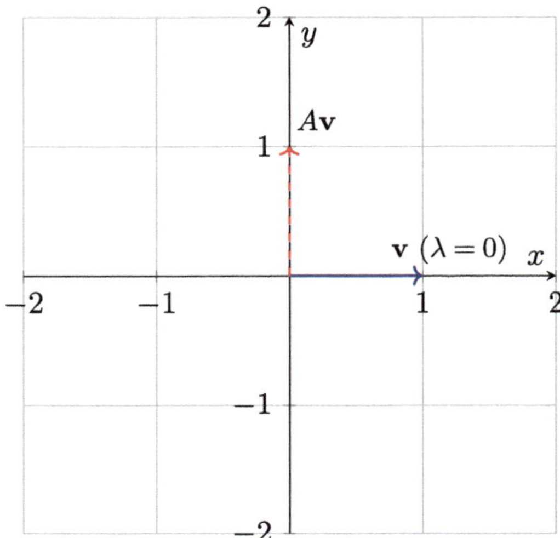

Check eigenvectors for $\lambda = -1$:

$$(C + I) = \begin{bmatrix} -5 & -3 & 1 \\ 5 & 3 & -1 \\ 2 & 3 & -4 \end{bmatrix} \Rightarrow \text{Row reduce to get eigenvectors}$$

Check eigenvectors for $\lambda = -4$:

$$(C + 4I) = \begin{bmatrix} -2 & -3 & 1 \\ 5 & 6 & -1 \\ 2 & 3 & -1 \end{bmatrix} \Rightarrow \text{rank}(C + 4I) = 2 \Rightarrow \text{nullity} = 1$$

For $\lambda = -4$, geometric multiplicity is 1 (less than its algebraic multiplicity 2).

$$\Rightarrow C \text{ is not diagonalizable.}$$

Example 8.21

Let:

$$M = \begin{bmatrix} -3 & 2 & 1 \\ 3 & -4 & -3 \\ -8 & 8 & 6 \end{bmatrix}.$$

Compute characteristic polynomial:

$$\det(M - \lambda I_3) = -(\lambda - 2)(\lambda + 1)(\lambda + 2),$$

$$\Rightarrow \lambda_1 = 2, \quad \lambda_2 = -1, \quad \lambda_3 = -2$$

- All eigenvalues have algebraic multiplicity 1
- We expect a complete basis of eigenvectors

Since M has 3 distinct eigenvalues, it is diagonalizable.

Let's once again check the conditions for diagonalizability.
An $n \times n$ matrix A is diagonalizable if and only if **both** of the following conditions are satisfied:

1. The total number of eigenvalues (counting multiplicities) equals n.
2. For each eigenvalue λ, the dimension of the corresponding eigenspace equals the **algebraic multiplicity** of λ.

Mathematically, for each eigenvalue λ of A:

$$\underbrace{n - \operatorname{rank}(A - \lambda I_n)}_{\text{Geometric multiplicity}} = \underbrace{\text{Multiplicity of } \lambda}_{\text{Algebraic multiplicity}}.$$

Diagonalizability depends on whether there are enough linearly independent eigenvectors to form a basis of $\mathbb{R}^n$. This is guaranteed if and only if the geometric multiplicity equals the algebraic multiplicity for all eigenvalues.

Example 8.22 We are given the matrix:

$$A = \begin{bmatrix} 0 & 2 & 1 \\ -2 & 0 & -2 \\ 0 & 0 & -1 \end{bmatrix}.$$

We want to check whether this matrix is diagonalizable.
Step 1- Find the Characteristic Polynomial

We compute the determinant of $A - t I_3$ as follows:

$$\det(A - t I_3) = \det \begin{bmatrix} -t & 2 & 1 \\ -2 & -t & -2 \\ 0 & 0 & -1-t \end{bmatrix}.$$

Expanding along the third row:

$$= (-1 - t) \cdot \det \begin{bmatrix} -t & 2 \\ -2 & -t \end{bmatrix} = (-1 - t)\,[(-t)(-t) - (2)(-2)] = (-1 - t)(t^2 + 4).$$

Hence, the characteristic polynomial is:

$$p(t) = -(t + 1)(t^2 + 4).$$

Step 2- Find Eigenvalues and Their Multiplicities

From the factorized characteristic polynomial, we identify the eigenvalues:

$$\lambda_1 = -1, \quad \lambda_2 = 2i, \quad \lambda_3 = -2i.$$

Each eigenvalue has algebraic multiplicity 1.
Step 3- Check Diagonalizability

A matrix is diagonalizable if:

- The sum of algebraic multiplicities equals the size of the matrix ($n = 3$),
- For each eigenvalue, the geometric multiplicity equals the algebraic multiplicity.

Since we have three distinct eigenvalues (even though some are complex) and each has algebraic multiplicity 1, and the corresponding eigenspaces are one-dimensional, A is diagonalizable over $\mathbb{C}$.
Step 4- Diagonalization

Let P be the matrix whose columns are eigenvectors of A, and D the diagonal matrix of corresponding eigenvalues:

$$P = \begin{bmatrix} -1 & i & -i \\ 0 & 1 & 1 \\ 1 & 0 & 0 \end{bmatrix}, \quad D = \begin{bmatrix} -1 & 0 & 0 \\ 0 & -2i & 0 \\ 0 & 0 & 2i \end{bmatrix}, \quad P^{-1} = \begin{bmatrix} 0 & 0 & 1 \\ -0.5i & 0.5 & -0.5i \\ 0.5i & 0.5 & 0.5i \end{bmatrix}.$$

Then, we have:

$$A = PDP^{-1}$$

Example 8.23 We want to check the Diagonalizability of Matrix B

$$B = \begin{bmatrix} -7 & -3 & -6 \\ 0 & -4 & 0 \\ 3 & 3 & 2 \end{bmatrix}.$$

Step 1- Find the Characteristic Polynomial

We compute the determinant of $B - tI_3$:

$$\det(B - tI_3) = \det \begin{bmatrix} -7-t & -3 & -6 \\ 0 & -4-t & 0 \\ 3 & 3 & 2-t \end{bmatrix}.$$

Using cofactor expansion (e.g., along the second row or simplifying block structures), the determinant simplifies to:

$$\det(B - t I_3) = -(t + 1)(t + 4)^2.$$

Step 2- Eigenvalues and Their Multiplicities

From the factorization of the characteristic polynomial:

$$\lambda_1 = -1 \quad \text{(alg. mult.1)},$$

$$\lambda_2 = -4 \quad \text{(alg. mult.2)}$$

Step 3- Check Geometric Multiplicity

To be diagonalizable, the dimension of the eigenspace corresponding to each eigenvalue must equal its algebraic multiplicity.
We compute:

$$\text{nullity}(B + 4I_3) = 3 - \text{rank}(B + 4I_3).$$

After calculating $B + 4I_3$, its rank is 1, hence:

$$\dim(\text{Null}(B + 4I_3)) = 3 - 1 = 2.$$

So, for eigenvalue -4, the geometric multiplicity is 2, equal to its algebraic multiplicity.
Step 4- Conclusion and Diagonalization

Since:

- The total algebraic multiplicities sum to 3 (the size of the matrix),
- Each eigenvalue's geometric multiplicity equals its algebraic multiplicity,

the matrix B is diagonalizable.
Step 5- Diagonal Form and Eigenbasis

The matrix of eigenvectors is:

$$P = \begin{bmatrix} -1 & -1 & -2 \\ 0 & 1 & 0 \\ 1 & 0 & 1 \end{bmatrix} \quad \text{and} \quad D = \begin{bmatrix} -1 & 0 & 0 \\ 0 & -4 & 0 \\ 0 & 0 & -4 \end{bmatrix}.$$

Therefore:

$$B = PDP^{-1}.$$

B is diagonalizable.

Example 8.24 Geometric multiplicities: 1 ($\lambda = 3$), 2 ($\lambda = -1$). Since these match the algebraic multiplicities, A is diagonalizable.

$$P = \begin{bmatrix} 0 & 1 & 0 \\ 1 & 0 & -1 \\ 1 & 0 & 1 \end{bmatrix}, \quad D = \begin{bmatrix} 3 & 0 & 0 \\ 0 & -1 & 0 \\ 0 & 0 & -1 \end{bmatrix}.$$

8.3   Linear Operators and Eigenvalues

A **linear operator** is a special type of transformation that preserves the structure of vector addition and scalar multiplication. Formally, a linear operator is a function:

$$T : \mathbb{R}^n \to \mathbb{R}^n,$$

such that for all vectors $\mathbf{u}, \mathbf{v} \in \mathbb{R}^n$ and all scalars $c \in \mathbb{R}$, the following two properties hold:

- **Additivity (Linearity over vector addition):**

$$T(\mathbf{u} + \mathbf{v}) = T(\mathbf{u}) + T(\mathbf{v}).$$

- **Homogeneity (Linearity over scalar multiplication):**

$$T(c\mathbf{u}) = cT(\mathbf{u}).$$

These two properties together define what it means for a transformation to be linear. Any transformation that fails to satisfy either property is considered non-linear and cannot be analyzed using standard linear algebra tools.

Matrix Representation: Every linear operator on $\mathbb{R}^n$ can be represented by a matrix $A \in \mathbb{R}^{n \times n}$, such that for any vector $\mathbf{v} \in \mathbb{R}^n$, the action of the operator is equivalent to matrix-vector multiplication:

$$T(\mathbf{v}) = A\mathbf{v}.$$

This representation allows us to study the behavior of the transformation using the algebraic and geometric properties of matrices.

Linear operators are functions that preserve the structure of vector spaces and can be studied through their matrix representations. The eigenvalues and eigenvectors of these matrices give powerful insights into the internal behavior of the operator, revealing invariant directions and scaling effects. These tools are especially critical in PCA, system stability, and neural network dynamics.

Example 8.25 For $T(x_1, x_2) = (-2x_2, -3x_1 + x_2)$:

$$A = \begin{bmatrix} 0 & -2 \\ -3 & 1 \end{bmatrix}.$$

$$\det(A - \lambda I) = (-\lambda)(1 - \lambda) - (-2)(-3) = \lambda^2 - \lambda - 6 = (\lambda - 3)(\lambda + 2) = 0.$$

Eigenvalues: $\lambda = 3, -2$. For $\lambda = 3$:

$$A - 3I = \begin{bmatrix} -3 & -2 \\ -3 & -2 \end{bmatrix}, \quad R_2 \to R_2 - R_1 : \begin{bmatrix} -3 & -2 \\ 0 & 0 \end{bmatrix},$$

$$x_1 = -\frac{2}{3}x_2 \implies \mathbf{v}_1 = t \begin{bmatrix} -2 \\ 3 \end{bmatrix}.$$

For $\lambda = -2$:

$$A + 2I = \begin{bmatrix} 2 & -2 \\ -3 & 3 \end{bmatrix}, \quad R_2 \to R_2 + \frac{3}{2}R_1 : \begin{bmatrix} 2 & -2 \\ 0 & 0 \end{bmatrix}, \quad x_1 = x_2 \implies \mathbf{v}_2 = t \begin{bmatrix} 1 \\ 1 \end{bmatrix}.$$

Python Example: Eigenvalues and Eigenvectors Computation

```python
import numpy as np
A = np.array([[0, -2], [-3, 1]])
eigenvalues, eigenvectors = np.linalg.eig(A)
print("Eigenvalues:", eigenvalues)
print("Eigenvectors:\n", eigenvectors)
```

Output:

```
Eigenvalues: [-2.  3.]
Eigenvectors:
 [[-0.71  0.55]
 [-0.71 -0.83]]
```

8.4 Eigenvalues in ML

Eigenvalues and eigenvectors are pivotal in ML, enabling efficient data analysis, transformation, and model optimization.

1. **PCA**: Eigenvectors of the covariance matrix $C = \frac{1}{n} X^{\mathrm{T}} X$ are principal components, capturing maximal variance. Eigenvalues indicate variance magnitude, guiding dimensionality reduction.
2. **Spectral Clustering**: Eigenvectors of the Laplacian matrix partition data into clusters, capturing nonlinear structures.
3. **Neural Network Optimization**: Eigenvalues of the Hessian matrix inform convergence in second-order optimization methods like Newton's method.
4. **Kernel Methods**: Eigenvectors of kernel matrices define feature spaces for nonlinear transformations in Support Vector Machines (SVMs).
5. **Manifold Learning**: Eigenvectors of similarity matrices reveal low-dimensional structures in high-dimensional data.
6. **Graph Neural Networks**: Eigenvalues of adjacency matrices influence message passing and graph convolutions.

Chapter Summary

This chapter provided an in-depth exploration of eigenvalues and eigenvectors, revealing them as key to understanding the intrinsic structure of linear transformations and matrices, with profound implications for ML algorithms like PCA and spectral clustering. It commenced with the definition in Sect. 8.1: for a square matrix $A \in \mathbb{R}^{n \times n}$, an eigenvector $\mathbf{v} \neq \mathbf{0}$ satisfies $A\mathbf{v} = \lambda\mathbf{v}$, where λ is the eigenvalue, representing scaling along invariant directions, as illustrated in Fig. 8.1 for a diagonal matrix stretching axes by $\lambda = 2, 3$. The eigenspace of λ formed the null space of $A - \lambda I$, a subspace containing all such vectors, with geometric interpretations emphasizing fixed directions under transformations.

Section 8.2 detailed computation via the characteristic equation $\det(A - \lambda I) = 0$, yielding the characteristic polynomial whose roots are eigenvalues, potentially real or complex. For each λ, eigenvectors were found by solving $(A - \lambda I)\mathbf{v} = \mathbf{0}$ through row reduction.

The characteristic polynomial and diagonalization in Sect. 8.2.5 extended this, with algebraic multiplicity as root counts and geometric multiplicity as eigenspace dimensions, requiring equality for diagonalizability $A = PDP^{-1}$ where P's columns are eigenvectors and D diagonalizes eigenvalues. Examples confirmed diagonalizability when multiplicities matched, like distinct eigenvalues ensuring independence, and non-diagonalizability when geometric multiplicity falls short of algebraic multiplicity, as in the case of repeated eigenvalue $\lambda = -4$ with geometric multiplicity 1 versus algebraic multiplicity 2.

Section 8.4 highlighted applications: PCA used covariance eigenvectors for variance maximization, spectral clustering Laplacian eigenvectors for partitioning, neural optimization Hessian eigenvalues for curvature, and kernel methods for feature spaces. Python examples with NumPy's `eig` and SymPy visualized eigenvectors (Fig. 8.4), computed polynomials, and verified diagonalization $A = PDP^{-1}$.

Takeaways

Key takeaways from the chapter were outlined as:

- Eigenvalues λ and eigenvectors $\mathbf{v}$ satisfying $A\mathbf{v} = \lambda\mathbf{v}$, defining invariant directions scaled by λ, with eigenspaces as null spaces of $A - \lambda I$.
- Computation through characteristic polynomial $\det(A - \lambda I) = 0$, solved for roots, followed by solving homogeneous systems for eigenvectors, simplified for triangular or block forms.
- Algebraic (root multiplicity) vs. geometric (eigenspace dimension) multiplicities, with equality ensuring diagonalizability $A = PDP^{-1}$ for orthogonal bases in symmetric cases.
- Linear operators as matrix-represented transformations, with eigenvalues revealing scaling, orthogonality for distinct λ, and complex values for rotations.
- ML integrations: PCA for data decorrelation via eigenvectors, clustering with spectral graphs, optimization using Hessian spectra, and kernels for nonlinear mappings.

Through these concepts, readers gained proficiency in eigenvalue analysis, enabling decomposition of transformations and application to dimensionality reduction and model interpretability in ML.

Exercises

1. Compute the eigenvalues and eigenvectors of $A = \begin{bmatrix} 1 & 1 \\ 0 & 2 \end{bmatrix}$.

2. For $A = \begin{bmatrix} 0 & 1 \\ -1 & 0 \end{bmatrix}$, find the eigenvalues and explain why they are complex.

3. Find the eigenvectors of A and verify that they are orthogonal if $A = \begin{bmatrix} 3 & 1 \\ 1 & 3 \end{bmatrix}$.

4. Find the eigenspace for $\lambda = -1$ of $A = \begin{bmatrix} -1 & 0 & 0 \\ 0 & 1 & 2 \\ 0 & 2 & 1 \end{bmatrix}$.

5. Determine if $\lambda = 0$ is an eigenvalue of $A = \begin{bmatrix} 1 & 2 \\ 3 & 4 \end{bmatrix}$.

6. Compute eigenvalues and eigenspaces for $A = \begin{bmatrix} 2 & 0 & 0 \\ 0 & 3 & 4 \\ 0 & 4 & 9 \end{bmatrix}$.

7. For $A = \begin{bmatrix} 1 & 2 \\ 0 & 1 \end{bmatrix}$, find eigenvalues, eigenvectors, and comment on diagonalizability.

8. If $A = \begin{bmatrix} 4 & 1 \\ 2 & 3 \end{bmatrix}$ then show that A has real eigenvalues (e.g., via the characteristic polynomial) and compute the eigenspaces.

9. If $A = \begin{bmatrix} 2 & 0 \\ 0 & 5 \end{bmatrix}$, show that A is already diagonal and identify P and D such that $A = PDP^{-1}$.

10. For $A = \begin{bmatrix} 0 & 1 & 0 \\ 1 & 0 & 0 \\ 0 & 0 & 3 \end{bmatrix}$, find eigenvalues, eigenvectors, and check diagonalizability.

11. Compute the characteristic polynomial of $A = \begin{bmatrix} 2 & 1 \\ 1 & 2 \end{bmatrix}$ and find its eigenvalues.

12. Verify that $\mathbf{v} = \begin{bmatrix} 1 \\ -1 \end{bmatrix}$ is an eigenvector of $A = \begin{bmatrix} 3 & 2 \\ 2 & 3 \end{bmatrix}$.

13. For $A = \begin{bmatrix} 1 & 0 & 0 \\ 3 & -2 & 0 \\ 2 & 3 & 4 \end{bmatrix}$, use the fact that A is triangular to confirm that the eigenvalues are the diagonal entries, and find eigenvectors.

14. Compute the eigenspace for $\lambda = 0$ of $A = \begin{bmatrix} 1 & 2 \\ 2 & 4 \end{bmatrix}$.

15. Show that $A = \begin{bmatrix} -2 & 1 \\ 1 & -2 \end{bmatrix}$ has orthogonal eigenvectors.

16. For the linear operator $T : \mathbb{R}^2 \to \mathbb{R}^2$ defined by $T(x_1, x_2) = (x_2, -x_1)$, find its matrix representation in the standard basis, then compute eigenvalues and eigenvectors.

17. Compute eigenvalues and eigenvectors of $A = \begin{bmatrix} 5 & -1 \\ -1 & 5 \end{bmatrix}$ and verify orthogonality.

18. Write a Python script to compute and visualize the eigenvectors as invariant directions of the transformation of $A = \begin{bmatrix} 1 & 2 \\ 2 & 1 \end{bmatrix}$.

19. Determine if $A = \begin{bmatrix} 1 & 1 \\ 1 & 1 \end{bmatrix}$ is diagonalizable and justify.

20. For $A = \begin{bmatrix} 2 & 0 & 1 \\ 0 & 3 & 0 \\ 1 & 0 & 2 \end{bmatrix}$, compute eigenvalues and eigenspaces.

Chapter 9
Vector Spaces and Subspaces

Introduction

Vector spaces and their subspaces form the backbone of linear algebra, providing the mathematical framework to represent and manipulate data in high-dimensional settings. A vector space is a collection of vectors closed under addition and scalar multiplication. Understanding these concepts, along with bases, coordinate systems, and orthogonality, is crucial for simplifying complex problems in linear algebra and its applications, such as ML.

This chapter offers a comprehensive exploration of vector spaces, subspaces, and orthogonality, enriched with geometric interpretations, computational methods, and Python implementations using NumPy and Matplotlib. We begin with the definition and properties of vector spaces, followed by an in-depth study of subspaces, including null, column, and row spaces. Visualizations clarify geometric interpretations, while Python code bridges theory to practice. A dedicated section on ML applications ties these concepts to real-world problems, ensuring relevance for data science enthusiasts. Extensive exercises reinforce both theoretical and computational skills.

Topics Covered

This chapter is organized as follows:

- **Section** 9.1 **Vector Spaces; Definition and Basics**: Defines vector spaces and their properties, with examples in $\mathbb{R}^n$.
- **Section** 9.2 **Subspaces**: Explores subspaces, including null, column, and row spaces, with examples and visualizations.
- **Section** 9.3 **Null Space**: Delves into the null space of a matrix, defining it as the set of vectors that map to the zero vector, and explores its properties as a subspace

© The Author(s), under exclusive license to Springer Nature Singapore Pte Ltd. 2026   263
Md. Jalil Piran, *Linear Algebra with Applications in Machine Learning*,
https://doi.org/10.1007/978-981-95-5167-5_9

and its connection to the Rank-Nullity Theorem, and introduces the column and row spaces as complementary perspectives on the range and constraint structure of a linear transformation.

- **Section** 9.4 **Basis and Dimension**: Defines bases as linearly independent spanning sets, introduces the dimension of a subspace, and presents the nullity of a matrix along with the Dimension Theorem.
- **Section** 9.5 **Coordinate Systems**: Introduces the concept of coordinate systems, explaining how a basis provides a unique coordinate representation for any vector within a vector space, with a focus on change-of-basis.
- **Section** 9.6 **ML Applications**: Discusses the role of subspaces and orthogonality in PCA, SVMs, and optimization.
- **Summary and Exercises**: Recaps key concepts and provides a comprehensive set of problems to deepen understanding and computational proficiency.

All code accompanying the chapter is publicly available at https://github.com/jalil-piran/Linear-Algebra-with-Applications-in-Machine-Learning/blob/main/Chapter_09_Vector_Spaces.ipynb.

By blending theory, computation, and visualization, this chapter equips readers with the tools to navigate vector spaces and their applications, laying a foundation for advanced topics like eigenvalues and SVD.

Common Acronyms and Their Definitions

ML	Machine Learning
PCA	Principal Component Analysis
SVD	Singular Value Decomposition
SVMs	Support Vector Machines
REF	Row Echelon Form
RREF	Reduced Row Echelon Form
LR	Linear Regression

9.1   Vector Spaces: Definition and Basics

A **vector space** over the real numbers $\mathbb{R}$ is a set V equipped with vector addition and scalar multiplication satisfying eight axioms: commutativity, associativity, identity, and inverses for addition; compatibility, distributivity, and identity for scalar multiplication. Common examples include $\mathbb{R}^n$, the set of polynomials of degree at most n, and the set of $m \times n$ matrices.

Definition

A set V with operations $+$ (vector addition) and $\cdot$ (scalar multiplication by $\mathbb{R}$) is a **vector space** if:

- **Addition commutativity**: $\mathbf{u} + \mathbf{v} = \mathbf{v} + \mathbf{u}$.
- **Addition associativity**: $(\mathbf{u} + \mathbf{v}) + \mathbf{w} = \mathbf{u} + (\mathbf{v} + \mathbf{w})$.
- **Additive identity**: There exists $\mathbf{0} \in V$ such that $\mathbf{v} + \mathbf{0} = \mathbf{v}$.
- **Additive inverses**: For each $\mathbf{v} \in V$, there exists $-\mathbf{v} \in V$ such that $\mathbf{v} + (-\mathbf{v}) = \mathbf{0}$.
- **Scalar multiplication compatibility**: $a(b\mathbf{v}) = (ab)\mathbf{v}$.
- **Scalar multiplication identity**: $1 \cdot \mathbf{v} = \mathbf{v}$.
- **Distributivity over vector addition**: $a(\mathbf{u} + \mathbf{v}) = a\mathbf{u} + a\mathbf{v}$.
- **Distributivity over scalar addition**: $(a + b)\mathbf{v} = a\mathbf{v} + b\mathbf{v}$.

Example 9.1 The set $\mathbb{R}^3$, consisting of all ordered triples of real numbers, forms a vector space when equipped with standard vector addition and scalar multiplication. Each element in $\mathbb{R}^3$ can be written as a column vector

$$\mathbf{u} = \begin{bmatrix} u_1 \\ u_2 \\ u_3 \end{bmatrix}, \quad \mathbf{v} = \begin{bmatrix} v_1 \\ v_2 \\ v_3 \end{bmatrix}, \quad c \in \mathbb{R}.$$

- **Addition**:

$$\mathbf{u} + \mathbf{v} = \begin{bmatrix} u_1 + v_1 \\ u_2 + v_2 \\ u_3 + v_3 \end{bmatrix}.$$

This operation adds vectors component-wise. For instance, adding $\begin{bmatrix} 1 \\ 2 \\ 3 \end{bmatrix} + \begin{bmatrix} 4 \\ 5 \\ 6 \end{bmatrix} = \begin{bmatrix} 5 \\ 7 \\ 9 \end{bmatrix}$.

- **Scalar multiplication**:

$$c\mathbf{u} = \begin{bmatrix} cu_1 \\ cu_2 \\ cu_3 \end{bmatrix}.$$

Each component is scaled by the real number c. For example, $2 \begin{bmatrix} 1 \\ -3 \\ 4 \end{bmatrix} = \begin{bmatrix} 2 \\ -6 \\ 8 \end{bmatrix}$.

- **Zero vector**:

$$\mathbf{0} = \begin{bmatrix} 0 \\ 0 \\ 0 \end{bmatrix}.$$

Adding the zero vector leaves any vector unchanged, i.e., $\mathbf{u} + \mathbf{0} = \mathbf{u}$.

- **Additive inverse**:

$$-\mathbf{u} = \begin{bmatrix} -u_1 \\ -u_2 \\ -u_3 \end{bmatrix}.$$

Adding a vector to its inverse yields the zero vector: $\mathbf{u} + (-\mathbf{u}) = \mathbf{0}$.

Beyond these, other axioms such as associativity, commutativity of addition, and distributive properties also hold. Therefore, $\mathbb{R}^3$ satisfies all of the vector space axioms and is indeed a vector space. Geometrically, vectors in $\mathbb{R}^3$ can be interpreted as points or arrows in three-dimensional space, making this example particularly intuitive.

Example 9.2 The set of polynomials $P_2 = \{a_0 + a_1 x + a_2 x^2 \mid a_0, a_1, a_2 \in \mathbb{R}\}$ forms a vector space. Addition and scalar multiplication are defined as:

$$(a_0 + a_1 x + a_2 x^2) + (b_0 + b_1 x + b_2 x^2) = (a_0 + b_0) + (a_1 + b_1)x + (a_2 + b_2)x^2.$$

$$c(a_0 + a_1 x + a_2 x^2) = ca_0 + ca_1 x + ca_2 x^2.$$

The zero polynomial is $p(x) = 0$, and the inverse of $p(x)$ is $-p(x)$. All axioms are satisfied.

9.2   Subspaces

A **subspace** is a subset of a vector space (e.g., $\mathbb{R}^n$) that is itself a vector space under the same operations of vector addition and scalar multiplication. This means a subspace must satisfy three criteria:

- It must include the zero vector.
- It must be closed under vector addition.
- It must be closed under scalar multiplication.

Subspaces are foundational in ML because they often represent constrained or reduced feature spaces. For instance, SVMs define decision boundaries as subspaces, while techniques like PCA find optimal subspaces to represent data with minimal loss.

Definition

A subset $V \subseteq \mathbb{R}^n$ is a **subspace** if:

1. **Zero vector**: $\mathbf{0} \in V$.
2. **Closure under addition**: For any $\mathbf{u}, \mathbf{v} \in V$, $\mathbf{u} + \mathbf{v} \in V$.
3. **Closure under scalar multiplication**: For any scalar $c \in \mathbb{R}$ and $\mathbf{u} \in V$, $c\mathbf{u} \in V$.

Example 9.3

Consider:

$$U = \left\{ \begin{bmatrix} 2x - y \\ 3x \\ x + y \end{bmatrix} \mid x, y \in \mathbb{R} \right\}.$$

We want to verify if U is a subspace of $\mathbb{R}^3$.

- **Zero vector**: Set $x = 0$, $y = 0$:

$$\mathbf{u} = \begin{bmatrix} 2(0) - 0 \\ 3(0) \\ 0 + 0 \end{bmatrix} = \begin{bmatrix} 0 \\ 0 \\ 0 \end{bmatrix} \in U.$$

- **Addition**: Let $\mathbf{u}_1 = \begin{bmatrix} 2x_1 - y_1 \\ 3x_1 \\ x_1 + y_1 \end{bmatrix}$, $\mathbf{u}_2 = \begin{bmatrix} 2x_2 - y_2 \\ 3x_2 \\ x_2 + y_2 \end{bmatrix}$.

$$\mathbf{u}_1 + \mathbf{u}_2 = \begin{bmatrix} 2(x_1 + x_2) - (y_1 + y_2) \\ 3(x_1 + x_2) \\ (x_1 + x_2) + (y_1 + y_2) \end{bmatrix} \in U.$$

- **Scalar multiplication**: For $\mathbf{u} = \begin{bmatrix} 2x - y \\ 3x \\ x + y \end{bmatrix}$, scalar c:

$$c\mathbf{u} = \begin{bmatrix} 2(cx) - (cy) \\ 3(cx) \\ (cx) + (cy) \end{bmatrix} \in U.$$

Thus, U is a subspace, expressible as:

$$U = \mathrm{Span} \left\{ \begin{bmatrix} 2 \\ 3 \\ 1 \end{bmatrix}, \begin{bmatrix} -1 \\ 0 \\ 1 \end{bmatrix} \right\}$$

Example 9.4 Consider:

$$S = \left\{ \begin{bmatrix} x \\ y \end{bmatrix} \in \mathbb{R}^2 \mid x \geq 0 \right\}.$$

– Zero vector: $\begin{bmatrix} 0 \\ 0 \end{bmatrix} \in S.$

– Addition: For $\mathbf{u} = \begin{bmatrix} a \\ b \end{bmatrix}, \mathbf{v} = \begin{bmatrix} c \\ d \end{bmatrix}, a, c \geq 0, \mathbf{u} + \mathbf{v} = \begin{bmatrix} a+c \\ b+d \end{bmatrix}, a+c \geq 0.$

– Scalar multiplication: For $\mathbf{u} = \begin{bmatrix} a \\ b \end{bmatrix}, a > 0, c = -1$:

$$c\mathbf{u} = \begin{bmatrix} -a \\ -b \end{bmatrix}, \quad -a < 0 \notin S.$$

S is not a subspace due to failure of closure under scalar multiplication.

Python Example: Visualizing a Subspace Plane

```python
import numpy as np
import matplotlib.pyplot as plt
from mpl_toolkits.mplot3d import Axes3D

w2, w3 = np.meshgrid(np.linspace(-2, 2, 20),
    np.linspace(-2, 2, 20))
w1 = (5 * w2 - 4 * w3) / 6

fig = plt.figure()
ax = fig.add_subplot(111, projection='3d')
ax.plot_surface(w2, w3, w1, alpha=0.5,
    cmap='viridis')
ax.set_xlabel('$w_2$')
ax.set_ylabel('$w_3$')
ax.set_zlabel('$w_1$')
plt.title('Subspace: $6w_1 - 5w_2 + 4w_3 = 0$')
ax.view_init(elev=35, azim=45)
plt.show()
```

Output:

See Figure 9.1.

Fig. 9.1 Subspace in $\mathbb{R}^3$:
Plane defined by
$6w_1 - 5w_2 + 4w_3 = 0$

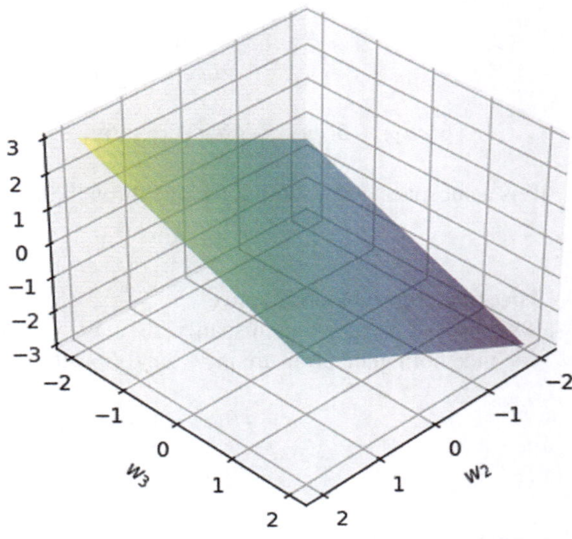

Example 9.5 Consider:

$$W = \left\{ \begin{bmatrix} w_1 \\ w_2 \\ w_3 \end{bmatrix} \in \mathbb{R}^3 \mid 6w_1 - 5w_2 + 4w_3 = 0 \right\} .$$

This defines a plane through the origin in $\mathbb{R}^3$. Check the conditions:

– Zero vector:
$$6(0) - 5(0) + 4(0) = 0 \quad \Rightarrow \quad \mathbf{0} \in W.$$

– Addition: Let $\mathbf{u} = \begin{bmatrix} u_1 \\ u_2 \\ u_3 \end{bmatrix}, \mathbf{v} = \begin{bmatrix} v_1 \\ v_2 \\ v_3 \end{bmatrix} \in W$, so $6u_1 - 5u_2 + 4u_3 = 0, 6v_1 - 5v_2 + 4v_3 = 0.$

$$\mathbf{u} + \mathbf{v} = \begin{bmatrix} u_1 + v_1 \\ u_2 + v_2 \\ u_3 + v_3 \end{bmatrix},$$

$$6(u_1 + v_1) - 5(u_2 + v_2) + 4(u_3 + v_3)$$
$$= (6u_1 - 5u_2 + 4u_3) + (6v_1 - 5v_2 + 4v_3) = 0 + 0 = 0.$$

– Scalar multiplication: For scalar c,

$$c\mathbf{u} = \begin{bmatrix} cu_1 \\ cu_2 \\ cu_3 \end{bmatrix},$$

$$6(cu_1) - 5(cu_2) + 4(cu_3) = c(6u_1 - 5u_2 + 4u_3) = c \cdot 0 = 0.$$

W is a subspace, representing a plane through the origin, visualized in Fig. 9.1.

Theorem: Span Is a Subspace

If $\mathbf{v}_1, \mathbf{v}_2, \ldots, \mathbf{v}_n \in \mathbb{R}^3$, then $\text{span}\{\mathbf{v}_1, \ldots, \mathbf{v}_n\}$ is a subspace of $\mathbb{R}^3$.

Since both generators are in $\mathbb{R}^3$, and U is their span, we conclude:

$$U \text{ is a subspace of } \mathbb{R}^3.$$

Example 9.6 Let us consider Example 9.3 again, we want to apply the subspace axioms.

Let:

$$\mathbf{u} = \begin{bmatrix} 2x_1 - y_1 \\ 3x_1 \\ x_1 + y_1 \end{bmatrix}, \quad \mathbf{v} = \begin{bmatrix} 2x_2 - y_2 \\ 3x_2 \\ x_2 + y_2 \end{bmatrix}.$$

(a) **Closure under addition**:

$$\mathbf{u} + \mathbf{v} = \begin{bmatrix} 2x_1 - y_1 \\ 3x_1 \\ x_1 + y_1 \end{bmatrix} + \begin{bmatrix} 2x_2 - y_2 \\ 3x_2 \\ x_2 + y_2 \end{bmatrix}$$

$$= \begin{bmatrix} 2(x_1 + x_2) - (y_1 + y_2) \\ 3(x_1 + x_2) \\ (x_1 + x_2) + (y_1 + y_2) \end{bmatrix},$$

which is again of the form $\begin{bmatrix} 2x - y \\ 3x \\ x + y \end{bmatrix}$, so closed under addition.

(b) **Closure under scalar multiplication**:

Let $c \in \mathbb{R}$,

$$c \cdot \mathbf{u} = c \cdot \begin{bmatrix} 2x - y \\ 3x \\ x + y \end{bmatrix} = \begin{bmatrix} 2(cx) - (cy) \\ 3(cx) \\ (cx) + (cy) \end{bmatrix},$$

which again has the same form, hence closed under scalar multiplication.

(c) **Zero vector in U**

We solve:

$$\begin{bmatrix} 2x - y \\ 3x \\ x + y \end{bmatrix} = \begin{bmatrix} 0 \\ 0 \\ 0 \end{bmatrix}.$$

This gives:

$$3x = 0 \Rightarrow x = 0,$$

$$2x - y = 0 \Rightarrow -y = 0,$$

$$x + y = 0 \Rightarrow y = 0.$$

Thus $x = y = 0 \Rightarrow \mathbf{0} \in U$.

All three subspace properties are satisfied. Hence:

$$\boxed{U \text{ is a subspace of } \mathbb{R}^3}$$

9.3 Null Space

The **null space** (also called the *kernel*) of a matrix $A \in \mathbb{R}^{m \times n}$ is the set of all input vectors $\mathbf{x} \in \mathbb{R}^n$ that are mapped to the zero vector in $\mathbb{R}^m$ under the linear transformation defined by A. In other words, it consists of all solutions to the homogeneous system:

$$A\mathbf{x} = \mathbf{0}.$$

Definition

$$\text{Null}(A) = \left\{ \mathbf{x} \in \mathbb{R}^n \mid A\mathbf{x} = \mathbf{0} \right\}.$$

This set is called the ***null space*** (or kernel) of A, and it consists of all vectors that are mapped to the zero vector under the linear transformation defined by A.

The null space represents the set of directions in the input space $\mathbb{R}^n$ that are completely *flattened* (i.e., collapsed to zero) by the transformation. These are the vectors that lie entirely in the kernel of the transformation and are "invisible" in the output space. If $A\mathbf{x} = \mathbf{0}$, then $\mathbf{x}$ contributes nothing to the range of the transformation.

Null Space as a Subspace: The null space $\text{Null}(A)$ is always a subspace of the domain $\mathbb{R}^n$, and it satisfies the three criteria for subspaces:

- **Contains the zero vector**: Since $A\mathbf{0} = \mathbf{0}$, the null space always includes the zero vector.

- **Closed under addition**: If $\mathbf{u}, \mathbf{v} \in \text{Null}(A)$, then $A(\mathbf{u} + \mathbf{v}) = A\mathbf{u} + A\mathbf{v} = \mathbf{0} + \mathbf{0} = \mathbf{0}$, so $\mathbf{u} + \mathbf{v} \in \text{Null}(A)$.
- **Closed under scalar multiplication**: If $\mathbf{v} \in \text{Null}(A)$ and $c \in \mathbb{R}$, then $A(c\mathbf{v}) = cA\mathbf{v} = c\mathbf{0} = \mathbf{0}$, so $c\mathbf{v} \in \text{Null}(A)$.

Solving the homogeneous equation $A\mathbf{x} = \mathbf{0}$ reveals the linear dependencies among the columns of A. If the null space contains only the zero vector, then the columns of A are linearly independent. If it contains non-zero vectors, then the columns are linearly dependent.

> **Definition**
>
> **Dimension and Nullity**: The **nullity** of a matrix A, denoted nullity(A), is the dimension of its null space. It indicates how many degrees of freedom exist in the homogeneous solution.

This leads to the **Rank-Nullity Theorem**, which provides a fundamental relationship between the number of columns in A, the rank, and the nullity.

> **Theorem: Rank-Nullity Theorem**
>
> $$\text{rank}(A) + \text{nullity}(A) = n$$
>
> where n is the number of columns of A, and rank(A) is the number of pivot columns in its row-reduced echelon form.

This theorem tells us that the total number of variables (columns) is partitioned into:

- Pivot variables (contributing to rank).
- Free variables (contributing to nullity).

We can summarize the above topics as follows:

- The null space of a matrix is the set of input vectors mapped to zero.
- It is always a subspace of $\mathbb{R}^n$.
- The dimension of the null space is the nullity, related to rank via the Rank-Nullity Theorem.
- It provides insight into the solution structure of linear systems and the linear dependence of matrix columns.

Example 9.7 Let:

$$A = \begin{bmatrix} 3 & 6 & 6 & 3 & 9 \\ 6 & 12 & 13 & 0 & 3 \end{bmatrix}$$

We want to find all $\mathbf{x} \in \mathbb{R}^5$ such that:

$$Ax = 0 \Rightarrow \begin{bmatrix} 3 & 6 & 6 & 3 & 9 \\ 6 & 12 & 13 & 0 & 3 \end{bmatrix} \begin{bmatrix} x_1 \\ x_2 \\ x_3 \\ x_4 \\ x_5 \end{bmatrix} = \begin{bmatrix} 0 \\ 0 \end{bmatrix}$$

This gives the system:

$$3x_1 + 6x_2 + 6x_3 + 3x_4 + 9x_5 = 0$$
$$6x_1 + 12x_2 + 13x_3 + 0x_4 + 3x_5 = 0$$

Step 1- Form the augmented matrix

$$\begin{bmatrix} 3 & 6 & 6 & 3 & 9 & | & 0 \\ 6 & 12 & 13 & 0 & 3 & | & 0 \end{bmatrix}$$

Step 2- RREF:

- $R_1 \leftarrow R_1/3$:

$$\begin{bmatrix} 1 & 2 & 2 & 1 & 3 & | & 0 \\ 6 & 12 & 13 & 0 & 3 & | & 0 \end{bmatrix}$$

- $R_2 \leftarrow R_2 - 6R_1$:

$$\begin{bmatrix} 1 & 2 & 2 & 1 & 3 & | & 0 \\ 0 & 0 & 1 & -6 & -15 & | & 0 \end{bmatrix}$$

- $R_1 \leftarrow R_1 - 2R_2$

$$\begin{bmatrix} 1 & 2 & 0 & 13 & 33 & | & 0 \\ 0 & 0 & 1 & -6 & -15 & | & 0 \end{bmatrix}$$

Step 3- Solve the system

$$x_2 = t, \quad x_4 = s, \quad x_5 = u,$$
$$x_1 = -2t - 13s - 33u,$$
$$x_3 = 6s + 15u.$$

General solution:

$$\mathbf{x} = \begin{bmatrix} x_1 \\ x_2 \\ x_3 \\ x_4 \\ x_5 \end{bmatrix} = t \begin{bmatrix} -2 \\ 1 \\ 0 \\ 0 \\ 0 \end{bmatrix} + s \begin{bmatrix} -13 \\ 0 \\ 6 \\ 1 \\ 0 \end{bmatrix} + u \begin{bmatrix} -33 \\ 0 \\ 15 \\ 0 \\ 1 \end{bmatrix}$$

The basis for the null space is:

$$\text{Null}(A) = \text{Span} \left\{ \begin{bmatrix} -2 \\ 1 \\ 0 \\ 0 \\ 0 \end{bmatrix}, \begin{bmatrix} -13 \\ 0 \\ 6 \\ 1 \\ 0 \end{bmatrix}, \begin{bmatrix} -33 \\ 0 \\ 15 \\ 0 \\ 1 \end{bmatrix} \right\}$$

Dimension: 3   (nullity = 3)

Since the null space is a span of vectors in $\mathbb{R}^5$, and it satisfies closure under addition and scalar multiplication, we confirm:

$$\boxed{\text{The null space of } A \text{ is a subspace of } \mathbb{R}^5.}$$

Example 9.8

We want to check null space membership for:

$$A = \begin{bmatrix} 2 & 5 & 1 \\ -1 & -7 & -5 \\ 3 & 4 & -2 \end{bmatrix}, \quad \mathbf{u} = \begin{bmatrix} 1 \\ 0 \\ 4 \end{bmatrix}, \quad \mathbf{v} = \begin{bmatrix} 2 \\ -1 \\ 1 \end{bmatrix}.$$

Compute:

$$A\mathbf{u} = \begin{bmatrix} 2(1) + 5(0) + 1(4) \\ -1(1) - 7(0) - 5(4) \\ 3(1) + 4(0) - 2(4) \end{bmatrix} = \begin{bmatrix} 6 \\ -21 \\ -5 \end{bmatrix} \neq \mathbf{0}.$$

$$A\mathbf{v} = \begin{bmatrix} 2(2) + 5(-1) + 1(1) \\ -1(2) - 7(-1) - 5(1) \\ 3(2) + 4(-1) - 2(1) \end{bmatrix} = \begin{bmatrix} 4 - 5 + 1 \\ -2 + 7 - 5 \\ 6 - 4 - 2 \end{bmatrix} = \begin{bmatrix} 0 \\ 0 \\ 0 \end{bmatrix}.$$

$$\mathbf{v} \in \text{Null}(A)$$

but

$$\mathbf{u} \notin \text{Null}(A)$$

Example 9.9

For the following matrix, we want to compute the null space via RREF.

$$A = \begin{bmatrix} 1 & 2 & 1 & -1 \\ 2 & 4 & 0 & -8 \\ 0 & 0 & 2 & 6 \end{bmatrix}.$$

Perform row reduction: 1. $R_2 \to R_2 - 2R_1$:

$$\begin{bmatrix} 1 & 2 & 1 & -1 \\ 0 & 0 & -2 & -6 \\ 0 & 0 & 2 & 6 \end{bmatrix}.$$

2. $R_3 \to R_3 + R_2$:

$$\begin{bmatrix} 1 & 2 & 1 & -1 \\ 0 & 0 & -2 & -6 \\ 0 & 0 & 0 & 0 \end{bmatrix}.$$

3. Scale $R_2 \to -\frac{1}{2}R_2$:

$$\begin{bmatrix} 1 & 2 & 1 & -1 \\ 0 & 0 & 1 & 3 \\ 0 & 0 & 0 & 0 \end{bmatrix}.$$

4. $R_1 \to R_1 - R_2$:

$$R = \begin{bmatrix} 1 & 2 & 0 & -4 \\ 0 & 0 & 1 & 3 \\ 0 & 0 & 0 & 0 \end{bmatrix}.$$

- Pivots: Columns 1, 3 (x_1, x_3).
- Free variables: $x_2 = s$, $x_4 = t$.
- Solve:

$$x_1 + 2x_2 - 4x_4 = 0 \quad \Rightarrow \quad x_1 = -2s + 4t,$$

$$x_3 + 3x_4 = 0 \quad \Rightarrow \quad x_3 = -3t.$$

$$\mathbf{x} = \begin{bmatrix} x_1 \\ x_2 \\ x_3 \\ x_4 \end{bmatrix} = s \begin{bmatrix} -2 \\ 1 \\ 0 \\ 0 \end{bmatrix} + t \begin{bmatrix} 4 \\ 0 \\ -3 \\ 1 \end{bmatrix}.$$

$$\text{Null}(A) = \text{Span} \left\{ \begin{bmatrix} -2 \\ 1 \\ 0 \\ 0 \end{bmatrix}, \begin{bmatrix} 4 \\ 0 \\ -3 \\ 1 \end{bmatrix} \right\}.$$

The null space has dimension 2 (two free variables).

The null space represents directions in $\mathbb{R}^n$ that are "squashed" to zero by A. If $\text{Null}(A) = \{\mathbf{0}\}$, all columns of A are linearly independent, and no non-trivial input is lost.

Python Example: Null Space

```python
import numpy as np
from scipy.linalg import null_space

A = np.array([[1, 2, 1, -1], [2, 4, 0, -8], [0,
    0, 2, 6]])
ns = null_space(A)
print("Null space basis:\n", ns)
```

Output:

```
Null space basis:
 [[ 0.15  0.91]
 [ 0.47 -0.41]
 [-0.82 -0.07]
 [ 0.27  0.02]]
```

9.3.1 Column and Row Spaces

The **column space** and **row space** of a matrix provide two complementary perspectives on the linear transformation associated with a matrix $A \in \mathbb{R}^{m \times n}$. They reveal, respectively, the set of possible outputs and the system of constraints embedded within the transformation.

The column space of a matrix A is the subspace spanned by its column vectors. It consists of all possible linear combinations of these columns.

Example 9.10 Consider a matrix $A = \begin{bmatrix} 1 & 2 & 0 \\ 2 & 4 & 0 \\ 0 & 0 & 1 \end{bmatrix}$ with 3 columns. The column space is spanned by the first two columns (which are linearly dependent) and the third column, giving a dimension of 2. The null space consists of vectors x such that $Ax = 0$, which yields $x_1 + 2x_2 = 0$ and $x_3 = 0$, leading to a dimension of 1 (one free variable). Thus, the dimension of the column space (2) plus the dimension of the null space (1) equals the number of columns (3), satisfying the Counting Theorem.

The Counting Theorem states that for any $m \times n$ matrix A, the dimension of the column space plus the dimension of the null space equals n (the number of columns), i.e., rank(A) + nullity(A) = n, which is precisely the Rank-Nullity Theorem already stated formally in Sect. 9.3 of this chapter.

- **Column Space**

  The **column space** of a matrix A, denoted $\mathrm{Col}(A)$, is the subspace of $\mathbb{R}^m$ spanned by the columns of A. Formally,

$$\mathrm{Col}(A) = \left\{ A\mathbf{x} \in \mathbb{R}^m \mid \mathbf{x} \in \mathbb{R}^n \right\} = \mathrm{Span}(\text{columns of } A).$$

  This space contains all possible output vectors $A\mathbf{x}$ that can result from applying the linear transformation A to any vector $\mathbf{x} \in \mathbb{R}^n$. Geometrically, the column space represents the image or range of the transformation A, and thus lies in $\mathbb{R}^m$, the codomain.

  - If $\mathrm{Col}(A) = \mathbb{R}^m$, the transformation is **surjective** (onto).
  - The dimension of the column space is equal to the **rank** of A, i.e., the number of linearly independent columns.

  Understanding the column space helps determine whether a system $A\mathbf{x} = \mathbf{b}$ has a solution: the system is consistent if and only if $\mathbf{b} \in \mathrm{Col}(A)$.

- **Row Space**

  The **row space** of a matrix $A \in \mathbb{R}^{m \times n}$, denoted $\mathrm{Row}(A)$, is the subspace of $\mathbb{R}^n$ spanned by the rows of A, viewed as vectors in $\mathbb{R}^n$. Formally,

$$\mathrm{Row}(A) = \mathrm{Col}(A^{\mathrm{T}}) = \mathrm{Span}(\text{rows of } A).$$

  Each row in the matrix A represents a linear equation (or constraint) in the variables $x_1, x_2, \ldots, x_n$. The row space therefore captures the set of all linear combinations of these constraints.

  - The row space is a subspace of the domain $\mathbb{R}^n$.
  - Its dimension equals the rank of the matrix, just like the column space.
  - The row space is preserved under elementary row operations, which is why Gaussian elimination helps identify a basis for it.

  The row space provides insight into the relationships between the variables in a linear system. It helps identify the independent constraints and is used when solving systems via row reduction.

  In summary:

- **Column Space**: Output space of the transformation $A\mathbf{x}$; subspace of $\mathbb{R}^m$; span of columns of A.
- **Row Space**: Constraint space; subspace of $\mathbb{R}^n$; span of rows of A; same dimension as the column space (i.e., **rank**).

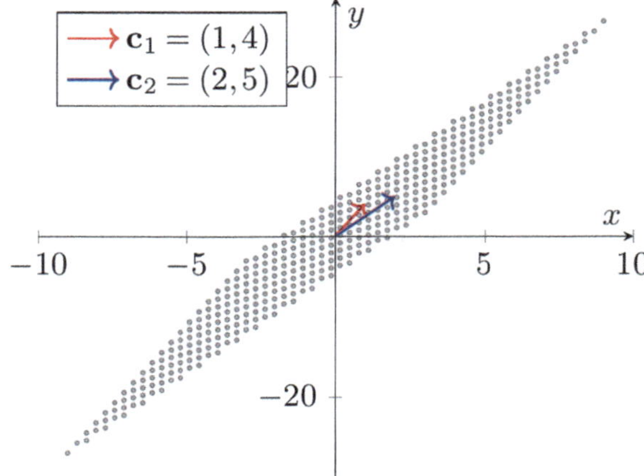

Fig. 9.2 Column space as the span of two column vectors in $\mathbb{R}^2$

- **Rank**: Dimension of both the row space and the column space; number of linearly independent rows (or columns).

Example 9.11 For:

$$A = \begin{bmatrix} 1 & 2 & 3 \\ 4 & 5 & 6 \end{bmatrix}.$$

RREF:

$$\begin{bmatrix} 1 & 0 & -1 \\ 0 & 1 & 2 \end{bmatrix}.$$

– Column space: Pivot columns (1, 2):

$$\mathrm{Col}(A) = \mathrm{Span}\left\{ \begin{bmatrix} 1 \\ 4 \end{bmatrix}, \begin{bmatrix} 2 \\ 5 \end{bmatrix} \right\}.$$

– Row space: Non-zero rows of RREF:

$$\mathrm{Row}(A) = \mathrm{Span}\left\{ \begin{bmatrix} 1 & 0 & -1 \end{bmatrix}, \begin{bmatrix} 0 & 1 & 2 \end{bmatrix} \right\}.$$

We can see its visualization in Fig. 9.2.

We now state the Rank Theorem, which establishes that for any matrix, the dimension of its column space is equal to the dimension of its row space. This common dimension is called the *rank* of the matrix. It shows that the number of linearly independent columns equals the number of linearly independent rows, providing a fundamental property of matrices in linear algebra.

Theorem: Rank Theorem
Dimension of column space = dimension of row space. This is the rank.

Example 9.12 Consider the matrix

$$A = \begin{bmatrix} 1 & 2 & 3 \\ 4 & 5 & 6 \end{bmatrix}.$$

The column space of A is spanned by its linearly independent columns. Here, the first two columns are linearly independent, while the third column can be written as a combination of the first two:

$$\begin{bmatrix} 3 \\ 6 \end{bmatrix} = -1 \cdot \begin{bmatrix} 1 \\ 4 \end{bmatrix} + 2 \cdot \begin{bmatrix} 2 \\ 5 \end{bmatrix}$$

Thus, $\dim(\mathrm{Col}(A)) = 2$.
The row space is spanned by its linearly independent rows. Here, the two rows are also linearly independent, so $\dim(\mathrm{Row}(A)) = 2$.
This confirms the **Rank Theorem**:

$$\boxed{\dim(\mathrm{Col}(A)) = \dim(\mathrm{Row}(A)) = 2}$$

9.4 Basis and Dimension

A **basis** is a set of vectors that is linearly independent and spans a subspace $V \subseteq \mathbb{R}^n$. It's like the minimal set of "directions" needed to reach any point in V.

Definition
A set $\{\mathbf{b}_1, \ldots, \mathbf{b}_k\} \subseteq V$ is a basis for V if:

1. **Linearly independent**: $c_1 \mathbf{b}_1 + \cdots + c_k \mathbf{b}_k = \mathbf{0} \implies c_1 = \cdots = c_k = 0$.
2. **Spanning**: Every $\mathbf{v} \in V$ can be written as $\mathbf{v} = c_1 \mathbf{b}_1 + \cdots + c_k \mathbf{b}_k$.

Python Example: Basis Verification

```python
import numpy as np
basis = np.array([[-1, 1, 0], [-1, 0, 1]]).T
rank = np.linalg.matrix_rank(basis)
```

```
4 print("Rank (dimension):", rank)
```

Output:

```
Rank (dimension): 2
```

Example 9.13

For:

$$V = \left\{ \begin{bmatrix} x_1 \\ x_2 \\ x_3 \end{bmatrix} \in \mathbb{R}^3 : x_1 + x_2 + x_3 = 0 \right\}.$$

Solve:

$$x_1 = -x_2 - x_3, \quad \mathbf{x} = \begin{bmatrix} -x_2 - x_3 \\ x_2 \\ x_3 \end{bmatrix} = x_2 \begin{bmatrix} -1 \\ 1 \\ 0 \end{bmatrix} + x_3 \begin{bmatrix} -1 \\ 0 \\ 1 \end{bmatrix}.$$

Basis:

$$\left\{ \begin{bmatrix} -1 \\ 1 \\ 0 \end{bmatrix}, \begin{bmatrix} -1 \\ 0 \\ 1 \end{bmatrix} \right\}.$$

– Linear Independence:

$$a_1 \begin{bmatrix} -1 \\ 1 \\ 0 \end{bmatrix} + a_2 \begin{bmatrix} -1 \\ 0 \\ 1 \end{bmatrix} = \begin{bmatrix} -a_1 - a_2 \\ a_1 \\ a_2 \end{bmatrix} = \begin{bmatrix} 0 \\ 0 \\ 0 \end{bmatrix} \implies a_1 = a_2 = 0.$$

– Spanning: The vectors span V, as shown above.
– Dimension: $\dim(V) = 2$.

The basis vectors define the directions of a plane in $\mathbb{R}^3$. Any vector in V is a unique combination of these directions.

> **Theorem: Dimension Theorem**
> All bases for a vector space have the same number of vectors.

Example 9.14 Consider the vector space $\mathbb{R}^2$. One basis is $\{(1, 0), (0, 1)\}$, which has 2 vectors. Another basis could be $\{(1, 1), (1, -1)\}$, also with 2 vectors. According to the Dimension Theorem, both bases must have the same number of vectors, confirming the dimension of $\mathbb{R}^2$ is 2.

Nullity of a Matrix

The **nullity** of a matrix A is the dimension of its null space:

$$\text{nullity}(A) = \dim(\text{Null}(A)).$$

The null space is the set of all solutions to the homogeneous system:

$$A\mathbf{x} = \mathbf{0}.$$

The nullity tells us how many *free variables* (parameters) exist in the general solution of the system.

Example 9.15 Let:

$$A = \begin{bmatrix} 1 & 2 & 3 & 4 \\ 2 & 4 & 6 & 8 \\ 3 & 6 & 9 & 12 \end{bmatrix}.$$

We want to find:

- The null space of A
- A basis for the null space
- The nullity

Step 1- Form the augmented matrix

$$\begin{bmatrix} 1 & 2 & 3 & 4 & | & 0 \\ 2 & 4 & 6 & 8 & | & 0 \\ 3 & 6 & 9 & 12 & | & 0 \end{bmatrix}.$$

Step 2- Row reduction to RREF

- $R_2 \leftarrow R_2 - 2R_1$
- $R_3 \leftarrow R_3 - 3R_1$

$$\begin{bmatrix} 1 & 2 & 3 & 4 & | & 0 \\ 0 & 0 & 0 & 0 & | & 0 \\ 0 & 0 & 0 & 0 & | & 0 \end{bmatrix}.$$

Step 3- Back-substitution

From the first row:

$$x_1 + 2x_2 + 3x_3 + 4x_4 = 0 \Rightarrow x_1 = -2x_2 - 3x_3 - 4x_4.$$

Let:

$$x_2 = s, \quad x_3 = t, \quad x_4 = u \quad \text{(free variables)}.$$

Then:

$$\mathbf{x} = \begin{bmatrix} x_1 \\ x_2 \\ x_3 \\ x_4 \end{bmatrix} = \begin{bmatrix} -2s - 3t - 4u \\ s \\ t \\ u \end{bmatrix} = s \begin{bmatrix} -2 \\ 1 \\ 0 \\ 0 \end{bmatrix} + t \begin{bmatrix} -3 \\ 0 \\ 1 \\ 0 \end{bmatrix} + u \begin{bmatrix} -4 \\ 0 \\ 0 \\ 1 \end{bmatrix}.$$

The null space is the span of the following vectors:

$$\left\{ \begin{bmatrix} -2 \\ 1 \\ 0 \\ 0 \end{bmatrix}, \begin{bmatrix} -3 \\ 0 \\ 1 \\ 0 \end{bmatrix}, \begin{bmatrix} -4 \\ 0 \\ 0 \\ 1 \end{bmatrix} \right\}.$$

So:

$$\text{nullity}(A) = \dim(\text{Null}(A)) = 3.$$

Basis for the null space:

$$\left\{ \begin{bmatrix} -2 \\ 1 \\ 0 \\ 0 \end{bmatrix}, \begin{bmatrix} -3 \\ 0 \\ 1 \\ 0 \end{bmatrix}, \begin{bmatrix} -4 \\ 0 \\ 0 \\ 1 \end{bmatrix} \right\}.$$

Verification by Rank-Nullity Theorem

Matrix A is 3×4, so $n = 4$. From RREF, $\text{rank}(A) = 1$. By the rank-nullity theorem:

$$\boxed{\text{nullity}(A) = n - \text{rank}(A) = 4 - 1 = 3}$$

9.5   Coordinate Systems

A **coordinate system** assigns a unique set of coordinates to each vector in a subspace relative to a basis $\mathcal{B} = \{\mathbf{b}_1, \ldots, \mathbf{b}_k\}$:

$$\mathbf{v} = c_1 \mathbf{b}_1 + \cdots + c_k \mathbf{b}_k, \quad [\mathbf{v}]_{\mathcal{B}} = \begin{bmatrix} c_1 \\ \vdots \\ c_k \end{bmatrix}.$$

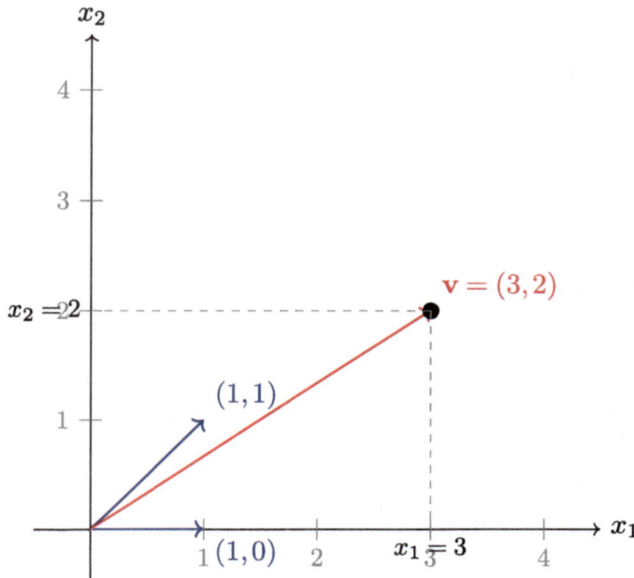

Fig. 9.3 Vector $\mathbf{v} = (3, 2)$ expressed in basis $\mathcal{B}$, with projections onto the x_1 and x_2 axes shown as dashed gray lines

Example 9.16 For $\mathcal{B} = \left\{ \begin{bmatrix} 1 \\ 0 \end{bmatrix}, \begin{bmatrix} 1 \\ 1 \end{bmatrix} \right\}$, find $[\mathbf{v}]_{\mathcal{B}}$ for $\mathbf{v} = \begin{bmatrix} 3 \\ 2 \end{bmatrix}$:

$$c_1 \begin{bmatrix} 1 \\ 0 \end{bmatrix} + c_2 \begin{bmatrix} 1 \\ 1 \end{bmatrix} = \begin{bmatrix} c_1 + c_2 \\ c_2 \end{bmatrix} = \begin{bmatrix} 3 \\ 2 \end{bmatrix}.$$

Solve:

$$c_2 = 2, \quad c_1 + 2 = 3 \implies c_1 = 1.$$

$$[\mathbf{v}]_{\mathcal{B}} = \begin{bmatrix} 1 \\ 2 \end{bmatrix}.$$

We can verify as shown in Fig. 9.3:

$$1 \cdot \begin{bmatrix} 1 \\ 0 \end{bmatrix} + 2 \cdot \begin{bmatrix} 1 \\ 1 \end{bmatrix} = \begin{bmatrix} 1 + 2 \\ 0 + 2 \end{bmatrix} = \begin{bmatrix} 3 \\ 2 \end{bmatrix}.$$

Example 9.17 For:

$$\mathcal{B} = \left\{ \begin{bmatrix} 1 \\ 1 \\ 1 \end{bmatrix}, \begin{bmatrix} 1 \\ -1 \\ 1 \end{bmatrix}, \begin{bmatrix} 1 \\ 2 \\ 2 \end{bmatrix} \right\}, \quad [\mathbf{v}]_{\mathcal{B}} = \begin{bmatrix} 3 \\ 6 \\ -2 \end{bmatrix}.$$

Compute:

$$\mathbf{v} = 3 \begin{bmatrix} 1 \\ 1 \\ 1 \end{bmatrix} + 6 \begin{bmatrix} 1 \\ -1 \\ 1 \end{bmatrix} - 2 \begin{bmatrix} 1 \\ 2 \\ 2 \end{bmatrix} = \begin{bmatrix} 3+6-2 \\ 3-6-4 \\ 3+6-4 \end{bmatrix} = \begin{bmatrix} 7 \\ -7 \\ 5 \end{bmatrix}.$$

A basis defines a new "lens" for viewing vectors, aligning coordinates with problem-specific directions (e.g., axes of maximum variance).

Example 9.18 Let the basis $\mathcal{B}$ for $\mathbb{R}^3$ be:

$$\mathcal{B} = \{\mathbf{b}_1 = (1, 1, 0), \ \mathbf{b}_2 = (0, 1, 1), \ \mathbf{b}_3 = (1, 0, 1)\}.$$

We are given the vector $\mathbf{v} = (0, 1, 3) \in \mathbb{R}^3$. We want to express $\mathbf{v}$ as a linear combination of the basis vectors:

$$\mathbf{v} = a_1 \mathbf{b}_1 + a_2 \mathbf{b}_2 + a_3 \mathbf{b}_3.$$

So,
$$(0, 1, 3) = a_1(1, 1, 0) + a_2(0, 1, 1) + a_3(1, 0, 1).$$

Compute the linear combination:

$$a_1(1, 1, 0) = (a_1, a_1, 0), \quad a_2(0, 1, 1) = (0, a_2, a_2), \quad a_3(1, 0, 1) = (a_3, 0, a_3).$$

Add the vectors:
$$\mathbf{v} = (a_1 + a_3, a_1 + a_2, a_2 + a_3).$$

Equating with $\mathbf{v} = (0, 1, 3)$, we get:

$$\begin{cases} a_1 + a_3 = 0, \\ a_1 + a_2 = 1, \\ a_2 + a_3 = 3, \end{cases}$$

Solving the system:

From the first equation: $a_3 = -a_1$

Substitute into the second: $a_1 + a_2 = 1 \Rightarrow a_2 = 1 - a_1$

Substitute both into the third: $(1 - a_1) + (-a_1) = 3 \Rightarrow 1 - 2a_1 = 3$

$\Rightarrow -2a_1 = 2 \Rightarrow a_1 = -1$

$\Rightarrow a_2 = 1 - (-1) = 2$

$\Rightarrow a_3 = -(-1) = 1$

Coordinate Vector of v in Basis $\mathcal{B}$

$$[\mathbf{v}]_{\mathcal{B}} = \begin{bmatrix} a_1 \\ a_2 \\ a_3 \end{bmatrix} = \begin{bmatrix} -1 \\ 2 \\ 1 \end{bmatrix}.$$

Example 9.19 Consider the basis:

$$B = \left\{ \mathbf{b}_1 = \begin{bmatrix} 1 \\ 1 \\ 0 \end{bmatrix}, \quad \mathbf{b}_2 = \begin{bmatrix} 0 \\ 1 \\ 1 \end{bmatrix}, \quad \mathbf{b}_3 = \begin{bmatrix} 1 \\ 0 \\ 1 \end{bmatrix} \right\}.$$

Let the vector:

$$\mathbf{v} = \begin{bmatrix} 0 \\ 1 \\ 3 \end{bmatrix}.$$

We want to find the coordinate vector of $\mathbf{v}$ relative to basis B, i.e., find scalars $a_1, a_2, a_3 \in \mathbb{R}$ such that:

$$\mathbf{v} = a_1\mathbf{b}_1 + a_2\mathbf{b}_2 + a_3\mathbf{b}_3.$$

Expanding the linear combination:

$$a_1 \begin{bmatrix} 1 \\ 1 \\ 0 \end{bmatrix} + a_2 \begin{bmatrix} 0 \\ 1 \\ 1 \end{bmatrix} + a_3 \begin{bmatrix} 1 \\ 0 \\ 1 \end{bmatrix} = \begin{bmatrix} a_1 \\ a_1 \\ 0 \end{bmatrix} + \begin{bmatrix} 0 \\ a_2 \\ a_2 \end{bmatrix} + \begin{bmatrix} a_3 \\ 0 \\ a_3 \end{bmatrix}$$

$$= \begin{bmatrix} a_1 + a_3 \\ a_1 + a_2 \\ a_2 + a_3 \end{bmatrix}.$$

We equate this to the vector $\mathbf{v}$:

$$\begin{bmatrix} a_1 + a_3 \\ a_1 + a_2 \\ a_2 + a_3 \end{bmatrix} = \begin{bmatrix} 0 \\ 1 \\ 3 \end{bmatrix}.$$

Solving the system:

$$\begin{cases} a_1 + a_3 = 0, \\ a_1 + a_2 = 1, \\ a_2 + a_3 = 3. \end{cases}$$

From the first equation: $a_3 = -a_1$ Substitute into the third equation:

$$a_2 + (-a_1) = 3 \Rightarrow a_2 = 3 + a_1.$$

Now substitute into the second equation:

$$a_1 + (3 + a_1) = 1 \Rightarrow 2a_1 + 3 = 1 \Rightarrow a_1 = -1.$$

Then:

$$a_2 = 3 + (-1) = 2, \quad a_3 = -(-1) = 1.$$

Therefore, the coordinate vector of $\mathbf{v}$ with respect to basis B is:

$$[\mathbf{v}]_B = \begin{bmatrix} -1 \\ 2 \\ 1 \end{bmatrix}.$$

9.6 Vector Spaces and Subspaces in ML

Subspaces, bases, and orthogonality are pivotal in ML, enabling efficient data representation and optimization.

1. **PCA**: Projects data onto a subspace spanned by eigenvectors of the covariance matrix (principal components), maximizing variance. The column space of the data matrix is approximated by a low-rank subspace, as computed in the PCA example below.
2. **SVMs**: Define decision boundaries as affine hyperplanes in feature space, maximizing the margin between classes.
3. **Feature Engineering**: A basis for the column space identifies independent features, reducing redundancy via RREF.
4. **Regularization**: In LR, if $\text{Null}(X) \neq \{\mathbf{0}\}$, regularization constrains solutions to a subspace, ensuring uniqueness.
5. **Sparse Coding**: Learns a sparse basis for efficient data representation.
6. **Autoencoders**: Map data to nonlinear latent subspaces, generalizing linear coordinate systems.
7. **Transformers**: Use attention mechanisms to project data into new coordinate systems, focusing on relevant features.
8. **Optimization**: Coordinate systems in weight spaces guide gradient-based optimization.

Chapter Summary

This chapter established vector spaces and subspaces as the foundational structures in linear algebra, providing the framework for representing and manipulating data in ML contexts like PCA and SVMs. It opened with the definition of a vector space in

Sect. 9.1, a set V closed under addition and scalar multiplication, satisfying axioms such as commutativity, associativity, and distributivity, with examples including $\mathbb{R}^3$ (standard operations yielding triples like $(1, 2, 3) + (4, 5, 6) = (5, 7, 9)$) and polynomials P_2 (addition and scaling preserving degree ≤ 2).

Subspaces in Sect. 9.2 were introduced as subsets inheriting vector space properties—containing zero, closed under addition and scaling—with examples verifying sets like $U = \{[2x - y, 3x, x + y]^T \mid x, y \in \mathbb{R}\}$ via span representation and axiom checks, contrasting non-subspaces like half-planes failing scalar closure. The null space in Sect. 9.3 was defined as $\text{Null}(A) = \{\mathbf{x} \mid A\mathbf{x} = \mathbf{0}\}$, always a subspace capturing flattened directions, with nullity as its dimension and the Rank-Nullity Theorem $\text{rank}(A) + \text{nullity}(A) = n$ partitioning columns into pivots and free variables, exemplified by row-reduced matrices yielding bases like $\text{Span}\{[-2, 1, 0, 0]^T, [-3, 0, 1, 0]^T, [-4, 0, 0, 1]^T\}$ (dimension 3).

Column and row spaces in Sect. 9.3.1 were examined: $\text{Col}(A)$ as the span of columns (output subspace of $\mathbb{R}^m$), $\text{Row}(A)$ as the span of rows (constraint subspace of $\mathbb{R}^n$), both of dimension $\text{rank}(A)$, with RREF identifying pivot-based bases. Coordinate systems in Sect. 9.5 used bases $\mathcal{B}$ for unique representations $[\mathbf{v}]_\mathcal{B} = [c_1, \ldots, c_k]^T$ where $\mathbf{v} = \sum c_i \mathbf{b}_i$, solved via systems like $a_1 + a_3 = 0$, yielding $[-1, 2, 1]^T$ for $\mathbf{v} = [0, 1, 3]^T$ in a 3D basis.

ML applications highlighted subspaces in PCA (projecting onto variance-maximizing column spaces of covariance), SVMs (hyperplane margins), and feature engineering (independent bases via rank), with Python code computing PCA bases and null spaces via `null_space`.

Takeaways

Key elements from the chapter are distilled into:

- Vector spaces as sets satisfying addition/scaling axioms, exemplified by $\mathbb{R}^n$ and polynomials, enabling structured data manipulation.
- Subspaces as zero-inclusive, closed subsets, verified through spans (e.g., planes like $x_1 + x_2 + x_3 = 0$) and contrasts with non-subspaces (e.g., half-planes).
- Null space $\text{Null}(A)$ as solution set to $A\mathbf{x} = \mathbf{0}$, a subspace with nullity $\dim(\text{Null}(A))$, linked by Rank-Nullity to rank for solution dimensionality.
- Column space $\text{Col}(A)$ (output span) and row space $\text{Row}(A)$ (constraint span), both rank-dimensional, with RREF yielding bases for consistency checks in systems.
- Coordinate systems transforming vectors to basis coefficients via linear solves, providing alternative "lenses" like principal axes in data analysis.
- Applications in ML: PCA for subspace projections, SVMs for boundary subspaces, and null spaces for dependence detection in features.

Through these explorations, readers mastered identifying and analyzing subspaces, equipping them to handle data constraints and bases in ML pipelines.

Exercises

1. Verify if $U = \left\{ \begin{bmatrix} x \\ y \\ z \end{bmatrix} \in \mathbb{R}^3 \mid x + y = 0 \right\}$ is a subspace of $\mathbb{R}^3$.

2. Is $V = \left\{ \begin{bmatrix} x \\ y \\ z \end{bmatrix} \in \mathbb{R}^3 \mid x^2 + y^2 = 1 \right\}$ a subspace? Explain.

3. Find a basis and dimension for $\text{Null}(A)$ where $A = \begin{bmatrix} 1 & 2 & 1 \\ 2 & 4 & 2 \end{bmatrix}$.

4. For matrix A in exercise 3, find a basis for $\text{Col}(A)$ by using the pivot columns identified from the RREF of (A), taken from the original matrix.

5. Find a basis for $\text{Row}(A)$ using the nonzero rows of the RREF of $A = \begin{bmatrix} 1 & 2 & 1 \\ 3 & 6 & 3 \end{bmatrix}$.

6. For $\mathcal{B} = \left\{ \begin{bmatrix} 1 \\ 0 \end{bmatrix}, \begin{bmatrix} 0 \\ 1 \end{bmatrix} \right\}$, find $[\mathbf{v}]_\mathcal{B}$ for $\mathbf{v} = \begin{bmatrix} 5 \\ 7 \end{bmatrix}$.

7. If $\mathcal{B} = \left\{ \begin{bmatrix} 1 \\ 0 \\ 1 \end{bmatrix}, \begin{bmatrix} 0 \\ 1 \\ 0 \end{bmatrix} \right\}$, determine whether the set is linearly independent, and hence whether it forms a basis for its span in $\mathbb{R}^3$.

8. Compute $\text{Null}(A)$ and its dimension for $A = \begin{bmatrix} 1 & 2 & 3 \\ 0 & 1 & 2 \end{bmatrix}$.

9. For $V = \{ \mathbf{x} \in \mathbb{R}^4 \mid x_1 + x_2 - x_4 = 0 \}$, find its dimension and basis.

10. Find a basis and dimension for the subspace of $\mathbb{R}^3$ spanned by $\begin{bmatrix} 1 \\ 2 \\ 3 \end{bmatrix}, \begin{bmatrix} 2 \\ 4 \\ 6 \end{bmatrix}, \begin{bmatrix} 1 \\ 1 \\ 1 \end{bmatrix}$.

11. Verify if $W = \left\{ \begin{bmatrix} x \\ y \\ z \end{bmatrix} \in \mathbb{R}^3 \mid 2x - y + z = 0 \right\}$ is a subspace.

12. Check if $\mathbf{u} = \begin{bmatrix} 1 \\ -1 \\ 2 \end{bmatrix}$ is in $\text{Null}(A)$ for $A = \begin{bmatrix} 1 & 1 & -1 \\ 2 & 0 & 2 \end{bmatrix}$.

13. Find the coordinate vector of $\mathbf{v} = \begin{bmatrix} 2 \\ 3 \end{bmatrix}$ with respect to $\mathcal{B} = \left\{ \begin{bmatrix} 2 \\ 1 \end{bmatrix}, \begin{bmatrix} -1 \\ 1 \end{bmatrix} \right\}$.

14. Verify orthogonality of $\begin{bmatrix} 1 \\ -1 \\ 0 \end{bmatrix}$ and $\begin{bmatrix} 1 \\ 1 \\ 0 \end{bmatrix}$.

15. Compute the nullity of $A = \begin{bmatrix} 1 & 2 & 3 & 4 \\ 2 & 4 & 6 & 8 \end{bmatrix}$ using RREF.

16. Find an orthonormal basis for the subspace spanned by $\begin{bmatrix} 1 \\ 2 \\ 3 \end{bmatrix}, \begin{bmatrix} 0 \\ 1 \\ 1 \end{bmatrix}$.

17. For $A = \begin{bmatrix} 1 & 0 & 1 \\ 0 & 1 & 1 \end{bmatrix}$, find bases for $\text{Col}(A)$ and $\text{Row}(A)$.

18. Write a Python script to compute the null space of $A = \begin{bmatrix} 1 & 2 \\ 2 & 4 \end{bmatrix}$ and verify its dimension.

Chapter 10
Orthogonality

Introduction

Orthogonality is a cornerstone of linear algebra, serving as a guiding principle for understanding and navigating high-dimensional vector spaces. Just as a compass provides direction, orthogonality allows us to measure lengths, angles, and distances, and to identify independent directions in a space. By analyzing how vectors relate to each other through concepts like the dot product and norms, we can uncover the structure of complex systems and simplify computations. This chapter introduces the fundamental ideas of orthogonality, including orthogonal complements, projections, and the decomposition of vectors into orthogonal components, providing a strong foundation for both theoretical understanding and practical implementation.

In addition to theory, this chapter emphasizes computational techniques and visual intuition. Step-by-step derivations are paired with Python (NumPy) examples to help readers connect abstract concepts with concrete calculations. Applications in ML, such as dimensionality reduction, PCA, and regression, are discussed in a dedicated section, illustrating how orthogonality underpins efficient algorithms and data analysis. Throughout, the focus remains on building geometric intuition and computational skills, equipping undergraduate and early graduate students with the tools to leverage orthogonality in both linear algebra and applied ML contexts.

Topics Covered

This chapter is organized as follows:

- **Section** 10.1 **Norm and Distance**: Introduces vector norms and distances in $\mathbb{R}^n$, with geometric and analytical interpretations.
- **Section** 10.2 **Dot Product and Orthogonality**: Defines the dot product, explores its algebraic and geometric significance, and introduces orthogonality and angle computations.

© The Author(s), under exclusive license to Springer Nature Singapore Pte Ltd. 2026   291
Md. Jalil Piran, *Linear Algebra with Applications in Machine Learning*,
https://doi.org/10.1007/978-981-95-5167-5_10

- **Section** 10.3 **Pythagorean Theorem and Geometric Insights**: Extends the Pythagorean theorem to vector spaces, with algebraic and geometric proofs.
- **Section** 10.4 **Cauchy–Schwarz and Triangle Inequalities**: Presents fundamental inequalities that bound vector interactions.
- **Section** 10.5 **Orthogonal Complement**: Discusses the definition and properties of orthogonal complements in vector spaces.
- **Section** 10.6 **Matrix and Subspace Orthogonality**: Relates orthogonality to row space, column space, and null space of a matrix.
- **Section** 10.7 **Dimension of a Subspace and Its Orthogonal Complement**: Establishes dimension formulas and decomposition of vectors.
- **Section** 10.8 **Orthogonal Projection**: Derives projection formulas and applies them to regression and approximation problems.
- **Section** 10.9 **Least Squares Approximation**: Formulates linear regression as an orthogonal projection problem, derives the normal equations for computing best-fit coefficients, and introduces the projection matrix for approximating vectors within a subspace.
- **Section** 10.10 **Best Quadratic Fit**: Introduces quadratic regression using orthogonal projection and least squares.
- **Section** 10.11 **Multivariable Least Squares Approximation**: Extends least squares to multiple independent variables.
- **Section** 10.12 **Orthogonality in ML**: Highlights applications in PCA, regression, clustering, cosine similarity, neural network optimization, and kernel methods.
- **Summary and Exercises**, Recaps key concepts and provides practice problems for reinforcement.

All code accompanying the chapter is publicly available at https://github.com/jalil-piran/Linear-Algebra-with-Applications-in-Machine-Learning/blob/main/Chapter_10_Orthogonality.ipynb.

Common Acronyms and Their Definitions

ML	Machine Learning
PCA	Principal Component Analysis
NLP	Natural Language Processing
SVMs	Support Vector Machines
REF	Row Echelon Form
RREF	Reduced Row Echelon Form
LR	Linear Regression

10.1 Norm and Distance

The **norm** of a vector measures its "length" in Euclidean space, while the **distance** between vectors quantifies their separation, like measuring the gap between two points on a map.

Definition

- The **Euclidean norm** (ℓ_2-norm) of $\mathbf{v} = [v_1, v_2, \ldots, v_n]^{\mathrm{T}} \in \mathbb{R}^n$ is:

$$\|\mathbf{v}\| = \sqrt{v_1^2 + v_2^2 + \cdots + v_n^2}.$$

A vector with $\|\mathbf{v}\| = 1$ is a **unit vector**, representing pure direction.
- The **distance** between $\mathbf{u}, \mathbf{v} \in \mathbb{R}^n$ is:

$$\|\mathbf{v} - \mathbf{u}\| = \sqrt{(v_1 - u_1)^2 + \cdots + (v_n - u_n)^2}.$$

The norm is like the length of a rope from the origin to a point in space. The distance between vectors is the length of the straight line connecting their tips, generalizing the Pythagorean theorem to $\mathbb{R}^n$.

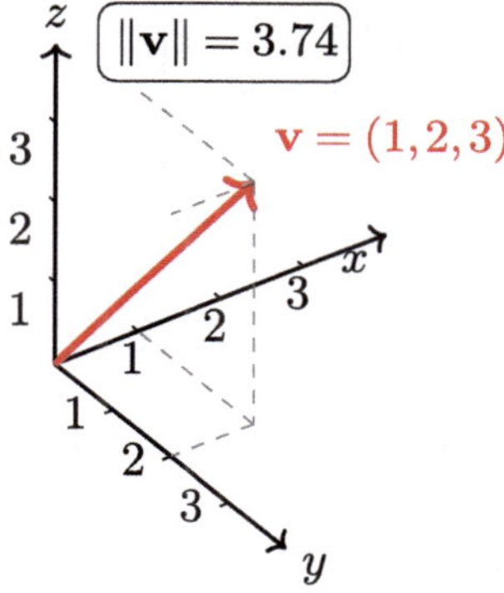

Fig. 10.1 3D visualization of the vector $\mathbf{v} = (1, 2, 3)$ (red arrow) in Cartesian space. Dashed gray lines indicate orthogonal projections onto the xy-, xz-, and yz-planes, illustrating the vector's components. The Euclidean norm $\|\mathbf{v}\| = \sqrt{1^2 + 2^2 + 3^2} = \sqrt{14} \approx 3.74$ is displayed in the upper right. Axes are labeled with ticks for clarity, and the viewpoint (azimuth 135°, inclination 25°) is chosen to avoid occlusion

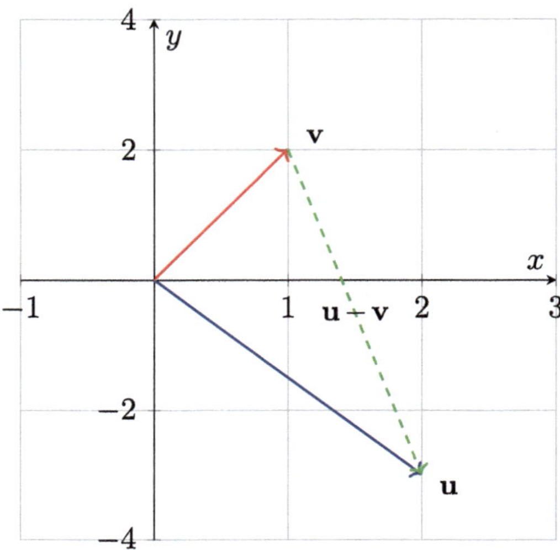

Fig. 10.2 Norm and distance between vectors **u** and **v** in 2D (projecting onto x–y plane)

Example 10.1 For:

$$
\mathbf{v} = \begin{bmatrix} 1 \\ 2 \\ 3 \end{bmatrix}, \quad \mathbf{u} = \begin{bmatrix} 2 \\ -3 \\ 0 \end{bmatrix}.
$$

– Norm of **v**:

$$
\|\mathbf{v}\| = \sqrt{1^2 + 2^2 + 3^2} = \sqrt{1 + 4 + 9} = \sqrt{14} \approx 3.74.
$$

– Figure 10.1 visualizes the vector and its norm. Norm of **u**:

$$
\|\mathbf{u}\| = \sqrt{2^2 + (-3)^2 + 0^2} = \sqrt{4 + 9 + 0} = \sqrt{13} \approx 3.606.
$$

– Distance:

$$
\mathbf{v} - \mathbf{u} = \begin{bmatrix} 1 - 2 \\ 2 - (-3) \\ 3 - 0 \end{bmatrix} = \begin{bmatrix} -1 \\ 5 \\ 3 \end{bmatrix},
$$

$$
\|\mathbf{v} - \mathbf{u}\| = \sqrt{(-1)^2 + 5^2 + 3^2} = \sqrt{1 + 25 + 9} = \sqrt{35} \approx 5.916.
$$

– Figure 10.2 shows the norm and distance between vectors v and u.

Python Example: Norm and Distance

```python
import numpy as np
v = np.array([1, 2, 3])
u = np.array([2, -3, 0])
norm_v = np.linalg.norm(v)
norm_u = np.linalg.norm(u)
distance = np.linalg.norm(v - u)
print(f"Norm of v: {norm_v:.2f}")
print(f"Norm of u: {norm_u:.2f}")
print(f"Distance: {distance:.2f}")
```

Output:

```
Norm of v: 3.74
Norm of u: 3.61
Distance: 5.92
```

10.2 Dot Product and Orthogonality

The **dot product** measures how much two vectors align, like checking if two arrows
point in similar directions. **Orthogonality** occurs when vectors are perpendicular,
forming a 90° angle, which implies they are independent in a geometric sense.

Definition

- The **dot product** of $\mathbf{u}, \mathbf{v} \in \mathbb{R}^n$ is:

$$\mathbf{u} \cdot \mathbf{v} = u_1 v_1 + u_2 v_2 + \cdots + u_n v_n = \mathbf{u}^{\mathsf{T}} \mathbf{v}.$$

  Geometrically: $\mathbf{u} \cdot \mathbf{v} = \|\mathbf{u}\| \|\mathbf{v}\| \cos\theta$.
- Vectors are **orthogonal** if:

$$\mathbf{u} \cdot \mathbf{v} = 0 \implies \theta = 90°.$$

- The zero vector is orthogonal to all vectors.

The dot product is a "conversation" between vectors: positive if they point in similar directions, negative if opposite, and zero if perpendicular. Orthogonal vectors are like walls meeting at a right angle, contributing independently to a structure.

Example 10.2 For:

$$\mathbf{u} = \begin{bmatrix} 2 \\ -1 \\ 3 \end{bmatrix}, \quad \mathbf{v} = \begin{bmatrix} 1 \\ 4 \\ -2 \end{bmatrix}, \quad \mathbf{w} = \begin{bmatrix} -8 \\ 3 \\ 2 \end{bmatrix}.$$

Compute dot products:

$$\mathbf{u} \cdot \mathbf{v} = 2 \cdot 1 + (-1) \cdot 4 + 3 \cdot (-2) = 2 - 4 - 6 = -8,$$

$$\mathbf{u} \cdot \mathbf{w} = 2 \cdot (-8) + (-1) \cdot 3 + 3 \cdot 2 = -16 - 3 + 6 = -13,$$

$$\mathbf{v} \cdot \mathbf{w} = 1 \cdot (-8) + 4 \cdot 3 + (-2) \cdot 2 = -8 + 12 - 4 = 0.$$

Thus, $\mathbf{v} \cdot \mathbf{w} = 0$, so $\mathbf{v}$ and $\mathbf{w}$ are orthogonal.

Python Example: Orthogonality Check

```python
import numpy as np
u = np.array([2, -1, 3])
v = np.array([1, 4, -2])
w = np.array([-8, 3, 2])
print("u dot  v =", np.dot(u, v))
print("u dot  w =", np.dot(u, w))
print("v dot  w =", np.dot(v, w))
```

Output:

```
u · v = -8
u · w = -13
v · w = 0
```

10.2.1   *Properties of the Dot Product*

- **Self Dot Product**:
$$\mathbf{u} \cdot \mathbf{u} = \|\mathbf{u}\|^2.$$

The dot product of a vector with itself equals the square of its norm (length). This follows from the definition $\mathbf{u} \cdot \mathbf{u} = u_1^2 + u_2^2 + \cdots + u_n^2$.

- **Zero Vector Condition**:

$$\mathbf{u} \cdot \mathbf{u} = 0 \iff \mathbf{u} = \mathbf{0}.$$

A vector has zero magnitude if and only if it is the zero vector. This is because the norm squared is a sum of squares, which can only be zero when all components are zero.

- **Commutativity**:

$$\mathbf{u} \cdot \mathbf{v} = \mathbf{v} \cdot \mathbf{u}.$$

The dot product is symmetric with respect to the order of operands due to the commutative property of scalar multiplication: $u_i v_i = v_i u_i$.

- **Distributivity Over Vector Addition**:

$$\mathbf{u} \cdot (\mathbf{v} + \mathbf{w}) = \mathbf{u} \cdot \mathbf{v} + \mathbf{u} \cdot \mathbf{w}.$$

The dot product distributes over vector addition, analogous to the distributive property of multiplication over addition in real numbers.

- **Compatibility with Scalar Multiplication**:

$$(c\mathbf{u}) \cdot \mathbf{v} = c(\mathbf{u} \cdot \mathbf{v}).$$

Scaling one vector by a scalar c scales the dot product by the same factor. This is due to the linearity of dot product with respect to scalar multiplication.

- **Norm Scaling**:

$$\|c\mathbf{u}\| = |c| \cdot \|\mathbf{u}\|.$$

The norm of a scalar multiple of a vector is the absolute value of the scalar times the norm of the vector. This preserves the positive magnitude and direction scaling.

- **Matrix Interaction with Transpose**:

$$A\mathbf{u} \cdot \mathbf{v} = \mathbf{u} \cdot (A^{\mathsf{T}}\mathbf{v}).$$

When applying a matrix A to $\mathbf{u}$, the dot product with another vector $\mathbf{v}$ can be transferred to the transpose of A acting on $\mathbf{v}$. This is essential in linear algebra, especially for deriving adjoints and gradients.

- **Geometric Insight**:
  The angle θ between vectors is:

$$\cos \theta = \frac{\mathbf{u} \cdot \mathbf{v}}{\|\mathbf{u}\| \|\mathbf{v}\|}.$$

Orthogonality ($\theta = 90°$) means no projection of one vector onto the other, ensuring independence.

10.3   Pythagorean Theorem and Geometric Insights

The **Pythagorean Theorem** extends to vectors: for orthogonal vectors $\mathbf{u}, \mathbf{v} \in \mathbb{R}^n$:

$$\|\mathbf{u} + \mathbf{v}\|^2 = \|\mathbf{u}\|^2 + \|\mathbf{v}\|^2.$$

This reflects that perpendicular vectors form a right triangle, with the sum vector as the hypotenuse.

Algebraic Proof

$$\|\mathbf{u} + \mathbf{v}\|^2 = (\mathbf{u} + \mathbf{v}) \cdot (\mathbf{u} + \mathbf{v}) = \mathbf{u} \cdot \mathbf{u} + 2\mathbf{u} \cdot \mathbf{v} + \mathbf{v} \cdot \mathbf{v}.$$

Since $\mathbf{u} \cdot \mathbf{v} = 0$:

$$\|\mathbf{u} + \mathbf{v}\|^2 = \|\mathbf{u}\|^2 + \|\mathbf{v}\|^2.$$

Example 10.3 For:

$$\mathbf{u} = \begin{bmatrix} 1 \\ 0 \end{bmatrix}, \quad \mathbf{v} = \begin{bmatrix} 0 \\ 1 \end{bmatrix}.$$

$$\mathbf{u} \cdot \mathbf{v} = 1 \cdot 0 + 0 \cdot 1 = 0 \text{ (orthogonal)},$$

$$\|\mathbf{u}\|^2 = 1^2 + 0^2 = 1, \quad \|\mathbf{v}\|^2 = 0^2 + 1^2 = 1,$$

$$\mathbf{u} + \mathbf{v} = \begin{bmatrix} 1 \\ 1 \end{bmatrix}, \quad \|\mathbf{u} + \mathbf{v}\|^2 = 1^2 + 1^2 = 2,$$

$$\|\mathbf{u}\|^2 + \|\mathbf{v}\|^2 = 1 + 1 = 2.$$

The theorem holds.

Geometric Application: Rhombus Condition

For vectors $\mathbf{u}, \mathbf{v}$, the parallelogram's diagonals are $\mathbf{u} + \mathbf{v}$ and $\mathbf{u} - \mathbf{v}$. If orthogonal:

$$(\mathbf{u} + \mathbf{v}) \cdot (\mathbf{u} - \mathbf{v}) = \|\mathbf{u}\|^2 - \|\mathbf{v}\|^2 = 0 \implies \|\mathbf{u}\| = \|\mathbf{v}\|.$$

This makes the parallelogram a rhombus (Fig. 10.3).

Fig. 10.3 Parallelogram formed by vectors **u**, **v**, and their sum **u** + **v**, with projections on axes

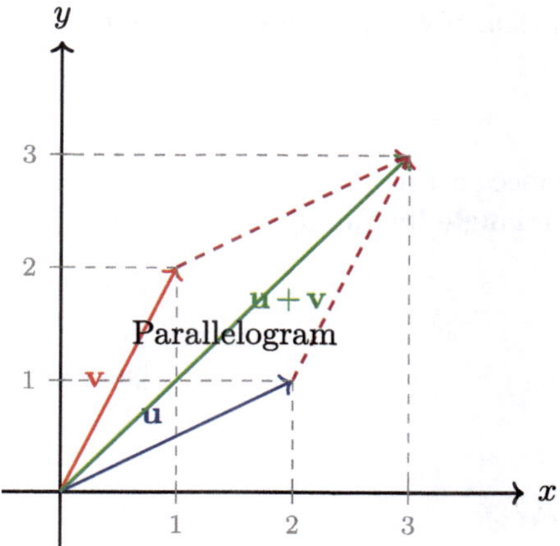

Python Example: Pythagorean Verification

```python
import numpy as np
u = np.array([1, 0])
v = np.array([0, 1])
lhs = np.linalg.norm(u + v)**2
rhs = np.linalg.norm(u)**2 + np.linalg.norm(v)**2
print(f"LHS: {lhs:.2f}, | RHS: {rhs:.2f} |")
```

Output:

```
LHS: 2.00, | RHS: 2.00 |
```

10.4 Cauchy-Schwarz and Triangle Inequalities

The **Cauchy-Schwarz** and **triangle inequalities** set bounds on vector interactions, ensuring geometric consistency in high-dimensional spaces.

Cauchy-Schwarz Inequality

$$|\mathbf{u} \cdot \mathbf{v}| \leq \|\mathbf{u}\| \cdot \|\mathbf{v}\|.$$

Equality holds if $\mathbf{u} = c\mathbf{v}$.

Proof: If $\mathbf{v} = \mathbf{0}$, both sides are zero. Otherwise:

$$|\mathbf{u} \cdot \mathbf{v}| = \|\mathbf{u}\|\|\mathbf{v}\||\cos\theta| \le \|\mathbf{u}\|\|\mathbf{v}\|,$$

since $|\cos\theta| \le 1$.

Triangle Inequality

$$\|\mathbf{u} + \mathbf{v}\| \le \|\mathbf{u}\| + \|\mathbf{v}\|.$$

Proof:

$$\|\mathbf{u} + \mathbf{v}\|^2 = \|\mathbf{u}\|^2 + 2\mathbf{u} \cdot \mathbf{v} + \|\mathbf{v}\|^2.$$

By Cauchy-Schwarz:

$$2\mathbf{u} \cdot \mathbf{v} \le 2\|\mathbf{u}\|\|\mathbf{v}\|.$$

$$\|\mathbf{u} + \mathbf{v}\|^2 \le \|\mathbf{u}\|^2 + 2\|\mathbf{u}\|\|\mathbf{v}\| + \|\mathbf{v}\|^2 = (\|\mathbf{u}\| + \|\mathbf{v}\|)^2.$$

Take square roots.

Example 10.4 For:

$$\mathbf{u} = \begin{bmatrix} 2 \\ -3 \\ 4 \end{bmatrix}, \quad \mathbf{v} = \begin{bmatrix} 1 \\ -2 \\ -5 \end{bmatrix}.$$

$$\mathbf{u} \cdot \mathbf{v} = 2 \cdot 1 + (-3) \cdot (-2) + 4 \cdot (-5) = 2 + 6 - 20 = -12, \quad |\mathbf{u} \cdot \mathbf{v}| = 12,$$

$$\|\mathbf{u}\| = \sqrt{2^2 + (-3)^2 + 4^2} = \sqrt{4 + 9 + 16} = \sqrt{29},$$

$$\|\mathbf{v}\| = \sqrt{1^2 + (-2)^2 + (-5)^2} = \sqrt{1 + 4 + 25} = \sqrt{30},$$

$$\|\mathbf{u}\| \cdot \|\mathbf{v}\| = \sqrt{29 \cdot 30} = \sqrt{870} \approx 29.5.$$

$$12 \le 29.5 \text{ (satisfied)}.$$

Python Example: Cauchy-Schwarz

```python
import numpy as np
u = np.array([2, -3, 4])
v = np.array([1, -2, -5])
dot_product = np.abs(np.dot(u, v))
```

```
5  norm_product = np.linalg.norm(u) *
       np.linalg.norm(v)
6  print("Dot product:", dot_product)
7  print(f"Norm product: {norm_product:.2f}")
```

Output:

```
Dot product: 12
Norm product: 29.50
```

Cauchy-Schwarz bounds the projection of one vector onto another, while the triangle inequality ensures the shortest path is the direct sum, like walking straight vs. along a triangle's sides.

10.5 Orthogonal Complement

The **orthogonal complement** of a subspace $S \subseteq \mathbb{R}^n$, denoted $S^\perp$, is:

$$S^\perp = \{\mathbf{v} \in \mathbb{R}^n \mid \mathbf{v} \cdot \mathbf{w} = 0 \text{ for all } \mathbf{w} \in S\}.$$

It contains all vectors perpendicular to S.
If S is a plane in $\mathbb{R}^3$, $S^\perp$ is the line perpendicular to it. This decomposition splits $\mathbb{R}^n$ into complementary directions.

10.5.1 Properties

- $S^\perp$ is a subspace.
  The set of all vectors in $\mathbb{R}^n$ that are orthogonal to every vector in a subspace S forms a subspace itself, satisfying closure under addition and scalar multiplication.
- $\mathbb{R}^n = S \oplus S^\perp$.
  Every vector in $\mathbb{R}^n$ can be uniquely written as the sum of one vector in S and one vector in $S^\perp$, forming a direct sum decomposition of the space.
- $(S^\perp)^\perp = S$.
  Taking the orthogonal complement twice returns the original subspace, assuming S is a subspace of $\mathbb{R}^n$.
- $\dim(S) + \dim(S^\perp) = n$.
  The dimensions of a subspace and its orthogonal complement always add up to the dimension of the entire space $\mathbb{R}^n$, reflecting the direct sum structure.

- $S \cap S^{\perp} = \{\mathbf{0}\}$.

  The only vector that lies in both a subspace and its orthogonal complement is the zero vector, since it is the only vector orthogonal to itself in both spaces.
- $\text{Row}(A)^{\perp} = \text{Null}(A)$, $\text{Col}(A)^{\perp} = \text{Null}(A^{\mathrm{T}})$.

  The orthogonal complement of the row space of a matrix A is its null space, and the orthogonal complement of its column space is the null space of A^{T}; this is a fundamental result in linear algebra.

Example 10.5 Let

$$S = \text{Span}\left\{ \begin{bmatrix} 1 \\ 1 \\ -1 \\ 4 \end{bmatrix}, \begin{bmatrix} 1 \\ -1 \\ 1 \\ 2 \end{bmatrix} \right\}.$$

We want to find a basis for the orthogonal complement $S^{\perp}$, which consists of all vectors $\mathbf{v} \in \mathbb{R}^4$ such that:

$$\mathbf{v} \cdot \mathbf{s}_1 = 0, \quad \mathbf{v} \cdot \mathbf{s}_2 = 0.$$

Let $\mathbf{v} = \begin{bmatrix} v_1 \\ v_2 \\ v_3 \\ v_4 \end{bmatrix}$. Then the orthogonality conditions yield the equations:

$$v_1 + v_2 - v_3 + 4v_4 = 0,$$

$$v_1 - v_2 + v_3 + 2v_4 = 0.$$

Step 1- Matrix Form
We represent this system as:

$$\begin{bmatrix} 1 & 1 & -1 & 4 \\ 1 & -1 & 1 & 2 \end{bmatrix} \begin{bmatrix} v_1 \\ v_2 \\ v_3 \\ v_4 \end{bmatrix} = \begin{bmatrix} 0 \\ 0 \end{bmatrix}.$$

Step 2- Row Reduction
Then, we apply elementary row operations:

$$\begin{bmatrix} 1 & 1 & -1 & 4 \\ 1 & -1 & 1 & 2 \end{bmatrix} \xrightarrow{R_2 \to R_2 - R_1} \begin{bmatrix} 1 & 1 & -1 & 4 \\ 0 & -2 & 2 & -2 \end{bmatrix}$$

$$\xrightarrow{R_2 \to -\frac{1}{2} R_2} \begin{bmatrix} 1 & 1 & -1 & 4 \\ 0 & 1 & -1 & 1 \end{bmatrix} \xrightarrow{R_1 \to R_1 - R_2} \begin{bmatrix} 1 & 0 & 0 & 3 \\ 0 & 1 & -1 & 1 \end{bmatrix}$$

From this we obtain:

$$v_1 = -3v_4,$$

$$v_2 = v_3 - v_4.$$

Step 3- Parametric Vector Form

Let $v_3 = t$, $v_4 = s$, then the general solution is:

$$\mathbf{v} = \begin{bmatrix} -3s \\ t-s \\ t \\ s \end{bmatrix} = t \begin{bmatrix} 0 \\ 1 \\ 1 \\ 0 \end{bmatrix} + s \begin{bmatrix} -3 \\ -1 \\ 0 \\ 1 \end{bmatrix}.$$

Basis for $S^{\perp}$

Therefore, a basis for the orthogonal complement is:

$$\left\{ \begin{bmatrix} 0 \\ 1 \\ 1 \\ 0 \end{bmatrix}, \begin{bmatrix} -3 \\ -1 \\ 0 \\ 1 \end{bmatrix} \right\}.$$

Python Example: Orthogonal Complement

```python
import numpy as np
from scipy.linalg import null_space
A = np.array([[1, 1], [1, -1], [-1, 1], [4,
    2]]).T
orthogonal_basis = null_space(A).T
print("Basis for S^{\perp}:\n", orthogonal_basis)
```

Output:

```
Basis for S^{\perp}:
[[ 0.01  0.71  0.71 -0.  ]
 [-0.93 -0.15  0.16  0.31]]
```

Example 10.6 Let $\mathbf{v} = \begin{bmatrix} 1 \\ 1 \\ -1 \end{bmatrix}$. We are interested in all vectors $\mathbf{x} = \begin{bmatrix} x_1 \\ x_2 \\ x_3 \end{bmatrix}$ such that:

$$\mathbf{v} \cdot \mathbf{x} = 1 \cdot x_1 + 1 \cdot x_2 + (-1) \cdot x_3 = 0.$$

This yields the condition:

$$x_1 + x_2 - x_3 = 0 \quad \Rightarrow \quad x_3 = x_1 + x_2.$$

General Form of Orthogonal Vectors

All vectors orthogonal to $\mathbf{v}$ lie on the plane defined by:

$$\mathbf{x} = \begin{bmatrix} x_1 \\ x_2 \\ x_1 + x_2 \end{bmatrix} = x_1 \begin{bmatrix} 1 \\ 0 \\ 1 \end{bmatrix} + x_2 \begin{bmatrix} 0 \\ 1 \\ 1 \end{bmatrix}.$$

So the set of all such vectors is the span of:

$$\text{Span} \left\{ \begin{bmatrix} 1 \\ 0 \\ 1 \end{bmatrix}, \begin{bmatrix} 0 \\ 1 \\ 1 \end{bmatrix} \right\}.$$

The set of all vectors orthogonal to $\mathbf{v}$ forms a plane in $\mathbb{R}^3$ that passes through the origin and is perpendicular to $\mathbf{v}$.

10.6 Matrix and Subspace Orthogonality

1. **Row Space Orthogonality**

Theorem

For any matrix $A \in \mathbb{R}^{m \times n}$,

$$(\text{Row}(A))^{\perp} = \text{Null}(A).$$

Proof: The row space of A is the subspace of $\mathbb{R}^n$ spanned by the rows of A. A vector $\mathbf{v} \in \mathbb{R}^n$ belongs to $(\text{Row}(A))^{\perp}$ if and only if it is orthogonal to every row of A. That is:

$$\mathbf{r}_i \cdot \mathbf{v} = 0, \quad \forall \, \mathbf{r}_i \in \text{rows of } A \Rightarrow A\mathbf{v} = \mathbf{0}.$$

Thus, $\mathbf{v} \in \text{Null}(A)$, proving the equality.

2. **Column Space Orthogonality**

Theorem

$$(\text{Col}(A))^{\perp} = \text{Null}(A^{\mathrm{T}}).$$

The orthogonal complement of the column space of a matrix A is exactly the null space of its transpose, linking orthogonality with solutions to homogeneous systems.

The column space of A is a subspace of $\mathbb{R}^m$, formed by the linear combinations of the columns of A. A vector $\mathbf{y} \in \mathbb{R}^m$ lies in $(\mathrm{Col}(A))^\perp$ if and only if it is orthogonal to every column of A, which means:

$$\mathbf{a}_j^\mathrm{T}\mathbf{y} = 0, \quad \forall\, \mathbf{a}_j \in \text{columns of } A \Rightarrow A^\mathrm{T}\mathbf{y} = \mathbf{0}.$$

Hence, $\mathbf{y} \in \mathrm{Null}(A^\mathrm{T})$, establishing that:

$$(\mathrm{Col}(A))^\perp = (\mathrm{Row}(A^\mathrm{T}))^\perp = \mathrm{Null}(A^\mathrm{T}).$$

3. **Dimension Relationship in Subspaces**

Theorem
For any subspace $W \subseteq \mathbb{R}^n$,

$$\dim(W) + \dim(W^\perp) = n$$

The dimension of a subspace and the dimension of its orthogonal complement always add up to the dimension of the entire space $\mathbb{R}^n$.

This relationship follows from the fundamental theorem of linear algebra. If W is a subspace of $\mathbb{R}^n$, then $W^\perp$ consists of all vectors orthogonal to W, and together they span the entire space:

- $\dim(W) = \mathrm{rank}(W)$, the number of linearly independent vectors in W.
- $\dim(W^\perp) = \mathrm{nullity}(W)$, the dimension of the orthogonal complement.

Thus,

$$\boxed{\mathrm{rank} + \mathrm{nullity} = n}$$

10.7 Dimension of a Subspace and Its Orthogonal Complement

Understanding the relationship between a subspace $W \subseteq \mathbb{R}^n$ and its orthogonal complement $W^\perp$ is fundamental in linear algebra. Together, these two subspaces divide the entire space into independent components, and their dimensions always add up to

Fig. 10.4 Unique
decomposition of $\mathbf{u} \in \mathbb{R}^n$ as
$\mathbf{w} \in W$ and $\mathbf{z} \in W^\perp$

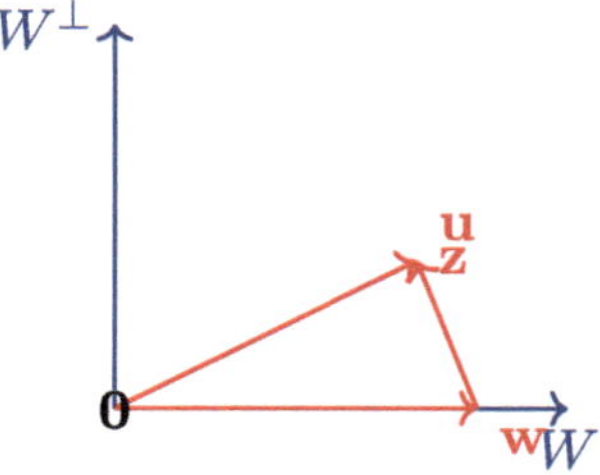

n, the dimension of the full space. This idea not only provides theoretical clarity but
also has important applications in projections, orthogonalization, and data analysis
methods such as PCA.

Basis Representations

- Basis for W: $\{\mathbf{w}_1, \mathbf{w}_2, \ldots, \mathbf{w}_k\}$, where $k = \dim(W)$.
- Basis for $W^\perp$: $\{\mathbf{z}_1, \mathbf{z}_2, \ldots, \mathbf{z}_{n-k}\}$, where $n - k = \dim(W^\perp)$.
- Together, the union of these bases forms a basis for $\mathbb{R}^n$.

Unique Decomposition of a Vector

Every vector $\mathbf{u} \in \mathbb{R}^n$ can be uniquely decomposed as (Fig. 10.4):

$$\mathbf{u} = \mathbf{w} + \mathbf{z},$$

where:

- $\mathbf{w} \in W$.
- $\mathbf{z} \in W^\perp$.

This decomposition is unique because $W \cap W^\perp = \{\mathbf{0}\}$, meaning they share only the
zero vector.

This geometric and algebraic property provides a fundamental framework in linear
algebra and ML, especially in techniques such as projections, orthogonalization (e.g.,
Gram-Schmidt), and dimensionality reduction (e.g., PCA).
In linear algebra, any vector $\mathbf{u} \in \mathbb{R}^n$ can be uniquely decomposed into two parts
relative to another non-zero vector $\mathbf{v} \in \mathbb{R}^n$:

$$\mathbf{u} = \mathbf{u}_\| + \mathbf{u}_\perp,$$

where:

- $\mathbf{u}_\|$ is the projection of $\mathbf{u}$ onto $\mathbf{v}$ (i.e., the part **parallel** to $\mathbf{v}$),
- $\mathbf{u}_\perp$ is the component of $\mathbf{u}$ that is **orthogonal** to $\mathbf{v}$.

This decomposition is unique (Fig. 10.5).

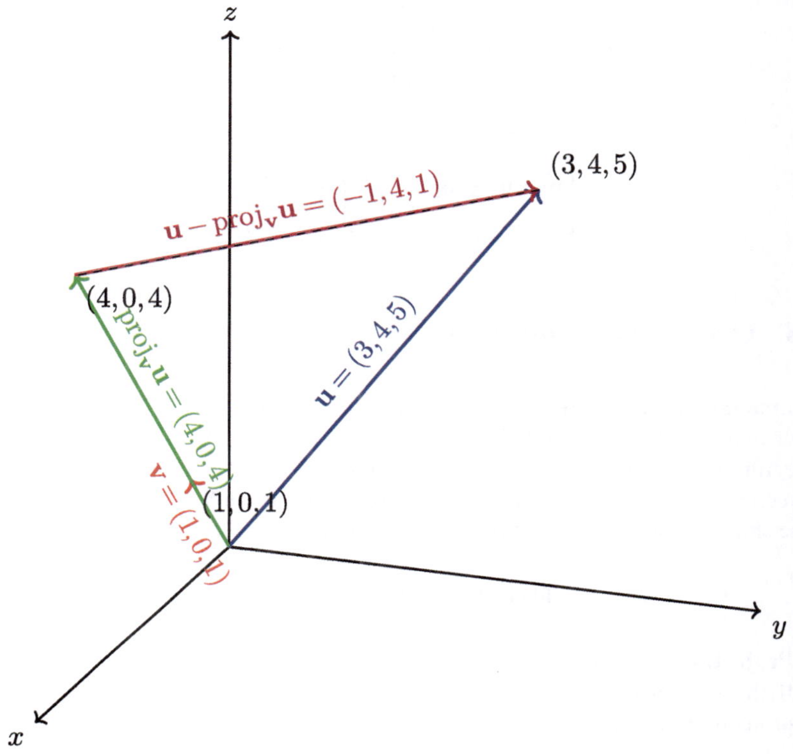

Fig. 10.5 Decomposition of $\mathbf{u} = (3, 4, 5)$ into components parallel and orthogonal to $\mathbf{v} = (1, 0, 1)$

Example 10.7 We want to decompose $\mathbf{u} = \begin{bmatrix} 3 \\ 4 \\ 5 \end{bmatrix}$ relative to $\mathbf{v} = \begin{bmatrix} 1 \\ 0 \\ 1 \end{bmatrix}$.

Step 1- Compute the projection of $\mathbf{u}$ onto $\mathbf{v}$:

$$\mathbf{u}_{\parallel} = \mathrm{proj}_{\mathbf{v}}\mathbf{u} = \frac{\mathbf{u} \cdot \mathbf{v}}{\mathbf{v} \cdot \mathbf{v}}\mathbf{v},$$

$$\mathbf{u} \cdot \mathbf{v} = 3 \cdot 1 + 4 \cdot 0 + 5 \cdot 1 = 8, \quad \mathbf{v} \cdot \mathbf{v} = 1^2 + 0^2 + 1^2 = 2,$$

$$\Rightarrow \mathbf{u}_{\parallel} = \frac{8}{2}\begin{bmatrix} 1 \\ 0 \\ 1 \end{bmatrix} = 4\begin{bmatrix} 1 \\ 0 \\ 1 \end{bmatrix} = \begin{bmatrix} 4 \\ 0 \\ 4 \end{bmatrix}.$$

Step 2- Find the orthogonal component:

$$\mathbf{u}_\perp = \mathbf{u} - \mathbf{u}_\| = \begin{bmatrix} 3 \\ 4 \\ 5 \end{bmatrix} - \begin{bmatrix} 4 \\ 0 \\ 4 \end{bmatrix} = \begin{bmatrix} -1 \\ 4 \\ 1 \end{bmatrix}.$$

$$\boxed{\mathbf{u} = \mathbf{u}_\| + \mathbf{u}_\perp = \begin{bmatrix} 4 \\ 0 \\ 4 \end{bmatrix} + \begin{bmatrix} -1 \\ 4 \\ 1 \end{bmatrix}}$$

10.8 Orthogonal Projection

The **orthogonal projection** of a vector $\mathbf{v} \in \mathbb{R}^n$ onto a subspace $S \subseteq \mathbb{R}^n$ is the unique vector in S that is closest to $\mathbf{v}$. This projected vector is denoted by $\mathrm{proj}_S(\mathbf{v})$. Projecting $\mathbf{v}$ onto S means finding a vector in S such that the difference (error vector) between $\mathbf{v}$ and its projection is orthogonal to every vector in S. This can be interpreted as the shadow that $\mathbf{v}$ casts onto S under perpendicular light.

$$\mathrm{proj}_S(\mathbf{v}) = \arg \min_{\mathbf{s} \in S} \|\mathbf{v} - \mathbf{s}\|_2.$$

(a) **Projection Using an Orthonormal Basis**

If the subspace S has an orthonormal basis $\{\mathbf{u}_1, \mathbf{u}_2, \ldots, \mathbf{u}_k\}$, the orthogonal projection of $\mathbf{v}$ onto S is:

$$\mathrm{proj}_S(\mathbf{v}) = \sum_{i=1}^{k} (\mathbf{v} \cdot \mathbf{u}_i) \mathbf{u}_i.$$

This formula exploits the fact that the dot product with an orthonormal basis gives the coefficients directly.

Orthogonal projections can be carried out either by using an orthonormal basis, which simplifies the computation by taking inner products, or by applying a matrix representation, which provides a compact algebraic form suitable for general transformations. An orthonormal basis is one in which every basis vector has unit norm ($\|u_i\| = 1$) and all basis vectors are mutually orthogonal ($u_i \cdot u_j = 0 \ for \ i \neq j$). If the basis $u_1, \ldots, u_k$ is orthogonal but not orthonormal, the projection formula generalizes to $\mathrm{proj}_S(v) = $ sum of $[(v \cdot u_i)/(u_i \cdot u_i)]u_i$.

(b) **Projection Using a Matrix Representation**

If $A \in \mathbb{R}^{n \times k}$ is a matrix whose columns form a basis (not necessarily orthonormal) for the subspace S, then the projection matrix P_S is:

$$P_S = A(A^{\mathrm{T}} A)^{-1} A^{\mathrm{T}},$$

$$\mathrm{proj}_S(\mathbf{v}) = P_S \mathbf{v}.$$

This generalizes the orthogonal projection to any basis and is especially useful in higher-dimensional computations (e.g., least-squares problems).

Example 10.8 Let:

$$
\mathbf{v} = \begin{bmatrix} 2 \\ 2 \\ 3 \end{bmatrix}, \quad \mathbf{u} = \begin{bmatrix} 1 \\ 0 \\ 1 \end{bmatrix}.
$$

We want to project $\mathbf{v}$ onto the line $S = \mathrm{Span}(\mathbf{u})$.
Step 1- We normalize $\mathbf{u}$:

$$
\hat{\mathbf{u}} = \frac{1}{\sqrt{1^2 + 0^2 + 1^2}} \begin{bmatrix} 1 \\ 0 \\ 1 \end{bmatrix} = \frac{1}{\sqrt{2}} \begin{bmatrix} 1 \\ 0 \\ 1 \end{bmatrix}.
$$

Step 2- We compute projection:

$$
\mathrm{proj}_S(\mathbf{v}) = (\mathbf{v} \cdot \hat{\mathbf{u}})\hat{\mathbf{u}} = \left(\frac{2 \cdot 1 + 2 \cdot 0 + 3 \cdot 1}{\sqrt{2}} \right) \cdot \frac{1}{\sqrt{2}} \begin{bmatrix} 1 \\ 0 \\ 1 \end{bmatrix} = \frac{5}{2} \begin{bmatrix} 1 \\ 0 \\ 1 \end{bmatrix} = \begin{bmatrix} 2.5 \\ 0 \\ 2.5 \end{bmatrix}.
$$

Step 3- Error vector:

$$
\mathbf{e} = \mathbf{v} - \mathrm{proj}_S(\mathbf{v}) = \begin{bmatrix} 2 \\ 2 \\ 3 \end{bmatrix} - \begin{bmatrix} 2.5 \\ 0 \\ 2.5 \end{bmatrix} = \begin{bmatrix} -0.5 \\ 2 \\ 0.5 \end{bmatrix}.
$$

10.8.1 *Orthogonal Projection of a Vector onto Coordinate Axes*

Let $\mathbf{v} = \begin{bmatrix} 3 \\ 2 \end{bmatrix}$. We can decompose it as:

$$
\mathbf{v} = \mathbf{w} + \mathbf{z},
$$

where:

- $\mathbf{w}$ is the orthogonal projection of $\mathbf{v}$ onto the x-axis: $\begin{bmatrix} 3 \\ 0 \end{bmatrix}$,

- $\mathbf{z} = \mathbf{v} - \mathbf{w} = \begin{bmatrix} 0 \\ 2 \end{bmatrix}$, which is orthogonal to $\mathbf{w}$ (Fig. 10.6).

Key Properties

- The projection $\mathrm{proj}_S(\mathbf{v}) \in S$.

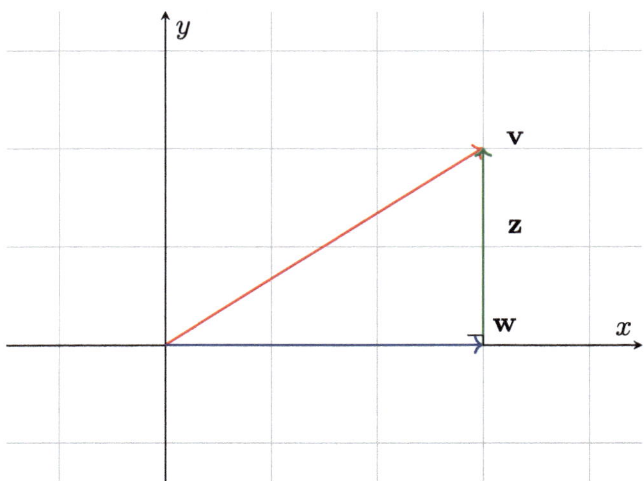

Fig. 10.6 Vector **v** is decomposed into its projection **w** onto the x-axis and the orthogonal component **z**

- The error vector $\mathbf{v} - \mathrm{proj}_S(\mathbf{v}) \perp S$.
- If P is a projection matrix, then $P^2 = P$ and $P^{\mathrm{T}} = P$ (symmetric idempotent).
- Projection minimizes Euclidean distance from **v** to the subspace S (Fig. 10.7).

Example 10.9 For $\mathbf{v} = \begin{bmatrix} 4 \\ 1 \end{bmatrix}$, $S = \mathrm{Span} \left\{ \begin{bmatrix} 2 \\ 1 \end{bmatrix} \right\}$:

$$\mathbf{u} = \begin{bmatrix} 2 \\ 1 \end{bmatrix}, \quad \mathbf{v} \cdot \mathbf{u} = 4 \cdot 2 + 1 \cdot 1 = 9,$$

$$\|\mathbf{u}\|^2 = 2^2 + 1^2 = 5,$$

$$\mathrm{proj}_S(\mathbf{v}) = \frac{9}{5} \begin{bmatrix} 2 \\ 1 \end{bmatrix} = \begin{bmatrix} 3.6 \\ 1.8 \end{bmatrix}.$$

Error:

$$\mathbf{e} = \mathbf{v} - \mathrm{proj}_S(\mathbf{v}) = \begin{bmatrix} 4 \\ 1 \end{bmatrix} - \begin{bmatrix} 3.6 \\ 1.8 \end{bmatrix} = \begin{bmatrix} 0.4 \\ -0.8 \end{bmatrix}.$$

$$\|\mathbf{e}\| = \sqrt{0.4^2 + (-0.8)^2} = \sqrt{0.16 + 0.64} = \sqrt{0.8} \approx 0.894.$$

Fig. 10.7 Projection of **v** onto $S = \mathrm{Span}\{\mathbf{u}\}$

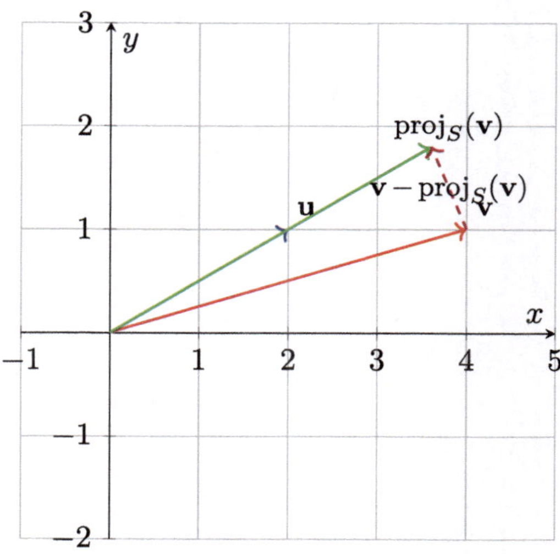

Python Example: Orthogonal Projection

```python
import numpy as np
v = np.array([4, 1])
u = np.array([2, 1])
proj_v = (np.dot(v, u) / np.dot(u, u)) * u
error = v - proj_v
print("Projection:", proj_v)
print("Error vector:", error)
print(f"Error norm: {np.linalg.norm(error):.2f}")
```

Output:

```
Projection: [3.6 1.8]
Error vector: [ 0.4 -0.8]
Error norm: 0.89
```

Example 10.10 Given:

$$\mathbf{v} = \begin{bmatrix} -6 \\ 4 \end{bmatrix}, \quad \mathbf{u} = \begin{bmatrix} 3 \\ 2 \end{bmatrix}.$$

We want to find the orthogonal projection and distance to a line.

Step 1- Projection of **v** onto **u**

$$\mathrm{proj}_{\mathbf{u}}(\mathbf{v}) = \frac{\mathbf{v} \cdot \mathbf{u}}{\mathbf{u} \cdot \mathbf{u}}\, \mathbf{u} = \frac{-10}{13}\begin{bmatrix} 3 \\ 2 \end{bmatrix} = \begin{bmatrix} -\frac{30}{13} \\ -\frac{20}{13} \end{bmatrix}.$$

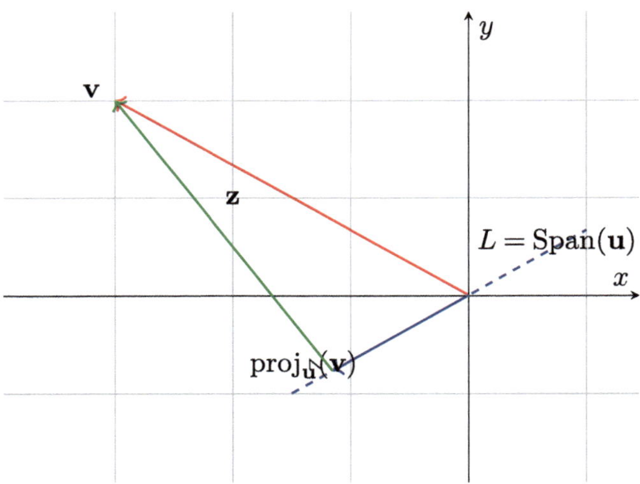

Fig. 10.8 The red vector $\mathbf{v}$ is projected onto the blue dashed line L. The green vector shows the orthogonal error (shortest distance) from $\mathbf{v}$ to L

Step 2–Orthogonal Vector (Error)

$$\mathbf{z} = \mathbf{v} - \mathrm{proj}_{\mathbf{u}}(\mathbf{v}) = \begin{bmatrix} -6 \\ 4 \end{bmatrix} - \begin{bmatrix} -\frac{30}{13} \\ -\frac{20}{13} \end{bmatrix} = \begin{bmatrix} -\frac{48}{13} \\ \frac{72}{13} \end{bmatrix}.$$

Step 3- Distance from $\mathbf{v}$ to Line L (Fig. 10.8)

$$\text{Distance} = \|\mathbf{z}\| = \frac{1}{13}\sqrt{48^2 + 72^2} = \frac{\sqrt{7488}}{13} \approx 6.655.$$

Example 10.11 Given:

$$\mathbf{x} = \begin{bmatrix} -2 \\ 3 \\ -1 \end{bmatrix}, \quad \mathbf{u} = \begin{bmatrix} -1 \\ 1 \\ 1 \end{bmatrix}, \quad L = \text{Span}(\mathbf{u}).$$

We want to decompose $\mathbf{x}$ into:

$$\mathbf{x} = \mathbf{x}_L + \mathbf{x}_L^{\perp},$$

where $\mathbf{x}_L$ is the projection of $\mathbf{x}$ onto the line L, and $\mathbf{x}_L^{\perp}$ is orthogonal to L.
Step 1- Compute the projection
The projection of $\mathbf{x}$ onto $\mathbf{u}$ is:

$$\mathbf{x}_L = \mathrm{proj}_{\mathbf{u}}(\mathbf{x}) = \frac{\mathbf{x} \cdot \mathbf{u}}{\mathbf{u} \cdot \mathbf{u}} \mathbf{u}$$

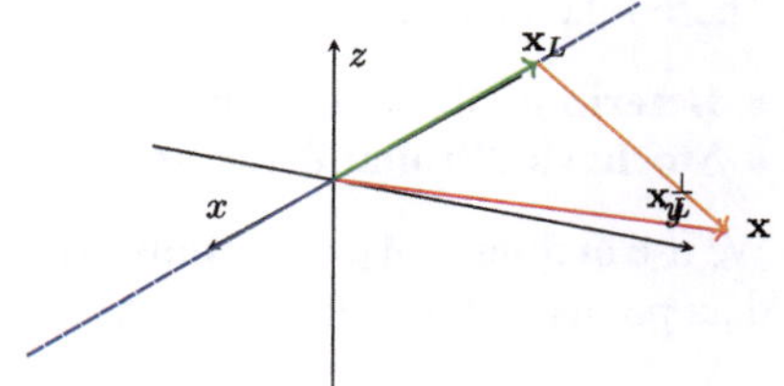

Fig. 10.9 Vector **x** decomposed into its projection onto line L (green) and the orthogonal component (orange)

Compute the dot products:

$$\mathbf{x}\cdot\mathbf{u} = (-2)(-1)+(3)(1)+(-1)(1) = 2+3-1 = 4,$$

$$\mathbf{u}\cdot\mathbf{u} = (-1)^2 + 1^2 + 1^2 = 1+1+1 = 3.$$

Therefore,

$$\mathbf{x}_L = \frac{4}{3}\begin{bmatrix} -1 \\ 1 \\ 1 \end{bmatrix} = \begin{bmatrix} -\frac{4}{3} \\ \frac{4}{3} \\ \frac{4}{3} \end{bmatrix}.$$

Step 2- Compute the orthogonal component

$$\mathbf{x}_L^{\perp} = \mathbf{x} - \mathbf{x}_L = \begin{bmatrix} -2 \\ 3 \\ -1 \end{bmatrix} - \begin{bmatrix} -\frac{4}{3} \\ \frac{4}{3} \\ \frac{4}{3} \end{bmatrix} = \begin{bmatrix} -\frac{2}{3} \\ \frac{5}{3} \\ -\frac{7}{3} \end{bmatrix}.$$

Final Decomposition as shown in the following figure

$$\mathbf{x} = \underbrace{\begin{bmatrix} -\frac{4}{3} \\ \frac{4}{3} \\ \frac{4}{3} \end{bmatrix}}_{\mathbf{x}_L} + \underbrace{\begin{bmatrix} -\frac{2}{3} \\ \frac{5}{3} \\ -\frac{7}{3} \end{bmatrix}}_{\mathbf{x}_L^{\perp}}$$

Figure 10.9 Vector **x** decomposed into its projection onto line L (green) and the orthogonal component (orange).

10.9 Least Squares Approximation

In many research problems, we are interested in finding simple mathematical relationships between variables. For instance, estimating the price of a car based on its age, model, and options. In this case, the price is the **dependent variable**, and the rest are **independent variables**.

These relationships can be:

- **Deterministic**: exact value determined from inputs.
- **Stochastic/Probabilistic**: includes variability or noise.

We use **orthogonal projections** to identify such relationships. The red dots represent data points and the blue line is a model fit (e.g., LR).

10.9.1 Error and Objective Function

The **error vector** e represents the difference between the observed and predicted values:

$$
e = \begin{bmatrix} y_1 - (a_0 + a_1 x_1) \\ y_2 - (a_0 + a_1 x_2) \\ \vdots \\ y_n - (a_0 + a_1 x_n) \end{bmatrix}.
$$

The objective is to minimize the squared error:

$$
E = \|e\|^2 = \sum_{i=1}^{n} [y_i - (a_0 + a_1 x_i)]^2 .
$$

10.9.2 Matrix Representation

Let vectors be defined as:

$$
\mathbf{v}_1 = \begin{bmatrix} 1 \\ 1 \\ \vdots \\ 1 \end{bmatrix}, \quad \mathbf{v}_2 = \begin{bmatrix} x_1 \\ x_2 \\ \vdots \\ x_n \end{bmatrix}, \quad \mathbf{y} = \begin{bmatrix} y_1 \\ y_2 \\ \vdots \\ y_n \end{bmatrix}, \quad C = [\mathbf{v}_1 \ \mathbf{v}_2].
$$

Then,

$$
E = \|\mathbf{y} - (a_0 \mathbf{v}_1 + a_1 \mathbf{v}_2)\|^2 = \|\mathbf{y} - C\mathbf{a}\|^2.
$$

Figure 10.10 illustrates a simple LR applied to a set of observed data points. The red markers denote the actual observations (x_i, y_i), and the blue line represents the fitted model

$$
y = a_0 + a_1 x.
$$

The green arrows correspond to the residuals, i.e., the vertical distances from each observed value to its predicted counterpart $(x_i, a_0 + a_1 x_i)$. These residuals measure

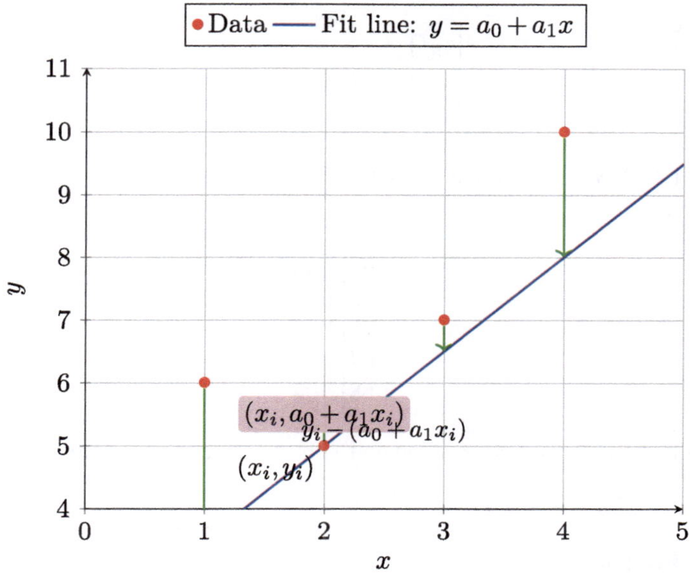

Fig. 10.10 Observed data, fitted line, residuals, and predicted point

the individual prediction errors and play a key role in minimizing the least squares
objective during model estimation.

10.9.3 Normal Equations

To minimize E, we solve:

$$C^{\mathrm{T}} C \begin{bmatrix} a_0 \\ a_1 \end{bmatrix} = C^{\mathrm{T}} \mathbf{y} \;\Rightarrow\; \begin{bmatrix} a_0 \\ a_1 \end{bmatrix} = (C^{\mathrm{T}} C)^{-1} C^{\mathrm{T}} \mathbf{y}.$$

When a linear fit is not sufficient, we extend the model:

$$y = a_0 + a_1 x + a_2 x^2.$$

$$\mathbf{v}_3 = \begin{bmatrix} x_1^2 \\ x_2^2 \\ \vdots \\ x_n^2 \end{bmatrix}, \quad C = [\mathbf{v}_1\ \mathbf{v}_2\ \mathbf{v}_3].$$

Then solve:

$$\begin{bmatrix} a_0 \\ a_1 \\ a_2 \end{bmatrix} = (C^{\mathrm{T}}C)^{-1}C^{\mathrm{T}}\mathbf{y}.$$

Example 10.12 We are given a dataset of 5 data points:

x_i	y_i
2.60	2.00
2.72	2.10
2.75	2.10
2.67	2.03
2.68	2.04

where, x_i is the independent variable and y_i is the dependent variable.

Matrix Formulation

We construct the design matrix C and observation vector $\mathbf{y}$:

$$C = \begin{bmatrix} 1 & 2.60 \\ 1 & 2.72 \\ 1 & 2.75 \\ 1 & 2.67 \\ 1 & 2.68 \end{bmatrix}, \quad \mathbf{y} = \begin{bmatrix} 2.00 \\ 2.10 \\ 2.10 \\ 2.03 \\ 2.04 \end{bmatrix}$$

The least squares solution minimizes the residual norm $\|C\mathbf{a} - \mathbf{y}\|$. The normal equations are:

$$C^{\mathrm{T}}C\mathbf{a} = C^{\mathrm{T}}\mathbf{y}.$$

Calculate:

$$C^{\mathrm{T}}C = \begin{bmatrix} 5 & 13.42 \\ 13.42 & 36.03 \end{bmatrix}, \quad C^{\mathrm{T}}\mathbf{y} = \begin{bmatrix} 10.27 \\ 27.57 \end{bmatrix}.$$

Solving:

$$\mathbf{a} = (C^{\mathrm{T}}C)^{-1}C^{\mathrm{T}}\mathbf{y} \approx \begin{bmatrix} 0.056 \\ 0.745 \end{bmatrix}.$$

Hence, the best-fit line is:

$$y = 0.056 + 0.745x.$$

- **Intercept (0.056)**: The predicted value of y when $x = 0$.

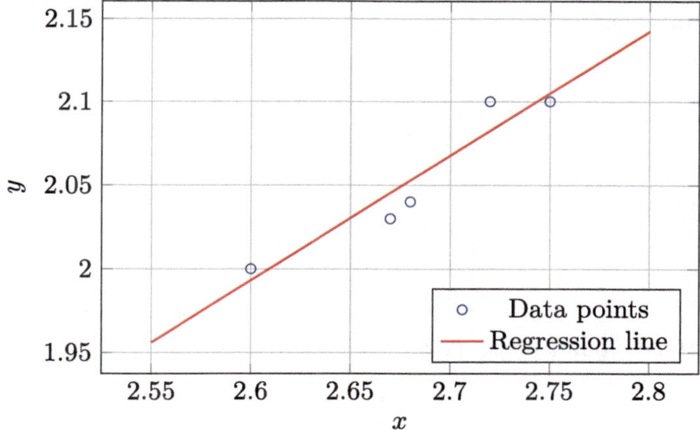

Fig. 10.11 Scatter plot of the given data points with the fitted regression line $y = 0.056 + 0.745x$

- **Slope (0.745)**: For every unit increase in x, y increases by approximately 0.745.

Figure 10.11 shows the distribution of the data points and how the regression line models the trend, indicating a positive linear relationship between x and y.
Once we built the mode, we can use it for prediction of a new input x. To predict y for a new input $x = 2.65$:

$$y = 0.056 + 0.745 \cdot 2.65 = 0.056 + 1.97625 = 2.03225 \approx \boxed{2.030}$$

Example 10.13 We are given data points:

$$(1, 62), \quad (3, 54), \quad (4, 50), \quad (5, 48), \quad (7, 40).$$

Our goal is to find a least-squares fit of the form:

$$y = a_0 + a_1 x + a_2 x^2.$$

The least-squares method minimizes the sum of squared errors:

$$E = \sum_{i=1}^{n} \left[y_i - (a_0 + a_1 x_i + a_2 x_i^2) \right]^2.$$

To formalize this, we define vectors:

$$\mathbf{v}_1 = \begin{bmatrix} 1 \\ 1 \\ \vdots \\ 1 \end{bmatrix}, \quad \mathbf{v}_2 = \begin{bmatrix} x_1 \\ x_2 \\ \vdots \\ x_n \end{bmatrix}, \quad \mathbf{v}_3 = \begin{bmatrix} x_1^2 \\ x_2^2 \\ \vdots \\ x_n^2 \end{bmatrix}, \quad \mathbf{y} = \begin{bmatrix} y_1 \\ y_2 \\ \vdots \\ y_n \end{bmatrix}.$$

Now construct matrix $C = [\mathbf{v}_1 \ \mathbf{v}_2 \ \mathbf{v}_3]$, a $n \times 3$ matrix. We solve the normal equation:

$$C^{\mathrm{T}}C \begin{bmatrix} a_0 \\ a_1 \\ a_2 \end{bmatrix} = C^{\mathrm{T}}\mathbf{y}.$$

The solution is:

$$\begin{bmatrix} a_0 \\ a_1 \\ a_2 \end{bmatrix} = (C^{\mathrm{T}}C)^{-1}C^{\mathrm{T}}\mathbf{y}.$$

Example 10.14 To fit a linear model $y = a_0 + a_1 x$, define:

$$C = \begin{bmatrix} 1 & 1 \\ 1 & 3 \\ 1 & 4 \\ 1 & 5 \\ 1 & 7 \end{bmatrix}, \quad \mathbf{y} = \begin{bmatrix} 62 \\ 54 \\ 50 \\ 48 \\ 40 \end{bmatrix}.$$

Solving:

$$\begin{bmatrix} a_0 \\ a_1 \end{bmatrix} = (C^{\mathrm{T}}C)^{-1}C^{\mathrm{T}}\mathbf{y} = \begin{bmatrix} 65.2 \\ -3.6 \end{bmatrix}.$$

Hence, the best fit line is:

$$y = 65.2 - 3.6x.$$

10.10   Best Quadratic Fit

In the previous section, we studied LR, where we attempted to find the best line that approximates a set of data points by minimizing the sum of squared errors. While linear equations model data with a constant rate of change, many real-world scenarios require more flexibility.

Quadratic regression fits a parabola to the data, enabling us to model acceleration, curvature, or non-linear phenomena.

General Model

We aim to find a quadratic polynomial:

$$y = a_0 + a_1 x + a_2 x^2,$$

to fit data points (x_1, y_1), (x_2, y_2), ..., (x_n, y_n). The error vector is defined as:

$$e = \begin{bmatrix} y_1 - (a_0 + a_1 x_1 + a_2 x_1^2) \\ y_2 - (a_0 + a_1 x_2 + a_2 x_2^2) \\ \vdots \\ y_n - (a_0 + a_1 x_n + a_2 x_n^2) \end{bmatrix}.$$

The objective is to minimize the total squared error:

$$E = \|e\|^2.$$

10.10.1   *Matrix Formulation*

Let

$$\mathbf{v}_1 = \begin{bmatrix} 1 \\ 1 \\ \vdots \\ 1 \end{bmatrix}, \quad \mathbf{v}_2 = \begin{bmatrix} x_1 \\ x_2 \\ \vdots \\ x_n \end{bmatrix}, \quad \mathbf{v}_3 = \begin{bmatrix} x_1^2 \\ x_2^2 \\ \vdots \\ x_n^2 \end{bmatrix}.$$

and define:

$$C = \begin{bmatrix} \mathbf{v}_1 \ \mathbf{v}_2 \ \mathbf{v}_3 \end{bmatrix}, \quad \mathbf{y} = \begin{bmatrix} y_1 \\ y_2 \\ \vdots \\ y_n \end{bmatrix}.$$

The normal equations are:

$$\begin{bmatrix} a_0 \\ a_1 \\ a_2 \end{bmatrix} = (C^\mathrm{T} C)^{-1} C^\mathrm{T} \mathbf{y}.$$

Figure 10.12 shows the result of fitting a quadratic model to a nonlinear dataset. The red points represent the observed data, and the blue curve represents the fitted quadratic polynomial of the form:

$$y = a_0 + a_1 x + a_2 x^2.$$

This model is derived using the method of least squares, minimizing the squared error between the predicted and observed values. Unlike LR, which fits a straight line, quadratic regression captures curvature in the data, making it more suitable for datasets that exhibit acceleration or deceleration patterns. The result is a best-fit curve that more accurately represents the underlying trend in the data.

Fig. 10.12 Polynomial regression of degree 2. The red dots represent sample data points, while the blue curve shows the fitted quadratic model $y = a_0 + a_1 x + a_2 x^2$. This illustrates how polynomial regression can capture nonlinear trends in the data

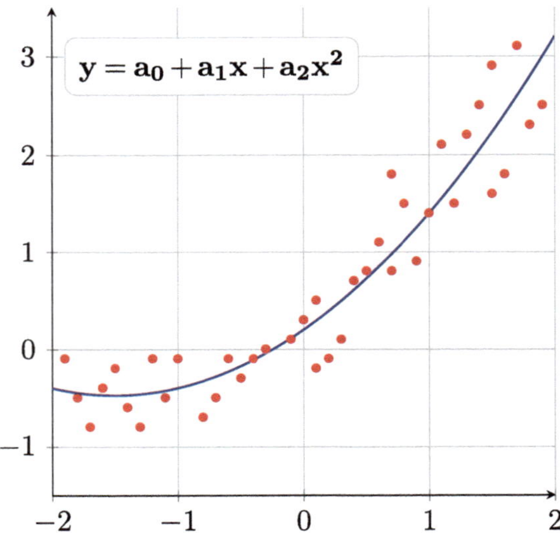

Example 10.15 A teacher collects data on how many hours five students studied and their corresponding exam scores. The data is shown in the table below:

Hours Studied (x)	Exam Score (y)
0	50
1	65
2	78
3	85
4	82

We aim to find a quadratic model of the form:

$$y = a_0 + a_1 x + a_2 x^2.$$

that best fits the data.

Step 1- Design Matrix C and Vector $\mathbf{y}$

We represent each data point using the matrix formulation. For a quadratic model, the design matrix includes a constant term, the input variable x, and the square of x:

$$C = \begin{bmatrix} 1 & 0 & 0 \\ 1 & 1 & 1 \\ 1 & 2 & 4 \\ 1 & 3 & 9 \\ 1 & 4 & 16 \end{bmatrix}, \quad \mathbf{y} = \begin{bmatrix} 50 \\ 65 \\ 78 \\ 85 \\ 82 \end{bmatrix}.$$

Step 2- Solve the Normal Equations

The coefficients $\mathbf{a} = \begin{bmatrix} a_0 & a_1 & a_2 \end{bmatrix}^{\mathrm{T}}$ can be computed using:

$$\mathbf{a} = (C^{\mathrm{T}}C)^{-1}C^{\mathrm{T}}\mathbf{y}.$$

We compute the matrices:

$$C^{\mathrm{T}}C = \begin{bmatrix} 5 & 10 & 30 \\ 10 & 30 & 100 \\ 30 & 100 & 354 \end{bmatrix}, \quad C^{\mathrm{T}}\mathbf{y} = \begin{bmatrix} 360 \\ 804 \\ 2454 \end{bmatrix}.$$

Now compute:

$$\mathbf{a} = (C^{\mathrm{T}}C)^{-1}C^{\mathrm{T}}\mathbf{y} \approx \begin{bmatrix} 49.2 \\ 20.39 \\ -3 \end{bmatrix}.$$

Step 3- Final Quadratic Model

So the best-fit quadratic equation is:

$$y = 49.2 + 20.39x - 3x^2.$$

- $a_0 = 49.2$ is the estimated base score with zero study hours.
- $a_1 = 20.39$ shows an initial steep improvement with more study.
- $a_2 = -3$ indicates diminishing returns, after a point, more study doesn't help and might even hurt performance.

Suppose a student studies for $x = 2.5$ h. We can predict their score:

$$y = 49.2 + 20.39(2.5) - 3(2.5)^2 = 81.425$$

So, the predicted score is 81.425.

The plot illustrates a quadratic regression model fitted to data representing the number of study hours and corresponding exam scores (Fig. 10.13).

This type of model is particularly useful for finding an optimal study duration that maximizes performance, reflecting a realistic non-linear trend common in learning and behavioral data.

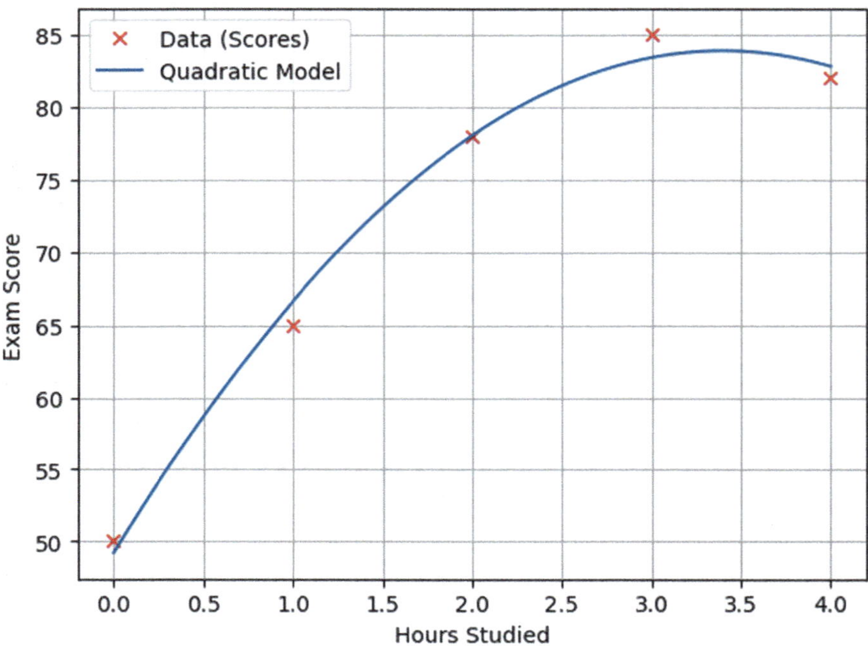

Fig. 10.13 Best quadratic fit for the relationship between study hours and exam scores

Python Example: Quadratic Regression

```python
import numpy as np

C = np.array([[1, 0, 0],
              [1, 1, 1],
              [1, 2, 4],
              [1, 3, 9],
              [1, 4, 16]])

y = np.array([50, 65, 78, 85, 82])

a = np.linalg.inv(C.T @ C) @ C.T @ y

print("Coefficients:", np.round(a, 2))
```

Output:

```
Coefficients: [49.2 20.4 -3.]
```

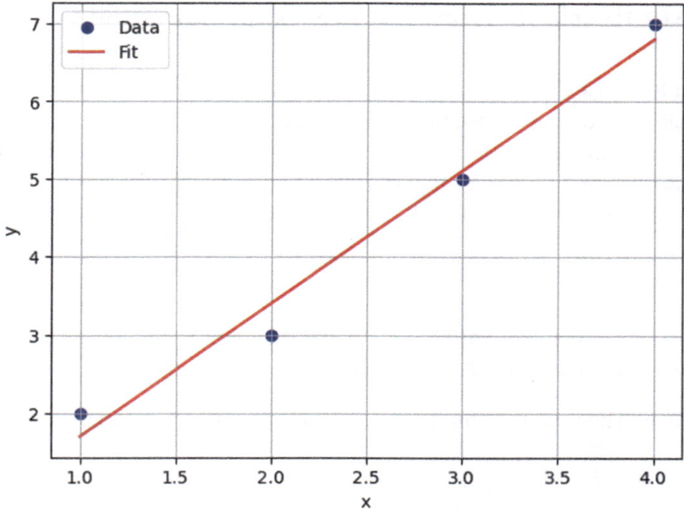

Fig. 10.14 A simple LR model fitted to a small dataset. The blue dots represent observed data points, while the red line shows the best-fit line obtained via least squares

---

Python Example: LR

```python
import numpy as np
import matplotlib.pyplot as plt
C = np.array([[1, 1], [1, 2], [1, 3], [1, 4]])
y = np.array([2, 3, 5, 7])
a = np.linalg.inv(C.T @ C) @ C.T @ y
x = np.array([1, 2, 3, 4])
plt.scatter(x, y, color='blue', label='Data')
plt.plot(x, a[0] + a[1] * x, 'r-', label='Fit')
plt.xlabel('x')
plt.ylabel('y')
plt.legend()
plt.grid(True)
plt.show()
```

Output:

```
See Figure 10.14.
```

Figure 10.14 illustrates the process of fitting a LR model to a set of data points. The goal is to minimize the residual error between the observed values (blue dots) and the predicted values from the linear model (red line). This visualization is fundamental in understanding how regression captures underlying trends in data through a best-fit approach.

10.11　Multivariable Least Squares Approximation

Multivariable Least Squares Approximation is a technique used to find the best-fitting linear relationship between multiple independent variables and a dependent variable by minimizing the sum of squared residuals. The residuals are the differences between the observed values y_A and the predicted values based on the model. The general form of the multiple LR model is given by:

$$y_A = a_0 + a_1 x_A + a_2 x_B,$$

where: y_A is the dependent variable, a_0 is the intercept term, and a_1 and a_2 are the coefficients for the independent variables x_A and x_B, respectively.

(a) Objective: The goal is to minimize the sum of squared residuals, defined as:

$$S = \sum_{i=1}^{n} (y_{A,i} - (a_0 + a_1 x_{A,i} + a_2 x_{B,i}))^2,$$

where n is the number of observations, and $y_{A,i}$, $x_{A,i}$, and $x_{B,i}$ are the observed values for the i-th data point.

(b) Solution: To find the optimal coefficients a_0, a_1, and a_2, we take the partial derivatives of S with respect to each coefficient and set them to zero:

$$\frac{\partial S}{\partial a_0} = -2 \sum_{i=1}^{n} (y_{A,i} - (a_0 + a_1 x_{A,i} + a_2 x_{B,i})) = 0,$$

$$\frac{\partial S}{\partial a_1} = -2 \sum_{i=1}^{n} (y_{A,i} - (a_0 + a_1 x_{A,i} + a_2 x_{B,i})) x_{A,i} = 0,$$

$$\frac{\partial S}{\partial a_2} = -2 \sum_{i=1}^{n} (y_{A,i} - (a_0 + a_1 x_{A,i} + a_2 x_{B,i})) x_{B,i} = 0.$$

These equations form a system of linear equations (normal equations) that can be solved to obtain the least squares estimates of a_0, a_1, and a_2. In matrix form, this can be expressed as:

$$\mathbf{X}^T \mathbf{X} \mathbf{a} = \mathbf{X}^T \mathbf{y},$$

where: $\mathbf{X}$ is the design matrix with rows $[1, x_{A,i}, x_{B,i}]$, $\mathbf{y}$ is the vector of observed $y_{A,i}$ values, and $\mathbf{a}$ is the vector of coefficients $[a_0, a_1, a_2]^T$.
The solution is:

$$\mathbf{a} = (\mathbf{X}^T \mathbf{X})^{-1} \mathbf{X}^T \mathbf{y}.$$

10.12 Orthogonality in ML

Orthogonality is pivotal in ML, enabling efficient data representations, decorrelation, and optimization.

Applications

- **PCA**: Projects data onto orthogonal principal components, maximizing variance and reducing dimensionality.
- **Clustering (k-means, hierarchical)**: Uses Euclidean distances and orthogonal projections to group data points based on similarity.
- **Cosine Similarity**: Measures similarity between vectors via dot products, widely used in NLP and recommendation systems.
- **LR**: Solves the least squares problem by minimizing the error vector orthogonal to the column space of the input matrix.
- **Neural Network Optimization**: Orthogonal weight initialization helps reduce neuron correlation, improving learning stability and convergence speed.
- **Kernel Methods**: Employ dot products in high-dimensional feature spaces to enable linear separation of non-linear data via kernel trick.

Python Example: PCA

```python
import numpy as np
X = np.array([[1, 2], [3, 4], [5, 6]])
C = np.cov(X.T, bias=False)
eigvals, eigvecs = np.linalg.eigh(C)
basis = eigvecs[:, np.argsort(eigvals)[::-1]]
print("PCA basis:\n", basis)
```

Output:

```
PCA basis:
 [[ 0.71 -0.71]
 [ 0.71  0.71]]
```

Summary: Nonsingular vs. Singular Matrices

In the previous chapters, we studied different topics, and here is a consolidated summary of what we have learned so far regarding the properties of square matrices.

A square matrix $A \in \mathbb{R}^{n \times n}$ can be classified as either **nonsingular** or **singular**, depending on its invertibility and the behavior of its algebraic, geometric, and numerical properties. These two categories are fundamental in linear algebra, as they determine the solvability of linear systems, the structure of vector spaces, the stability of numerical algorithms, and the spectral characteristics of A.

A matrix $A \in \mathbb{R}^{n \times n}$ is called **nonsingular** (or **invertible**) if there exists a matrix $A^{-1} \in \mathbb{R}^{n \times n}$ such that

$$AA^{-1} = A^{-1}A = I_n,$$

where I_n is the $n \times n$ identity matrix. Equivalently, A is nonsingular if and only if $\det(A) \neq 0$.

A matrix $A \in \mathbb{R}^{n \times n}$ is called **singular** if it is not invertible. Equivalently, A is singular if and only if $\det(A) = 0$, which implies that its rows or columns are linearly dependent, the rank of A is strictly less than n, and zero is an eigenvalue of A.

Property	Nonsingular Matrix	Singular Matrix
Invertibility	Invertible, A^{-1} exists, $AA^{-1} = I$.	Not invertible, A^{-1} does not exist.
Determinant	$\det(A) \neq 0$.	$\det(A) = 0$.
Linear Independence	Columns and rows are linearly independent.	Columns and rows are linearly dependent.
Solutions of $Ax = b$	Unique solution for all $b \in \mathbb{R}^n$.	Either no solution or infinitely many.
Solutions of $Ax = 0$	Only the trivial solution $x = 0$.	Infinitely many solutions.
Rank and Pivots	Full rank $r = n$, n nonzero pivots.	Rank $r < n$, fewer than n pivots.
RREF Form	$R = I$.	RREF has at least one zero row.
Null Space Dimension	Null space contains only $\{0\}$.	Null space has dimension $n - r > 0$.
Column Space	Spans $\mathbb{R}^n$.	Proper subspace of $\mathbb{R}^n$.
Row Space	Spans $\mathbb{R}^n$.	Proper subspace of $\mathbb{R}^n$.
Eigenvalues	All eigenvalues are nonzero.	Zero is an eigenvalue.
Singular Values	n strictly positive singular values.	Fewer than n positive singular values.
Condition Number	Finite condition number.	Condition number is infinite.
Quadratic Form $A^T A$	Symmetric and positive definite.	Positive semidefinite.
System Stability	Stable solutions, small perturbations in b give small changes in x.	Unstable solutions, small perturbations may cause large deviations.

- **A nonsingular matrix** guarantees full rank, invertibility, stability, and well-posed solutions to linear systems.

> - A **singular matrix** leads to rank deficiency, dependence in rows/columns, loss of invertibility, and potential non-uniqueness or inconsistency in solutions.

Chapter Summary

This chapter provided a comprehensive examination of orthogonality, a key concept in linear algebra that enables the measurement of lengths, angles, and distances, as well as the decomposition of vectors into independent components. It began with norms and distances in Sect. 10.1, defining the Euclidean norm $\|\mathbf{v}\| = \sqrt{\sum v_i^2}$ for vectors like $\mathbf{v} = [1, 2, 3]^T$ yielding $\sqrt{14} \approx 3.742$, and distance as $\|\mathbf{u} - \mathbf{v}\| \approx 5.916$ for $\mathbf{u} = [2, -3, 0]^T$, visualized in 3D projections (Fig. 10.1) and 2D illustrations (Fig. 10.2).

The dot product and orthogonality were introduced in Sect. 10.2, with $\mathbf{u} \cdot \mathbf{v} = \sum u_i v_i = \|\mathbf{u}\| \|\mathbf{v}\| \cos \theta$ measuring alignment, such as $\mathbf{v} \cdot \mathbf{w} = 0$ confirming perpendicularity for $\mathbf{v} = [1, 4, -2]^T, \mathbf{w} = [-8, 3, 2]^T$. Properties included self-dot equaling norm squared and commutativity, with Python verifying orthogonality.

Section 10.3 extended the Pythagorean theorem to orthogonal vectors ($\|\mathbf{u} + \mathbf{v}\|^2 = \|\mathbf{u}\|^2 + \|\mathbf{v}\|^2$), proven algebraically and applied to rhombus conditions for parallelograms (Fig. 10.3). Cauchy-Schwarz ($|\mathbf{u} \cdot \mathbf{v}| \leq \|\mathbf{u}\| \|\mathbf{v}\|$) and triangle inequalities in Sect. 10.4 bounded interactions, exemplified by $12 \leq \sqrt{29} \cdot \sqrt{30} \approx 29.5$.

Orthogonal complements in Sect. 10.5 were defined as $S^\perp = \{\mathbf{v} \mid \mathbf{v} \cdot \mathbf{w} = 0 \ \forall \ \mathbf{w} \in S\}$, with bases computed via null spaces for spans like Span$\{[1, 1, -1, 4]^T, [1, -1, 1, 2]^T\}$, yielding $\{[0, 1, 1, 0]^T, [-3, -1, 0, 1]^T\}$. Properties in Sect. 10.5.1 included $S^\perp$ as a subspace, $\mathbb{R}^n = S \oplus S^\perp, (S^\perp)^\perp = S, \dim(S) + \dim(S^\perp) = n, S \cap S^\perp = \{\mathbf{0}\}$, and matrix links (Row$(A)^\perp = $ Null(A), Col$(A)^\perp = $ Null(A^T)).

Orthogonal projections in Sect. 10.8 found the closest subspace vector, with formulas for orthonormal bases $\sum(\mathbf{v} \cdot \mathbf{u}_i)\mathbf{u}_i$ and matrices $P_S = A(A^T A)^{-1} A^T$, as in projecting $[4, 1]^T$ onto Span$\{[2, 1]^T\}$ yielding $[3.6, 1.8]^T$ (Fig. 10.7). Least squares in Sect. 10.9 minimized errors via normal equations $(C^T C)\mathbf{a} = C^T \mathbf{y}$, exemplified by linear fits like $y = 65.2 - 3.6x$ for study hours data and quadratic $y = 49.2 + 20.39x - 3x^2$ (Fig. 10.13). Multivariable extensions handled multiple variables with design matrices.

Applications in Sect. 10.12 showcased orthogonality in PCA (orthogonal eigenvector projections), clustering (Euclidean distances), cosine similarity, regression (error orthogonality), neural optimization, and kernels. An appendix summarized nonsingular vs. singular matrices, contrasting invertibility and rank.

Takeaways
Key insights were captured as:

- Norms and distances quantify vector magnitudes and separations, foundational for geometric computations.
- Dot products reveal alignment and orthogonality, with properties ensuring algebraic consistency.
- Orthogonal complements decompose spaces into $S \oplus S^{\perp}$, linking to null/row/column spaces via dimension formula $\dim(S) + \dim(S^{\perp}) = n$.
- Projections minimize distances to subspaces, enabling least squares for linear/quadratic/multivariable regression via normal equations.
- Applications in ML, from PCA decorrelation to Kernel Methods and cosine-based similarities.

This chapter empowered readers to leverage orthogonality for vector analysis, approximation, and data modeling in linear algebra and ML.

Exercises

1. Compute the Euclidean norm of the vector $\mathbf{v} = \begin{bmatrix} 1 \\ 2 \\ -3 \end{bmatrix}$ and verify using NumPy code.

2. Determine if the vectors $\mathbf{u} = \begin{bmatrix} 2 \\ -1 \\ 0 \end{bmatrix}$ and $\mathbf{v} = \begin{bmatrix} 0 \\ 3 \\ 4 \end{bmatrix}$ are orthogonal. Compute the angle between them.

3. For $\mathbf{u} = \begin{bmatrix} 1 \\ 1 \end{bmatrix}$ and $\mathbf{v} = \begin{bmatrix} 2 \\ 3 \end{bmatrix}$, verify the Cauchy-Schwarz inequality $|\mathbf{u} \cdot \mathbf{v}| \leq \|\mathbf{u}\| \|\mathbf{v}\|$.

4. Verify the Pythagorean theorem for the orthogonal vectors $\mathbf{u} = \begin{bmatrix} 1 \\ 0 \end{bmatrix}$ and $\mathbf{v} = \begin{bmatrix} 0 \\ 2 \end{bmatrix}$.

5. Compute the distance between $\mathbf{u} = \begin{bmatrix} 3 \\ 1 \end{bmatrix}$ and $\mathbf{v} = \begin{bmatrix} 0 \\ 4 \end{bmatrix}$, and visualize in 2D.

6. If $\mathbf{v} = \begin{bmatrix} 1 \\ -1 \\ 2 \end{bmatrix}$, find all vectors orthogonal to $\mathbf{v}$ by describing the solution set parametrically, and identify the geometric object.

7. Determine the basis for the orthogonal complement of $S = \text{Span}\left\{\begin{bmatrix} 1 \\ 0 \\ 1 \end{bmatrix}\right\}$ in $\mathbb{R}^3$.

8. Compute the orthogonal projection of $\mathbf{v} = \begin{bmatrix} 4 \\ 2 \\ 6 \end{bmatrix}$ onto the line spanned by $\mathbf{u} = \begin{bmatrix} 1 \\ 1 \\ 1 \end{bmatrix}$.

9. For the subspace $W = \text{Span}\left\{\begin{bmatrix} 1 \\ 1 \\ 0 \end{bmatrix}, \begin{bmatrix} 0 \\ 1 \\ 1 \end{bmatrix}\right\}$, in $\mathbb{R}^3$ with the standard inner product, verify that $\dim(W) + \dim(W^\perp) = 3$.

10. Decompose $\mathbf{x} = \begin{bmatrix} 3 \\ 1 \\ -2 \end{bmatrix}$ into components parallel and orthogonal to $\mathbf{u} = \begin{bmatrix} 1 \\ 2 \\ 3 \end{bmatrix}$.

11. Solve the least squares problem $A\mathbf{x} = \mathbf{b}$ where $A = \begin{bmatrix} 1 & 1 \\ 1 & 2 \\ 1 & 3 \end{bmatrix}$, $\mathbf{b} = \begin{bmatrix} 1 \\ 2 \\ 3 \end{bmatrix}$, and verify the error is orthogonal to the column space.

12. Fit a LR model $y = a_0 + a_1 x$ to the data points $(1, 2)$, $(2, 4)$, $(3, 5)$, $(4, 4)$ using normal equations.

13. For the quadratic model $y = a_0 + a_1 x + a_2 x^2$, find coefficients for data $(0, 1)$, $(1, 2)$, $(2, 1)$, $(3, 0)$ using least squares.

14. Prove that if $\mathbf{u} \cdot \mathbf{v} = 0$, then $\|\mathbf{u} + \mathbf{v}\|^2 = \|\mathbf{u}\|^2 + \|\mathbf{v}\|^2$.

15. Compute the projection matrix P_W for $W = \text{Span}\left\{\begin{bmatrix} 1 \\ 1 \end{bmatrix}\right\}$ and verify $P_W^2 = P_W$.

16. Find the distance from point $\mathbf{p} = \begin{bmatrix} 1 \\ 2 \\ 3 \end{bmatrix}$ to the plane $x + y + z = 0$.

17. Verify that for an orthogonal matrix Q, the columns satisfy $Q^T Q = I$ and the rows satisfy $Q Q^T = I$.

18. For the matrix $A = \begin{bmatrix} 1 & 2 & 3 \\ 4 & 5 & 6 \\ 7 & 8 & 9 \end{bmatrix}$, compute $\text{Col}(A)^\perp$ and verify it equals $\text{Null}(A^T)$.

19. In PCA context, center the data and compute the first principal component for the 2D dataset $[(1, 2), (3, 4), (5, 6)]$.

20. Prove the triangle inequality $\|\mathbf{u} + \mathbf{v}\| \leq \|\mathbf{u}\| + \|\mathbf{v}\|$ using the dot product.

21. Find the least squares solution for the overdetermined system with $A = \begin{bmatrix} 1 & 0 \\ 0 & 1 \\ 1 & 1 \end{bmatrix}$, $\mathbf{b} = \begin{bmatrix} 1 \\ 2 \\ 4 \end{bmatrix}$.

22. For $S = \text{Span} \left\{ \begin{bmatrix} 1 \\ 0 \\ 0 \\ 0 \end{bmatrix}, \begin{bmatrix} 0 \\ 1 \\ 0 \\ 0 \end{bmatrix} \right\}$, find a basis for $S^\perp$ in $\mathbb{R}^4$.

23. Verify that the error vector in least squares regression is orthogonal to each column of the design matrix.

24. Fit a multiple LR model $y = a_0 + a_1 x_1 + a_2 x_2$ to at least four data points and compute R^2.

25. Show that for any subspace W, $(W^\perp)^\perp = W$ using the definition of orthogonal complement.

Chapter 11
Matrix Decompositions: Factorization and SVD

Introduction

Matrix decompositions are fundamental tools in linear algebra that reveal the underlying structure of matrices and enable efficient solutions to a wide range of computational problems. This chapter explores key factorization techniques, beginning with essential concepts such as norms, dot products, and orthogonality, which form the building blocks for understanding more advanced decompositions. We study QR and LU, which are crucial for solving linear systems and least squares problems, and examine the spectral decomposition of symmetric matrices. The chapter culminates in a detailed treatment of the SVD, a versatile method that generalizes eigenvalue decomposition to non-square matrices and finds extensive applications in data analysis and ML.

By knowing these topics, readers will gain insights into how decompositions simplify complex operations, improve numerical stability, and facilitate dimensionality reduction. The chapter also highlights practical applications in ML, including PCA, regression, and feature selection, demonstrating the real-world relevance of these mathematical concepts. Through examples, proofs, and computational illustrations, we aim to provide a solid foundation for applying matrix decompositions in both theoretical and applied contexts.

Topics Covered

This chapter is organized as follows:

- **Section** 11.1 **Matrix Norms**: Defines different types of matrix norms, including Frobenius, ℓ_1, and spectral norms, with examples.
- **Section** 11.2 **Orthonormality**: Explores the concepts of orthogonal and orthonormal vectors and functions, and their geometric meaning.
- **Section** 11.3 **Orthogonal Matrices**: Defines orthogonal matrices and their key properties, including norm preservation, inverse-transpose equivalence $Q^{-1} =$

© The Author(s), under exclusive license to Springer Nature Singapore Pte Ltd. 2026   331
Md. Jalil Piran, *Linear Algebra with Applications in Machine Learning*,
https://doi.org/10.1007/978-981-95-5167-5_11

Q^T, and determinant constraints. Introduces QR factorization via the Gram-Schmidt process, with applications to least-squares problems and numerical stability.

- **Section** 11.4 **Orthogonal Operators and Their Properties**: Defines orthogonal linear operators and establishes their equivalence with orthogonal matrices. Presents key properties, including closure under composition and inversion, and demonstrates construction of orthogonal operators mapping specified vectors.
- **Section** 11.5 **Orthogonal Diagonalization of a Symmetric Matrix**: Demonstrates the procedure for orthogonally diagonalizing a symmetric matrix $A = PDP^T$ by computing eigenvalues, constructing orthonormal eigenvectors, and verifying the factorization.
- **Section** 11.6 **Symmetric Matrices and Spectral Decomposition**: States and proves the Spectral Theorem for real symmetric matrices, establishes orthogonality of eigenvectors corresponding to distinct eigenvalues, and develops the decomposition of symmetric matrices into sums of rank-1 projection matrices scaled by eigenvalues.
- **Section** 11.7 **LU Decomposition**: Details the factorization of a matrix into a unit lower triangular matrix L and an upper triangular matrix UUU using Gaussian elimination. Discusses partial pivoting $PA = LU$ for numerical stability.
- **Section** 11.8 **Singular Value Decomposition (SVD)**: Provides a comprehensive treatment of SVD ($A = \Sigma V^T$) for arbitrary matrices, covering singular values and vectors, economy and truncated SVD, low-rank approximation via the Eckart-Young-Mirsky theorem, and solving overdetermined, underdetermined, and rank-deficient linear systems using the pseudoinverse.
- **Section** 11.9 **ML Applications**: Highlights applications of matrix decompositions in machine learning, including PCA for dimensionality reduction, QR-based regression, SVD-driven feature selection, image compression, and latent semantic analysis.
- **Summary and Exercises**: Recaps key concepts and provides a set of problems to reinforce theoretical concepts and practical computations.

All code accompanying the chapter is publicly available at https://github.com/jalil-piran/Linear-Algebra-with-Applications-in-Machine-Learning/blob/main/Chapter_11_Decompositions.ipynb.

Acronyms

SVD　Singular Value Decomposition
ML　Machine Learning
QR　Q (orthogonal) and R (upper triangular) factors
LU　L (lower triangular) and U (upper triangular) factors
PCA　Principal Component Analysis
GD　Gradient Descent
LR　Linear Regression
LSA　Latent Semantic Analysis

NLP Natural Language Processing
RMSE Root Mean Square Error

11.1 Matrix Norms

A **matrix norm** quantifies the "size" or "magnitude" of a matrix, analogous to a vector's length. Norms measure how much a matrix transformation stretches or compresses vectors, crucial for stability analysis and error estimation.

Definition

For $A = [a_{ij}] \in \mathbb{R}^{m \times n}$:

- **Frobenius Norm**:

$$\|A\|_F = \sqrt{\sum_{i=1}^{m} \sum_{j=1}^{n} |a_{ij}|^2}.$$

  Treats A as a flattened vector.
- **ℓ_1 Norm**:

$$\|A\|_1 = \max_{1 \leq j \leq n} \sum_{i=1}^{m} |a_{ij}|.$$

  Sums absolute entries.
- **ℓ_∞ Norm**:

$$\|A\|_\infty = \max_{1 \leq i \leq m} \sum_{j=1}^{n} |a_{ij}|.$$

  Maximum absolute row sum.
- **$\ell_{2,1}$ Norm**:

$$\|A\|_{2,1} = \sum_{j=1}^{n} \sqrt{\sum_{i=1}^{m} a_{ij}^2}.$$

  Sum of column Euclidean norms.
- **Spectral Norm**:

$$\|A\|_2 = \sqrt{\lambda_{\max}(A^T A)}.$$

  Largest singular value, measuring maximum stretch.

- **Max Norm:**

$$\|A\|_{\max} = \max_{i,j} |a_{ij}|.$$

Largest absolute entry.

The Frobenius norm is like the Euclidean length of a matrix's entries, while the spectral norm measures the maximum "amplification" of a vector under A. The ℓ_1 and ℓ_∞ norms focus on entry-wise or row-wise magnitudes, respectively.

Example 11.1 Consider the matrix:

$$A = \begin{bmatrix} 1 & -2 \\ 3 & 4 \end{bmatrix}.$$

- **Frobenius Norm** ($\|A\|_F$): Measures the square root of the sum of the squares of all elements in the matrix (akin to the Euclidean norm for matrices).

$$\|A\|_F = \sqrt{1^2 + (-2)^2 + 3^2 + 4^2} = \sqrt{1 + 4 + 9 + 16} = \sqrt{30} \approx 5.477.$$

- ℓ_1 **Norm** ($\|A\|_1$): Often defined as the maximum absolute column sum of the matrix.

$$\|A\|_1 = \max\{|1| + |3|, |-2| + |4|\} = \max\{4, 6\} = 6.$$

- ℓ_∞ **Norm** ($\|A\|_\infty$): Defined as the maximum absolute row sum of the matrix.

$$\|A\|_\infty = \max\{|1| + |-2|, |3| + |4|\} = \max\{3, 7\} = 7.$$

- **Max Norm** ($\|A\|_{\max}$): The largest absolute value among all entries of the matrix.

$$\|A\|_{\max} = \max\{|1|, |-2|, |3|, |4|\} = 4.$$

- **Spectral Norm** ($\|A\|_2$): Equal to the square root of the largest eigenvalue of $A^T A$. Represents the largest singular value of the matrix.

$$A^T A = \begin{bmatrix} 1 & 3 \\ -2 & 4 \end{bmatrix} \begin{bmatrix} 1 & -2 \\ 3 & 4 \end{bmatrix} = \begin{bmatrix} 10 & 10 \\ 10 & 20 \end{bmatrix}$$

Eigenvalues via characteristic polynomial:

$$\det(A^T A - \lambda I) = \begin{vmatrix} 10 - \lambda & 10 \\ 10 & 20 - \lambda \end{vmatrix} = (10 - \lambda)(20 - \lambda) - 100$$

$$= \lambda^2 - 30\lambda + 100 \Rightarrow \lambda = \frac{30 \pm \sqrt{900 - 400}}{2} = 15 \pm 5\sqrt{5}.$$

Largest eigenvalue: $\lambda_{\max} = 15 + 5\sqrt{5}$

$$\|A\|_2 = \sqrt{15 + 5\sqrt{5}} \approx 5.11.$$

- $\ell_{2,1}$ **Norm** ($\|A\|_{2,1}$): It computes the ℓ_2 norm of each row, then sums them:

$$\|A\|_{2,1} = \sum_{j=1}^{m} \|\mathbf{a}_i\|_2,$$

where $\mathbf{a}_i$ is the ith row of A. This is used in ML and sparse coding. For this matrix:

$$\|\text{Row}_1\|_2 = \sqrt{1^2 + (-23)^2} = \sqrt{5},$$

$$\|\text{Row}_2\|_2 = \sqrt{-2^2 + 4^2} = \sqrt{20},$$

$$\|A\|_{2,1} = \sqrt{10} + \sqrt{20} = 3.162 + 4.472 = 7.634.$$

Python Example: Matrix Norms

```python
import numpy as np
A = np.array([[1, -2], [3, 4]])
frob_norm = np.linalg.norm(A, 'fro')
l1_norm = np.linalg.norm(A, 1)
linf_norm = np.linalg.norm(A, np.inf)
spectral_norm = np.linalg.norm(A, 2)
max_norm = np.max(np.abs(A))
print(f"Frobenius: {frob_norm:.2f}")
print(f"L1: {l1_norm:.2f}")
l21_norm = np.sum(np.linalg.norm(A, axis=0))
print(f"L-infinity: {linf_norm:.2f}")
print(f"L2,1: {l21_norm:.2f}")
print(f"Spectral: {spectral_norm:.2f}")
print(f"Max: {max_norm:.2f}")
```

Output:

```
Frobenius norm: 5.48
L1 norm: 6.00
```

```
L-infinity norm: 7.00
L2,1 norm: 7.63
Spectral norm: 5.12
Max norm: 4.00
```

Example 11.2 Given:

$$A = \begin{bmatrix} 0 & 1 & 2 \\ 1 & 0 & 1 \end{bmatrix}.$$

- Matrix 1-Norm $\|A\|_1$
  The matrix 1-norm is defined as the maximum absolute column sum:

$$\|A\|_1 = \max_{1 \leq j \leq n} \sum_{i=1}^{m} |a_{ij}|.$$

We compute:

$$\text{Col 1: } |0| + |1| = 1, \quad \text{Col 2: } |1| + |0| = 1, \quad \text{Col 3: } |2| + |1| = 3$$

$$\Rightarrow \|A\|_1 = \max(1, 1, 3) = \boxed{3}$$

- Infinity Norm $\|A\|_\infty$
  The infinity norm is the maximum absolute row sum:

$$\|A\|_\infty = \max_{1 \leq i \leq m} \sum_{j=1}^{n} |a_{ij}|$$

Compute:

$$\text{Row 1: } |0| + |1| + |2| = 3, \quad \text{Row 2: } |1| + |0| + |1| = 2$$

$$\Rightarrow \|A\|_\infty = \max(3, 2) = \boxed{3}$$

- Frobenius Norm $\|A\|_F$
  The Frobenius norm is the square root of the sum of the squares of all entries in a matrix.

$$\|A\|_F = \sqrt{\sum_{i=1}^{m} \sum_{j=1}^{n} |a_{ij}|^2} = \sqrt{0^2 + 1^2 + 2^2 + 1^2 + 0^2 + 1^2}$$

$$= \sqrt{0 + 1 + 4 + 1 + 0 + 1} = \sqrt{7}.$$

$$\Rightarrow \boxed{\|A\|_F = \sqrt{7} \approx 2.6458}$$

- **Mixed Norm** $\|A\|_{2,1}$
  This norm is the sum of Euclidean norms of columns:

$$\|A\|_{2,1} = \sum_{j=1}^{n} \left(\sqrt{\sum_{i=1}^{m} |a_{ij}|^2} \right).$$

We compute column norms:

$$\text{Col 1: } \sqrt{0^2 + 1^2} = \sqrt{1} = 1,$$

$$\text{Col 2: } \sqrt{1^2 + 0^2} = \sqrt{1} = 1,$$

$$\text{Col 3: } \sqrt{2^2 + 1^2} = \sqrt{4+1} = \sqrt{5}.$$

$$\Rightarrow \boxed{\|A\|_{2,1} = 1 + 1 + \sqrt{5} \approx 3 + 2.236 = 4.236}$$

- **Spectral Norm** $\|A\|_2$
  The spectral norm is the largest singular value of A. To compute it, first we find $A^T A$, then, we compute eigenvalues of $A^T A$, and finally the largest eigenvalue is the spectral norm.

1. Compute $A^T A$:

$$A^T A = \begin{bmatrix} 0 & 1 \\ 1 & 0 \\ 2 & 1 \end{bmatrix} \begin{bmatrix} 0 & 1 & 2 \\ 1 & 0 & 1 \end{bmatrix} = \begin{bmatrix} 1 & 0 & 1 \\ 0 & 1 & 2 \\ 1 & 2 & 5 \end{bmatrix}.$$

2. Compute eigenvalues of $A^T A$. Let's denote the largest eigenvalue of $A^T A$ as $\lambda_{\max}$. Using numerical methods or software, we find the largest eigenvalue:

$$\lambda_{\max} \approx 6 \Rightarrow \|A\|_2 = \sqrt{\lambda_{\max}} \approx \boxed{2.44}$$

Example 11.3 The **spectral norm** of a matrix $\mathbf{X} \in \mathbb{R}^{m \times n}$, denoted $\|\mathbf{X}\|_2$, is defined as the largest singular value of $\mathbf{X}$. That is,

$$\|\mathbf{X}\|_2 = \sigma_{\max}(\mathbf{X}),$$

where $\sigma_1, \sigma_2, \ldots, \sigma_{\min(m,n)}$ are the singular values of $\mathbf{X}$. Let

$$\mathbf{X} = \begin{bmatrix} 3 & 1 \\ 0 & 2 \end{bmatrix}.$$

To compute the spectral norm, we first compute the singular values of $\mathbf{X}$. This involves finding the eigenvalues of $\mathbf{X}^T\mathbf{X}$:

$$\mathbf{X}^T\mathbf{X} = \begin{bmatrix} 3 & 0 \\ 1 & 2 \end{bmatrix} \begin{bmatrix} 3 & 1 \\ 0 & 2 \end{bmatrix} = \begin{bmatrix} 9 & 3 \\ 3 & 5 \end{bmatrix}.$$

The eigenvalues of this matrix are:

$$\lambda_1 = 10.6056, \quad \lambda_2 = 3.3944.$$

Thus, the singular values of $\mathbf{X}$ are:

$$\sigma_1 = \sqrt{10.6056} \approx 3.257, \quad \sigma_2 = \sqrt{3.3944} \approx 1.843.$$

Therefore, the spectral norm of $\mathbf{X}$ is:

$$\|\mathbf{X}\|_2 = \sigma_1 \approx 3.257.$$

11.2 Orthonormality

A set of vectors is **orthonormal** if each vector has unit length and is orthogonal to all others, like coordinate axes in a perfectly aligned grid.

Unit Vector: A vector $\mathbf{u}$ is called a *unit vector* if its Euclidean norm is 1:

$$\|\mathbf{u}\|_2 = \sqrt{u_1^2 + u_2^2 + \cdots + u_n^2} = 1.$$

Such vectors represent direction without magnitude and are often used to define axes, directions of motion, or to normalize other vectors.

Definition

- **Orthonormal Vectors**: Two vectors $\mathbf{u}, \mathbf{v} \in \mathbb{R}^n$ are said to be *orthonormal* if:
$$\mathbf{u} \cdot \mathbf{v} = 0 \ \ (\text{orthogonal}), \quad \|\mathbf{u}\|_2 = \|\mathbf{v}\|_2 = 1 \ \ (\text{unit length}).$$

  Orthonormal vectors are both perpendicular and of unit length. Sets of orthonormal vectors form the foundation of orthonormal bases in linear algebra, simplifying projection, decomposition, and SVD.
- **Normalization**: Given any nonzero vector $\mathbf{v}$, it can be converted to a unit vector by dividing it by its norm:

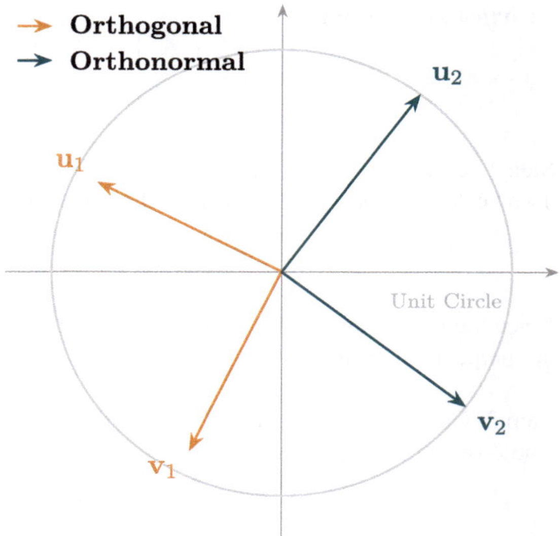

Fig. 11.1 Orthogonal vs. orthonormal vectors in $\mathbb{R}^2$

$$\hat{\mathbf{v}} = \frac{\mathbf{v}}{\|\mathbf{v}\|_2}.$$

This process, called normalization, preserves direction while standardizing length to 1. It is crucial in algorithms requiring standardized scales, such as cosine similarity and GD.

- **Orthonormal Matrix**: A matrix $Q \in \mathbb{R}^{n \times n}$ is called *orthonormal* (or orthogonal) if:

$$Q^T Q = Q Q^T = I_n,$$

where I_n is the identity matrix. This means the columns (and rows) of Q are orthonormal vectors. Orthonormal matrices preserve angles and lengths, making them fundamental in rotations, reflections, and SVD decompositions.

Orthonormal vectors are perpendicular and scaled to length 1, forming an ideal basis for projections and transformations without distortion.

Figure 11.1, compares the orthogonal vectors to orthonormal vectors in $\mathbb{R}^2$. The yellow pair $\mathbf{u}_1$, $\mathbf{v}_1$ represents orthogonal vectors, they meet at a right angle but do not have unit length. The teal pair $\mathbf{u}_2$, $\mathbf{v}_2$ illustrates orthonormal vectors, they are orthogonal and lie on the unit circle, hence also normalized. Orthonormal sets are particularly important in linear algebra, especially in the context of the SVD, where the columns of U and V matrices are orthonormal.

Example 11.4 Given:

$$\mathbf{u} = \begin{bmatrix} 4 \\ 2 \\ -1 \end{bmatrix}, \quad \mathbf{v} = \begin{bmatrix} 1 \\ -3 \\ -2 \end{bmatrix}.$$

Step 1- Check for Orthogonality:
Two vectors are orthogonal if their dot product is zero:

$$\mathbf{u} \cdot \mathbf{v} = 4 \cdot 1 + 2 \cdot (-3) + (-1) \cdot (-2) = 4 - 6 + 2 = 0.$$

Since the dot product is zero, the vectors $\mathbf{u}$ and $\mathbf{v}$ are orthogonal. This means they are perpendicular in $\mathbb{R}^3$.

Step 2- Compute ℓ_2-Norm (Euclidean Norm):
The ℓ_2-norm of a vector $\mathbf{x} = [x_1, x_2, x_3]^T$ is given by:

$$\|\mathbf{x}\|_2 = \sqrt{x_1^2 + x_2^2 + x_3^2}.$$

Compute for $\mathbf{u}$:

$$\|\mathbf{u}\|_2 = \sqrt{4^2 + 2^2 + (-1)^2} = \sqrt{16 + 4 + 1} = \sqrt{21}.$$

Compute for $\mathbf{v}$:

$$\|\mathbf{v}\|_2 = \sqrt{1^2 + (-3)^2 + (-2)^2} = \sqrt{1 + 9 + 4} = \sqrt{14}.$$

Step 3- Normalize the Vectors:
Normalization converts a vector to unit length by dividing it by its norm:

$$\hat{\mathbf{x}} = \frac{\mathbf{x}}{\|\mathbf{x}\|_2}.$$

For $\mathbf{u}$:

$$\hat{\mathbf{u}} = \frac{1}{\sqrt{21}} \begin{bmatrix} 4 \\ 2 \\ -1 \end{bmatrix}.$$

For $\mathbf{v}$:

$$\hat{\mathbf{v}} = \frac{1}{\sqrt{14}} \begin{bmatrix} 1 \\ -3 \\ -2 \end{bmatrix}.$$

As shown in Fig. 11.2, vectors $\mathbf{u}$ and $\mathbf{v}$ are orthogonal (i.e., their dot product is zero). Their norms measure the length or magnitude of each vector. The normalized vectors

$\hat{\mathbf{u}}$ and $\hat{\mathbf{v}}$ have unit length and preserve direction, which is essential for orthonormal bases in linear algebra and SVD.

Python Example: Orthonormalization

```python
import numpy as np
u = np.array([4, 2, -1])
v = np.array([1, -3, -2])
dot_product = np.dot(u, v)
u_norm = u / np.linalg.norm(u)
v_norm = v / np.linalg.norm(v)
print("Dot product:", dot_product)
print("Normalized u:", u_norm)
print("Normalized v:", v_norm)
```

Output:

```
Dot product: 0
Normalized u: [ 0.87  0.44 -0.22]
Normalized v: [ 0.27 -0.8  -0.53]
```

Orthogonal Subspaces

Two subspaces A and B of $\mathbb{R}^n$ are said to be **orthogonal** if:

$$\text{For every } \mathbf{a} \in A \text{ and every } \mathbf{b} \in B, \quad \mathbf{a} \cdot \mathbf{b} = 0.$$

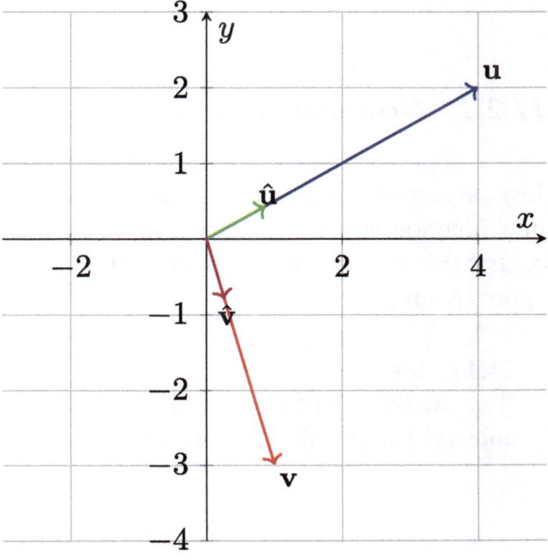

Fig. 11.2 Original and normalized orthogonal vectors (2D projection from R^3)

Example 11.5 Let subspace $A \subseteq \mathbb{R}^3$ be the set of all vectors of the form:

$$\mathbf{a} = \begin{bmatrix} a_1 \\ 0 \\ 0 \end{bmatrix}.$$

This means A is a line along the x-axis.

Let subspace $B \subseteq \mathbb{R}^3$ be the set of all vectors of the form:

$$\mathbf{b} = \begin{bmatrix} 0 \\ b_2 \\ b_3 \end{bmatrix}.$$

This describes the yz-plane (i.e., all vectors perpendicular to the x-axis).

Dot Product: Now take any vectors $\mathbf{a} \in A$ and $\mathbf{b} \in B$, and compute their dot product:

$$\mathbf{a} \cdot \mathbf{b} = \begin{bmatrix} a_1 \\ 0 \\ 0 \end{bmatrix} \cdot \begin{bmatrix} 0 \\ b_2 \\ b_3 \end{bmatrix} = a_1 \cdot 0 + 0 \cdot b_2 + 0 \cdot b_3 = 0.$$

Since the dot product is zero for *any* vectors $\mathbf{a} \in A$ and $\mathbf{b} \in B$, we conclude that:

$$\boxed{\text{Subspaces } A \text{ and } B \text{ are orthogonal.}}$$

Orthogonal subspaces arise frequently in linear algebra and numerical analysis, particularly in the context of orthogonal complements, projections, and the fundamental subspaces of a matrix (e.g., row space vs. null space).

11.2.1 Orthogonal Functions and Inner Product

Just as vectors can be orthogonal in Euclidean space, functions can be orthogonal in a function space with respect to a defined inner product. This concept plays a central role in areas such as Fourier analysis, orthogonal polynomials, and functional approximation.

Definition
Two real-valued functions $f(x)$ and $g(x)$ are said to be **orthogonal** on an interval $[a, b]$ if their inner product is zero:

$$\langle f, g \rangle = \int_a^b f(x)g(x)\,dx = 0$$

This inner product is an extension of the dot product and captures the idea of "perpendicularity" between functions over an interval.

Example 11.6 Let:

$$f(x) = x, \quad g(x) = 1, \quad \text{interval: } [-1, 1].$$

Compute the inner product:

$$\langle f, g \rangle = \int_{-1}^1 x \cdot 1 \, dx = \int_{-1}^1 x \, dx = \left. \frac{x^2}{2} \right|_{-1}^1 = \frac{1^2}{2} - \frac{(-1)^2}{2} = \frac{1}{2} - \frac{1}{2} = 0.$$

Since the inner product is zero, $f(x) = x$ and $g(x) = 1$ are orthogonal functions on the interval $[-1, 1]$.

Orthogonality of functions is always with respect to a particular interval and inner product definition. Two functions that are orthogonal on one interval may not be orthogonal on another.

Orthogonal functions are fundamental in constructing orthonormal bases in function spaces (e.g., Legendre, Chebyshev, Fourier bases), which are used in:

- Solving differential equations,
- Signal processing and compression,
- Data approximation and ML (e.g., PCA in function space).

Example 11.7 Let:

$$f(x) = 1 + x, \quad g(x) = x - x^2.$$

We want to verify whether f and g are orthogonal on the interval $[-2, 2]$.

Step 1- Compute the Inner Product

$$\langle f, g \rangle = \int_{-2}^2 (1 + x)(x - x^2)\,dx.$$

Step 2- Expand the Product

$$(1 + x)(x - x^2) = x - x^2 + x^2 - x^3 = x - x^3.$$

Step 3- Integrate

$$\langle f, g \rangle = \int_{-2}^{2} (x - x^3)\, dx = \int_{-2}^{2} x\, dx - \int_{-2}^{2} x^3\, dx.$$

Step 4- Use Symmetry of Odd Functions
Both x and x^3 are odd functions, and their integrals over symmetric intervals cancel out:

$$\int_{-2}^{2} x\, dx = 0, \quad \int_{-2}^{2} x^3\, dx = 0.$$

$$\Rightarrow \langle f, g \rangle = 0 - 0 = \boxed{0}$$

Since $\langle f, g \rangle = 0$, the functions $f(x) = 1 + x$ and $g(x) = x - x^2$ are **orthogonal** on the interval $[-2, 2]$.

11.3   Orthogonal Matrices

Orthogonal matrices play a central role in linear algebra, particularly in preserving geometric structures such as length and angle under transformations. These matrices appear in many applications, including numerical methods, physics, and ML algorithms.

An $n \times n$ matrix Q is **orthogonal** if:

$$Q^T Q = Q Q^T = I_n.$$

This means Q preserves lengths and angles, acting like a rotation or reflection.

11.3.1   Key Properties of Orthogonal Matrices

- **Inverse**: For an orthogonal matrix Q, the inverse is equal to its transpose:

$$Q^{-1} = Q^T.$$

This follows from the definition $Q^T Q = I$, and it greatly simplifies computations, especially in solving linear systems and performing matrix decompositions like QR and SVD.

- **Norm Preservation**: Applying Q to any vector $\mathbf{u}$ preserves its Euclidean norm:

$$\|Q\mathbf{u}\|_2 = \|\mathbf{u}\|_2.$$

This means that orthogonal transformations are *isometries*; they do not stretch or shrink vectors but only rotate or reflect them in space.

- **Inner Product**: Orthogonal transformations preserve the dot product (and hence the angle) between any two vectors:

$$\mathbf{u} \cdot \mathbf{v} = (Q\mathbf{u}) \cdot (Q\mathbf{v}).$$

This is crucial in applications like PCA and SVD, where geometric relationships (like orthogonality or projections) must remain intact after transformation.

- **Determinant**: The determinant of an orthonormal matrix is either $+1$ (pure rotation) or -1 (rotation with reflection):

$$\det(Q) = \pm 1.$$

This indicates that the transformation preserves volume and orientation (if $+1$) or reverses orientation (if -1).

- **Stability**: The set of orthogonal matrices is closed under matrix multiplication and inversion:

$$Q_1, Q_2 \text{ orthogonal} \Rightarrow Q_1 Q_2 \text{ and } Q_1^{-1} \text{ are also orthogonal.}$$

This property is fundamental in numerical linear algebra for designing stable and efficient algorithms, especially for decompositions and iterative methods.

Orthogonal matrices are rigid transformations, like spinning a globe without stretching it. They preserve the geometry of space, making them ideal for coordinate changes.

Example 11.8 We want to verify orthogonal matrix
Given matrix:

$$A = \begin{bmatrix} \frac{1}{\sqrt{2}} & \frac{1}{\sqrt{2}} \\ \frac{1}{\sqrt{2}} & -\frac{1}{\sqrt{2}} \end{bmatrix}$$

Step 1- Check Orthonormality of Columns
Let:

$$\mathbf{a}_1 = \begin{bmatrix} \frac{1}{\sqrt{2}} \\ \frac{1}{\sqrt{2}} \end{bmatrix}, \quad \mathbf{a}_2 = \begin{bmatrix} \frac{1}{\sqrt{2}} \\ -\frac{1}{\sqrt{2}} \end{bmatrix}.$$

Check Orthogonality:

$$\mathbf{a}_1 \cdot \mathbf{a}_2 = \left(\frac{1}{\sqrt{2}}\right)\left(\frac{1}{\sqrt{2}}\right) + \left(\frac{1}{\sqrt{2}}\right)\left(-\frac{1}{\sqrt{2}}\right) = \frac{1}{2} - \frac{1}{2} = 0.$$

Check Normalization:

$$\|\mathbf{a}_1\| = \sqrt{\left(\frac{1}{\sqrt{2}}\right)^2 + \left(\frac{1}{\sqrt{2}}\right)^2} = \sqrt{\frac{1}{2} + \frac{1}{2}} = \sqrt{1} = 1,$$

$$\|\mathbf{a}_2\| = \sqrt{\left(\frac{1}{\sqrt{2}}\right)^2 + \left(-\frac{1}{\sqrt{2}}\right)^2} = \sqrt{\frac{1}{2} + \frac{1}{2}} = \sqrt{1} = 1.$$

Since the columns are orthogonal and have unit length, they are **orthonormal**.

Step 2- Verify Orthogonality of Matrix

A matrix is orthogonal if:

$$A^T A = I \quad \text{and} \quad A^{-1} = A^T.$$

We compute the transpose:

$$A^T = \begin{bmatrix} \frac{1}{\sqrt{2}} & \frac{1}{\sqrt{2}} \\ \frac{1}{\sqrt{2}} & -\frac{1}{\sqrt{2}} \end{bmatrix}^T = \begin{bmatrix} \frac{1}{\sqrt{2}} & \frac{1}{\sqrt{2}} \\ \frac{1}{\sqrt{2}} & -\frac{1}{\sqrt{2}} \end{bmatrix}.$$

Now we compute $A^T A$:

$$A^T A = \begin{bmatrix} \frac{1}{\sqrt{2}} & \frac{1}{\sqrt{2}} \\ \frac{1}{\sqrt{2}} & -\frac{1}{\sqrt{2}} \end{bmatrix} \begin{bmatrix} \frac{1}{\sqrt{2}} & \frac{1}{\sqrt{2}} \\ \frac{1}{\sqrt{2}} & -\frac{1}{\sqrt{2}} \end{bmatrix} = \begin{bmatrix} 1 & 0 \\ 0 & 1 \end{bmatrix} = I.$$

Since the columns of A are orthonormal and $A^T A = I$, we conclude:

$$\boxed{A \text{ is an orthogonal matrix} \quad \Rightarrow \quad A^{-1} = A^T}$$

Orthogonal matrices preserve length and angle, and they play a key role in transformations such as rotations and reflections in linear algebra, as well as in SVD and QR factorizations.

Another way to determine whether a matrix is orthogonal is by checking the product of the matrix and its transpose.

Definition

A matrix $A \in \mathbb{R}^{n \times n}$ is **orthogonal** if:

$$A^T A = A A^T = I_n$$

This implies that:

$$A^{-1} = A^T$$

Orthogonal matrices preserve vector norms and angles, making them fundamental in numerical linear algebra and transformations.

Example 11.9 Let:

$$A = \begin{bmatrix} 0.7 & -0.3 \\ 0.3 & 0.7 \end{bmatrix}.$$

Compute the transpose:

$$A^T = \begin{bmatrix} 0.7 & 0.3 \\ -0.3 & 0.7 \end{bmatrix}.$$

Now compute the product $A A^T$:

$$
\begin{aligned}
A A^T &= \begin{bmatrix} 0.7 & -0.3 \\ 0.3 & 0.7 \end{bmatrix} \begin{bmatrix} 0.7 & 0.3 \\ -0.3 & 0.7 \end{bmatrix} \\
&= \begin{bmatrix} 0.7 \cdot 0.7 + (-0.3) \cdot (-0.3) & 0.7 \cdot 0.3 + (-0.3) \cdot 0.7 \\ 0.3 \cdot 0.7 + 0.7 \cdot (-0.3) & 0.3 \cdot 0.3 + 0.7 \cdot 0.7 \end{bmatrix}. \\
&= \begin{bmatrix} 0.49 + 0.09 & 0.21 - 0.21 \\ 0.21 - 0.21 & 0.09 + 0.49 \end{bmatrix} = \begin{bmatrix} 0.58 & 0 \\ 0 & 0.58 \end{bmatrix} \neq I_2.
\end{aligned}
$$

Since $A A^T \neq I$, the matrix A is not orthogonal. While the columns may appear "almost" orthogonal, they are not unit-length, and the inner products do not produce exact orthonormality. This method is reliable for verifying orthogonality in practice, especially when working with floating-point approximations in numerical computing.

Python Example: Orthogonal Matrix

```python
import numpy as np
Q = np.array([[np.sqrt(2)/2, -np.sqrt(2)/2],
        [np.sqrt(2)/2, np.sqrt(2)/2]])
is_orthogonal = np.allclose(Q.T @ Q, np.eye(2))
det_Q = np.linalg.det(Q)
```

```
5  print("Is orthogonal:", is_orthogonal)
6  print("Determinant:", det_Q)
```

Output:

```
Is orthogonal: True
Determinant: 1.0
```

Solving systems of linear equations often requires stable numerical methods, especially when dealing with large matrices or systems that arise in practical applications like data fitting and signal processing. Traditional direct methods such as Gaussian elimination can be sensitive to rounding errors. To mitigate such issues, matrix factorizations are used. Among them, the **QR factorization** plays a pivotal role.

Definition

Let $A \in \mathbb{R}^{m \times n}$ be a matrix with linearly independent columns. A **QR factorization** of A is a decomposition:

$$A = QR$$

where:

- $Q \in \mathbb{R}^{m \times n}$ is a matrix with orthonormal columns, i.e., $Q^T Q = I_n$,
- $R \in \mathbb{R}^{n \times n}$ is an upper triangular matrix with nonzero diagonal entries.

If A is square (i.e., $m = n$), then Q becomes an orthogonal matrix satisfying:

$$Q^T Q = QQ^T = I_n.$$

The structure of the QR factorization is as follows.

- **Q-Factor**:

  - Size: $m \times n$
  - Columns form an orthonormal basis in $\mathbb{R}^m$
  - $Q^T Q = I$

- **R-Factor**:

  - Size: $n \times n$
  - Upper triangular with nonzero diagonal
  - Nonsingular, provided the columns of A are linearly independent

Applications:

- Solving linear systems $A\mathbf{x} = \mathbf{b}$
- Least squares approximation when A is tall (i.e., $m > n$)
- Eigenvalue computation using the QR algorithm.

11.3.2 Gram-Schmidt Process for QR Factorization

The **Gram-Schmidt process** is an algorithm to transform a set of linearly independent vectors into an orthonormal set. This is foundational in constructing the Q
matrix of a QR factorization. Given $A = [\mathbf{a}_1, \mathbf{a}_2, \ldots, \mathbf{a}_n]$, the algorithm proceeds as
follows:
Step-by-Step Procedure:

1. Set $\mathbf{q}_1 = \mathbf{a}_1 / \|\mathbf{a}_1\|$
2. For $k = 2$ to n:

   - Project $\mathbf{a}_k$ onto the space orthogonal to $\{\mathbf{q}_1, \ldots, \mathbf{q}_{k-1}\}$:

$$\mathbf{v}_k = \mathbf{a}_k - \sum_{j=1}^{k-1} (\mathbf{q}_j^T \mathbf{a}_k)\mathbf{q}_j$$

   - Normalize: $\mathbf{q}_k = \mathbf{v}_k / \|\mathbf{v}_k\|$.

The entries of R are:

$$r_{ij} = \mathbf{q}_i^T \mathbf{a}_j, \quad i \le j.$$

Geometric Interpretation and Stability

   Geometrically, the Gram-Schmidt process orthogonalizes the vectors by remov-
ing their projections onto previously constructed directions. This aligns the basis
with standard axes, simplifying computations.

Numerical Stability: Classical Gram-Schmidt is known to be numerically unstable
in finite precision. Modified Gram-Schmidt and Householder reflections are often
preferred in practice.

   QR factorization, especially via the Gram-Schmidt process, is an indispensable
tool in numerical linear algebra. It offers an intuitive geometric transformation and a
stable method for solving overdetermined systems and improving algorithmic robustness in ML tasks. Understanding its derivation and properties enhances one's ability
to design and analyze modern computational systems.

Example 11.10 Let us consider the matrix:

$$A = \begin{bmatrix} 1 & 1 & 0 \\ 1 & 0 & 1 \\ 0 & 1 & 1 \end{bmatrix}.$$

We denote its column vectors as:

$$\mathbf{a}_1 = \begin{bmatrix} 1 \\ 1 \\ 0 \end{bmatrix}, \quad \mathbf{a}_2 = \begin{bmatrix} 1 \\ 0 \\ 1 \end{bmatrix}, \quad \mathbf{a}_3 = \begin{bmatrix} 0 \\ 1 \\ 1 \end{bmatrix}.$$

Step 1- First Orthogonal Vector

$$\mathbf{q}_1' = \mathbf{a}_1 = \begin{bmatrix} 1 \\ 1 \\ 0 \end{bmatrix}, \quad r_{11} = \|\mathbf{q}_1'\| = \sqrt{1^2 + 1^2 + 0^2} = \sqrt{2}.$$

$$\mathbf{q}_1 = \frac{1}{\sqrt{2}} \begin{bmatrix} 1 \\ 1 \\ 0 \end{bmatrix}.$$

Step 2- Orthogonalize $\mathbf{a}_2$

$$r_{12} = \mathbf{q}_1^T \mathbf{a}_2 = \begin{bmatrix} \frac{1}{\sqrt{2}} & \frac{1}{\sqrt{2}} & 0 \end{bmatrix} \begin{bmatrix} 1 \\ 0 \\ 1 \end{bmatrix} = \frac{1}{\sqrt{2}}.$$

$$\mathbf{q}_2' = \mathbf{a}_2 - r_{12}\mathbf{q}_1 = \begin{bmatrix} 1 \\ 0 \\ 1 \end{bmatrix} - \frac{1}{2} \begin{bmatrix} 1 \\ 1 \\ 0 \end{bmatrix} = \begin{bmatrix} \frac{1}{2} \\ -\frac{1}{2} \\ 1 \end{bmatrix}$$

$$r_{22} = \|\mathbf{q}_2'\| = \sqrt{\frac{1}{4} + \frac{1}{4} + 1} = \sqrt{\frac{3}{2}}, \quad \mathbf{q}_2 = \frac{1}{\sqrt{\frac{3}{2}}} \begin{bmatrix} \frac{1}{2} \\ -\frac{1}{2} \\ 1 \end{bmatrix} = \begin{bmatrix} \frac{1}{\sqrt{6}} \\ -\frac{1}{\sqrt{6}} \\ \frac{2}{\sqrt{6}} \end{bmatrix}.$$

Step 3- Orthogonalize $\mathbf{a}_3$

$$r_{13} = \mathbf{q}_1^T \mathbf{a}_3 = \begin{bmatrix} \frac{1}{\sqrt{2}} & \frac{1}{\sqrt{2}} & 0 \end{bmatrix} \begin{bmatrix} 0 \\ 1 \\ 1 \end{bmatrix} = \frac{1}{\sqrt{2}}, \quad r_{23} = \mathbf{q}_2^T \mathbf{a}_3 = \begin{bmatrix} \frac{1}{\sqrt{6}} & -\frac{1}{\sqrt{6}} & \frac{2}{\sqrt{6}} \end{bmatrix} \begin{bmatrix} 0 \\ 1 \\ 1 \end{bmatrix} = \frac{1}{\sqrt{6}}$$

$$\mathbf{q}_3' = \mathbf{a}_3 - r_{13}\mathbf{q}_1 - r_{23}\mathbf{q}_2 = \begin{bmatrix} 0 \\ 1 \\ 1 \end{bmatrix} - \frac{1}{2}\begin{bmatrix} 1 \\ 1 \\ 0 \end{bmatrix} - \frac{1}{\sqrt{6}}\begin{bmatrix} \frac{1}{\sqrt{6}} \\ -\frac{1}{\sqrt{6}} \\ \frac{2}{\sqrt{6}} \end{bmatrix} = \begin{bmatrix} -\frac{2}{3} \\ \frac{2}{3} \\ \frac{2}{3} \end{bmatrix}$$

$$r_{33} = \|\mathbf{q}_3'\| = \sqrt{\left(\frac{2}{3}\right)^2 + \left(\frac{2}{3}\right)^2 + \left(\frac{2}{3}\right)^2} = \sqrt{\frac{4}{3}}, \quad \mathbf{q}_3 = \frac{1}{\sqrt{\frac{4}{3}}}\begin{bmatrix} -\frac{2}{3} \\ \frac{2}{3} \\ \frac{2}{3} \end{bmatrix} = \begin{bmatrix} -\frac{1}{\sqrt{3}} \\ \frac{1}{\sqrt{3}} \\ \frac{1}{\sqrt{3}} \end{bmatrix}.$$

Final QR Factorization

$$Q = \begin{bmatrix} \frac{1}{\sqrt{2}} & \frac{1}{\sqrt{6}} & -\frac{1}{\sqrt{3}} \\ \frac{1}{\sqrt{2}} & -\frac{1}{\sqrt{6}} & \frac{1}{\sqrt{3}} \\ 0 & \frac{2}{\sqrt{6}} & \frac{1}{\sqrt{3}} \end{bmatrix}, \quad R = \begin{bmatrix} \sqrt{2} & \frac{1}{\sqrt{2}} & \frac{1}{\sqrt{2}} \\ 0 & \sqrt{\frac{3}{2}} & \frac{1}{\sqrt{6}} \\ 0 & 0 & \sqrt{\frac{4}{3}} \end{bmatrix}.$$

To verify the result, we compute QR:

$$QR = A = \begin{bmatrix} 1 & 1 & 0 \\ 1 & 0 & 1 \\ 0 & 1 & 1 \end{bmatrix}.$$

The Gram-Schmidt process successfully orthonormalized the column vectors of A, resulting in matrix Q with orthonormal columns and an upper triangular matrix R. The product QR reconstructs the original matrix A, verifying the correctness of the QR. Figure 11.3 shows two vectors a_1 a_2 and their orthogonalized vectors u_1, u_2.

Fig. 11.3 Original vectors a_1, a_2 and orthogonalized vectors u_1, u_2

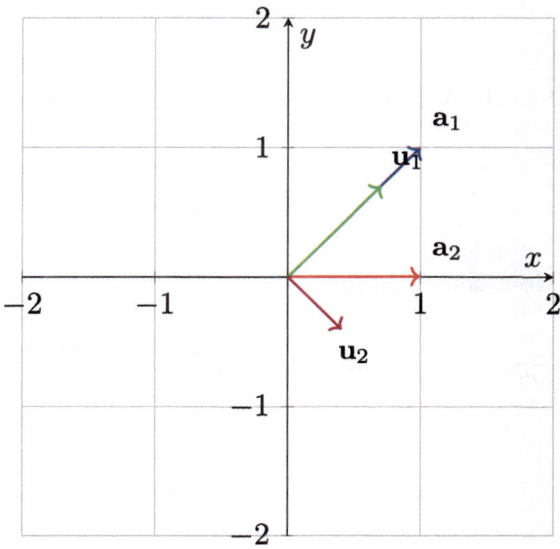

Python Example: QR Factorization

```python
import numpy as np
A = np.array([[1, 1, 0], [1, 0, 1], [0, 1, 1]])
Q, R = np.linalg.qr(A)
print("Q:\n", Q)
print("R:\n", R)
```

Output:

```
Q:
 [[-0.71  0.41 -0.58]
  [-0.71 -0.41  0.58]
  [-0.    0.82  0.58]]
R:
 [[-1.41 -0.71 -0.71]
  [ 0.    1.22  0.41]
  [ 0.    0.    1.15]]
```

Example 11.11 Let the matrix A be:

$$A = \begin{bmatrix} -1 & 3 \\ 1 & 5 \end{bmatrix}.$$

We denote the columns of A as:

$$v_1 = \begin{bmatrix} -1 \\ 1 \end{bmatrix}, \quad v_2 = \begin{bmatrix} 3 \\ 5 \end{bmatrix}.$$

Step 1- Orthonormalize the columns (Gram-Schmidt) Let $u_1 = v_1$, then normalize:

$$e_1 = \frac{u_1}{\|u_1\|} = \frac{1}{\sqrt{2}} \begin{bmatrix} -1 \\ 1 \end{bmatrix} = \begin{bmatrix} -\frac{\sqrt{2}}{2} \\ \frac{\sqrt{2}}{2} \end{bmatrix}.$$

Now, compute projection:

$$\text{proj}_{u_1}(v_2) = \frac{v_2 \cdot u_1}{\|u_1\|^2} u_1 = \frac{(3)(-1) + (5)(1)}{2} u_1 = \frac{-3 + 5}{2} u_1 = 1 \cdot u_1 = \begin{bmatrix} -1 \\ 1 \end{bmatrix}.$$

$$u_2 = v_2 - \text{proj}_{u_1}(v_2) = \begin{bmatrix} 3 \\ 5 \end{bmatrix} - \begin{bmatrix} -1 \\ 1 \end{bmatrix} = \begin{bmatrix} 4 \\ 4 \end{bmatrix}.$$

Normalize:

$$e_2 = \frac{u_2}{\|u_2\|} = \frac{1}{\sqrt{32}} \begin{bmatrix} 4 \\ 4 \end{bmatrix} = \frac{1}{4\sqrt{2}} \begin{bmatrix} 4 \\ 4 \end{bmatrix} = \begin{bmatrix} \frac{\sqrt{2}}{2} \\ \frac{\sqrt{2}}{2} \end{bmatrix}.$$

Step 2- Construct Q and R

$$Q = \begin{bmatrix} -\frac{\sqrt{2}}{2} & \frac{\sqrt{2}}{2} \\ \frac{\sqrt{2}}{2} & \frac{\sqrt{2}}{2} \end{bmatrix}.$$

$$R = Q^T A = \begin{bmatrix} \sqrt{2} & \sqrt{2} \\ 0 & 4\sqrt{2} \end{bmatrix}$$

Final Result

$$\boxed{A = QR \quad \text{where} \quad Q^T Q = I}$$

Example 11.12 Given matrix A, find the QR Factorization using Gram-Schmidt Process

$$A = \begin{bmatrix} 1 & 2 & 1 \\ 1 & 1 & 1 \\ 1 & 0 & 2 \\ 1 & 1 & 1 \end{bmatrix}.$$

We denote the columns of A as:

$$\mathbf{a}_1 = \begin{bmatrix} 1 \\ 1 \\ 1 \\ 1 \end{bmatrix}, \quad \mathbf{a}_2 = \begin{bmatrix} 2 \\ 1 \\ 0 \\ 1 \end{bmatrix}, \quad \mathbf{a}_3 = \begin{bmatrix} 1 \\ 1 \\ 2 \\ 1 \end{bmatrix}.$$

We will apply the Gram-Schmidt process to generate an orthonormal set $\{\mathbf{q}_1, \mathbf{q}_2, \mathbf{q}_3\}$.

Step 1- Compute $\mathbf{q}_1$

$$\mathbf{q}_1 = \frac{\mathbf{a}_1}{\|\mathbf{a}_1\|} = \frac{1}{\sqrt{1^2 + 1^2 + 1^2 + 1^2}} \begin{bmatrix} 1 \\ 1 \\ 1 \\ 1 \end{bmatrix} = \frac{1}{2} \begin{bmatrix} 1 \\ 1 \\ 1 \\ 1 \end{bmatrix}.$$

Step 2- Compute $\mathbf{q}_2$
Project $\mathbf{a}_2$ onto $\mathbf{q}_1$:

$$\text{proj}_{\mathbf{q}_1}(\mathbf{a}_2) = (\mathbf{q}_1^{\mathsf{T}}\mathbf{a}_2)\mathbf{q}_1 = \left(\frac{1}{2}[1, 1, 1, 1] \cdot [2, 1, 0, 1]\right) \cdot \frac{1}{2}\begin{bmatrix} 1 \\ 1 \\ 1 \\ 1 \end{bmatrix}.$$

$$= \frac{4}{2} \cdot \frac{1}{2}\begin{bmatrix} 1 \\ 1 \\ 1 \\ 1 \end{bmatrix} = 1 \cdot \begin{bmatrix} 1 \\ 1 \\ 1 \\ 1 \end{bmatrix}.$$

$$\mathbf{u}_2 = \mathbf{a}_2 - \text{proj}_{\mathbf{q}_1}(\mathbf{a}_2) = \begin{bmatrix} 2 \\ 1 \\ 0 \\ 1 \end{bmatrix} - \begin{bmatrix} 1 \\ 1 \\ 1 \\ 1 \end{bmatrix} = \begin{bmatrix} 1 \\ 0 \\ -1 \\ 0 \end{bmatrix}$$

$$\mathbf{q}_2 = \frac{\mathbf{u}_2}{\|\mathbf{u}_2\|} = \frac{1}{\sqrt{1^2 + 0^2 + (-1)^2 + 0^2}}\begin{bmatrix} 1 \\ 0 \\ -1 \\ 0 \end{bmatrix} = \frac{1}{\sqrt{2}}\begin{bmatrix} 1 \\ 0 \\ -1 \\ 0 \end{bmatrix}.$$

Step 3- Compute $\mathbf{q}_3$

$$\text{proj}_{\mathbf{q}_1}(\mathbf{a}_3) = (\mathbf{q}_1^{\mathsf{T}}\mathbf{a}_3)\mathbf{q}_1 = \left(\frac{1}{2}[1, 1, 1, 1] \cdot [1, 1, 2, 1]\right) \cdot \frac{1}{2}\begin{bmatrix} 1 \\ 1 \\ 1 \\ 1 \end{bmatrix}.$$

$$= \frac{5}{2} \cdot \frac{1}{2}\begin{bmatrix} 1 \\ 1 \\ 1 \\ 1 \end{bmatrix} = \frac{5}{4}\begin{bmatrix} 1 \\ 1 \\ 1 \\ 1 \end{bmatrix}.$$

$$\text{proj}_{\mathbf{q}_2}(\mathbf{a}_3) = (\mathbf{q}_2^{\mathsf{T}}\mathbf{a}_3)\mathbf{q}_2 = \left(\frac{1}{\sqrt{2}}[1, 0, -1, 0] \cdot [1, 1, 2, 1]\right) \cdot \frac{1}{\sqrt{2}}\begin{bmatrix} 1 \\ 0 \\ -1 \\ 0 \end{bmatrix}.$$

$$= \frac{-1}{2} \cdot \begin{bmatrix} 1 \\ 0 \\ -1 \\ 0 \end{bmatrix} = \frac{-1}{2}\begin{bmatrix} 1 \\ 0 \\ -1 \\ 0 \end{bmatrix}.$$

$$\mathbf{u}_3 = \mathbf{a}_3 - \text{proj}_{\mathbf{q}_1}(\mathbf{a}_3) - \text{proj}_{\mathbf{q}_2}(\mathbf{a}_3) = \begin{bmatrix} 1 \\ 1 \\ 2 \\ 1 \end{bmatrix} - \frac{5}{4}\begin{bmatrix} 1 \\ 1 \\ 1 \\ 1 \end{bmatrix} + \frac{1}{2}\begin{bmatrix} 1 \\ 0 \\ -1 \\ 0 \end{bmatrix}.$$

$$= \begin{bmatrix} 1 - \frac{5}{4} + \frac{1}{2} \\ 1 - \frac{5}{4} + 0 \\ 2 - \frac{5}{4} - \frac{1}{2} \\ 1 - \frac{5}{4} + 0 \end{bmatrix} = \begin{bmatrix} \frac{1}{4} \\ -\frac{1}{4} \\ \frac{1}{4} \\ -\frac{1}{4} \end{bmatrix} \Rightarrow \mathbf{q}_3 = \frac{1}{\frac{1}{2}}\begin{bmatrix} \frac{1}{4} \\ -\frac{1}{4} \\ \frac{1}{4} \\ -\frac{1}{4} \end{bmatrix} = \begin{bmatrix} \frac{1}{2} \\ -\frac{1}{2} \\ \frac{1}{2} \\ -\frac{1}{2} \end{bmatrix}.$$

QR Factorization Result

Let $Q = [\mathbf{q}_1 \ \mathbf{q}_2 \ \mathbf{q}_3]$, then:

$$Q = \begin{bmatrix} \frac{1}{2} & \frac{1}{\sqrt{2}} & \frac{1}{2} \\ \frac{1}{2} & 0 & -\frac{1}{2} \\ \frac{1}{2} & -\frac{1}{\sqrt{2}} & \frac{1}{2} \\ \frac{1}{2} & 0 & -\frac{1}{2} \end{bmatrix}$$

To get R, compute $R = Q^T A$.

$$R = \begin{bmatrix} \|\mathbf{a}_1\| & \mathbf{q}_1^\top \mathbf{a}_2 & \mathbf{q}_1^\top \mathbf{a}_3 \\ 0 & \|\mathbf{u}_2\| & \mathbf{q}_2^\top \mathbf{a}_3 \\ 0 & 0 & \|\mathbf{u}_3\| \end{bmatrix} = \begin{bmatrix} 2 & 2 & \frac{5}{2} \\ 0 & \sqrt{2} & -\sqrt{2}/2 \\ 0 & 0 & \frac{1}{2} \end{bmatrix}$$

Thus, $A = QR$ is the desired factorization.

Gram-Schmidt Orthonormalization

The Gram-Schmidt process is a method used in linear algebra to convert a set of linearly independent vectors into an orthonormal set that spans the same subspace. Given a list of vectors $\{\mathbf{v}_1, \mathbf{v}_2, \ldots, \mathbf{v}_n\}$, the algorithm iteratively constructs an orthogonal basis $\{\mathbf{w}_1, \mathbf{w}_2, \ldots, \mathbf{w}_n\}$, where each vector $\mathbf{w}_k$ is obtained by subtracting from $\mathbf{v}_k$ its projections onto all previous $\mathbf{w}_i$ for $i < k$. Each $\mathbf{w}_k$ is then normalized to produce the orthonormal basis $\{\mathbf{u}_1, \mathbf{u}_2, \ldots, \mathbf{u}_n\}$. This orthonormal set satisfies both orthogonality (mutual perpendicularity) and unit length, making it particularly useful for simplifying problems in vector spaces, such as solving least squares or constructing orthogonal projections.

The following example demonstrates Gram-Schmidt orthonormalization independently of QR factorization.

Example 11.13 We are given:

$$\mathbf{v}_1 = \begin{bmatrix} 1 \\ 1 \\ 1 \end{bmatrix}, \quad \mathbf{v}_2 = \begin{bmatrix} 1 \\ 0 \\ 1 \end{bmatrix}, \quad \mathbf{v}_3 = \begin{bmatrix} 1 \\ 1 \\ 2 \end{bmatrix}$$

Our goal is to produce an orthonormal set $\{\mathbf{u}_1, \mathbf{u}_2, \mathbf{u}_3\}$ using the Gram-Schmidt process.

Step 1- Compute $\mathbf{u}_1$

$$\mathbf{u}_1 = \frac{\mathbf{v}_1}{\|\mathbf{v}_1\|} = \frac{1}{\sqrt{1^2 + 1^2 + 1^2}} \begin{bmatrix} 1 \\ 1 \\ 1 \end{bmatrix} = \frac{1}{\sqrt{3}} \begin{bmatrix} 1 \\ 1 \\ 1 \end{bmatrix}$$

Step 2- Compute $\mathbf{u}_2$
First, subtract the projection of $\mathbf{v}_2$ onto $\mathbf{u}_1$:

$$\text{proj}_{\mathbf{u}_1}(\mathbf{v}_2) = \left(\frac{\mathbf{v}_2 \cdot \mathbf{u}_1}{\|\mathbf{u}_1\|^2} \right) \mathbf{u}_1 = (\mathbf{v}_2 \cdot \mathbf{u}_1)\mathbf{u}_1,$$

$$\mathbf{v}_2 \cdot \mathbf{u}_1 = \begin{bmatrix} 1 \\ 0 \\ 1 \end{bmatrix} \cdot \frac{1}{\sqrt{3}} \begin{bmatrix} 1 \\ 1 \\ 1 \end{bmatrix} = \frac{1+0+1}{\sqrt{3}} = \frac{2}{\sqrt{3}},$$

$$\text{proj}_{\mathbf{u}_1}(\mathbf{v}_2) = \frac{2}{\sqrt{3}} \cdot \frac{1}{\sqrt{3}} \begin{bmatrix} 1 \\ 1 \\ 1 \end{bmatrix} = \frac{2}{3} \begin{bmatrix} 1 \\ 1 \\ 1 \end{bmatrix}.$$

Now subtract:

$$\mathbf{w}_2 = \mathbf{v}_2 - \text{proj}_{\mathbf{u}_1}(\mathbf{v}_2) = \begin{bmatrix} 1 \\ 0 \\ 1 \end{bmatrix} - \frac{2}{3} \begin{bmatrix} 1 \\ 1 \\ 1 \end{bmatrix} = \begin{bmatrix} 1 - \frac{2}{3} \\ 0 - \frac{2}{3} \\ 1 - \frac{2}{3} \end{bmatrix} = \begin{bmatrix} \frac{1}{3} \\ -\frac{2}{3} \\ \frac{1}{3} \end{bmatrix}.$$

Now normalize $\mathbf{w}_2$:

$$\|\mathbf{w}_2\| = \sqrt{\left(\frac{1}{3}\right)^2 + \left(-\frac{2}{3}\right)^2 + \left(\frac{1}{3}\right)^2} = \sqrt{\frac{1+4+1}{9}} = \sqrt{\frac{6}{9}} = \sqrt{\frac{2}{3}}.$$

$$\mathbf{u}_2 = \frac{\mathbf{w}_2}{\|\mathbf{w}_2\|} = \frac{1}{\sqrt{2/3}} \begin{bmatrix} \frac{1}{3} \\ -\frac{2}{3} \\ \frac{1}{3} \end{bmatrix} = \sqrt{\frac{3}{2}} \cdot \frac{1}{3} \begin{bmatrix} 1 \\ -2 \\ 1 \end{bmatrix} = \frac{1}{\sqrt{6}} \begin{bmatrix} 1 \\ -2 \\ 1 \end{bmatrix}.$$

Step 3- Compute $\mathbf{u}_3$
Orthogonalize:

$$\mathbf{w}_3 = \mathbf{v}_3 - \text{proj}_{\mathbf{u}_1}(\mathbf{v}_3) - \text{proj}_{\mathbf{u}_2}(\mathbf{v}_3).$$

First projection:

$$\mathbf{v}_3 \cdot \mathbf{u}_1 = \begin{bmatrix} 1 \\ 1 \\ 2 \end{bmatrix} \cdot \frac{1}{\sqrt{3}} \begin{bmatrix} 1 \\ 1 \\ 1 \end{bmatrix} = \frac{1+1+2}{\sqrt{3}} = \frac{4}{\sqrt{3}}.$$

$$\text{proj}_{\mathbf{u}_1}(\mathbf{v}_3) = \frac{4}{3} \begin{bmatrix} 1 \\ 1 \\ 1 \end{bmatrix}.$$

Second projection:

$$\mathbf{v}_3 \cdot \mathbf{u}_2 = \begin{bmatrix} 1 \\ 1 \\ 2 \end{bmatrix} \cdot \frac{1}{\sqrt{6}} \begin{bmatrix} 1 \\ -2 \\ 1 \end{bmatrix} = \frac{1-2+2}{\sqrt{6}} = \frac{1}{\sqrt{6}}.$$

$$\text{proj}_{\mathbf{u}_2}(\mathbf{v}_3) = \frac{1}{6} \begin{bmatrix} 1 \\ -2 \\ 1 \end{bmatrix}.$$

Now compute:

$$\mathbf{w}_3 = \mathbf{v}_3 - \frac{4}{3} \begin{bmatrix} 1 \\ 1 \\ 1 \end{bmatrix} - \frac{1}{6} \begin{bmatrix} 1 \\ -2 \\ 1 \end{bmatrix} = \begin{bmatrix} 1 \\ 1 \\ 2 \end{bmatrix} - \begin{bmatrix} \frac{4}{3} \\ \frac{4}{3} \\ \frac{4}{3} \end{bmatrix} - \begin{bmatrix} \frac{1}{6} \\ -\frac{1}{3} \\ \frac{1}{6} \end{bmatrix}.$$

$$\mathbf{w}_3 = \begin{bmatrix} 1 - \frac{4}{3} - \frac{1}{6} \\ 1 - \frac{4}{3} + \frac{1}{3} \\ 2 - \frac{4}{3} - \frac{1}{6} \end{bmatrix} = \begin{bmatrix} -\frac{3}{6} = -\frac{1}{2} \\ 0 \\ \frac{3}{6} = 2 \end{bmatrix}.$$

Normalize:

$$\|\mathbf{w}_3\| = \sqrt{\left(-\frac{1}{2}\right)^2 + 0^2 + \left(\frac{1}{2}\right)^2} = \sqrt{\frac{1}{4} + \frac{1}{4}} = \sqrt{\frac{1}{2}} = 1/\sqrt{2}.$$

$$\mathbf{u}_3 = \frac{\mathbf{w}_3}{\|\mathbf{w}_3\|} = \sqrt{2} \begin{bmatrix} -\frac{1}{2} \\ 0 \\ \frac{1}{2} \end{bmatrix} = \begin{bmatrix} -1/\sqrt{2} \\ 0 \\ 1/\sqrt{2} \end{bmatrix}.$$

Final Orthonormal Set

$$\mathbf{u}_1 = \frac{1}{\sqrt{3}} \begin{bmatrix} 1 \\ 1 \\ 1 \end{bmatrix}, \quad \mathbf{u}_2 = \frac{1}{\sqrt{6}} \begin{bmatrix} 1 \\ -2 \\ 1 \end{bmatrix}, \quad \mathbf{u}_3 = \begin{bmatrix} -1/\sqrt{2} \\ 0 \\ 1/\sqrt{2} \end{bmatrix}.$$

11.4 Orthogonal Operators

Orthogonal matrices play a central role in preserving the structure of Euclidean space under linear transformations. In particular, we now explore how the properties of orthogonal matrices naturally extend to **orthogonal linear operators**.

Definition: Orthogonal Operator

Let $T : \mathbb{R}^n \to \mathbb{R}^n$ be a linear operator. We say that T is an **orthogonal operator** if it preserves inner products and norms. That is, for all $\mathbf{u}, \mathbf{v} \in \mathbb{R}^n$, the following equivalent conditions hold:

1. $T(\mathbf{u}) \cdot T(\mathbf{v}) = \mathbf{u} \cdot \mathbf{v}$ (Preservation of inner product)
2. $\|T(\mathbf{u})\| = \|\mathbf{u}\|$ (Preservation of length/norm)
3. The standard matrix A of T satisfies $A^T A = I$; that is, A is an orthogonal matrix.

Orthogonal operators preserve geometric properties of vectors, namely, their angles and lengths. This makes them critical in applications such as computer graphics, robotics, and ML where rigid transformations are common.

11.4.1 Properties of Orthogonal Operators

Let T and U be orthogonal operators on $\mathbb{R}^n$. Then:

- The composition $T \circ U$ is also an orthogonal operator.
- The inverse T^{-1} is an orthogonal operator.
- The standard matrix of T is an orthogonal matrix, so its transpose equals its inverse: $T^{-1} = T^T$.
- The determinant of the standard matrix of T satisfies $\det(T) = \pm 1$.

Example 11.14 Let $T : \mathbb{R}^2 \to \mathbb{R}^2$ be defined by:

$$T(\mathbf{x}) = Q\mathbf{x}, \quad \text{where} \quad Q = \begin{bmatrix} \cos\theta & -\sin\theta \\ \sin\theta & \cos\theta \end{bmatrix}$$

This represents a rotation by angle θ. We verify orthogonality:

$$Q^T Q = \begin{bmatrix} \cos\theta & \sin\theta \\ -\sin\theta & \cos\theta \end{bmatrix} \begin{bmatrix} \cos\theta & -\sin\theta \\ \sin\theta & \cos\theta \end{bmatrix} = \begin{bmatrix} \cos^2\theta + \sin^2\theta & 0 \\ 0 & \cos^2\theta + \sin^2\theta \end{bmatrix} = I$$

So, T is an orthogonal operator.

11.4.2   Equivalences for Orthogonal Operators

> **Theorem: Orthogonal Operator Theorem**
> Let $T : \mathbb{R}^n \to \mathbb{R}^n$ be a linear operator. Then the following are equivalent:
>
> - T is an orthogonal operator.
> - $\|T(\mathbf{u})\| = \|\mathbf{u}\|$ for all $\mathbf{u} \in \mathbb{R}^n$.
> - $T(\mathbf{u}) \cdot T(\mathbf{v}) = \mathbf{u} \cdot \mathbf{v}$ for all $\mathbf{u}, \mathbf{v} \in \mathbb{R}^n$.
> - The matrix representation Q of T is orthogonal: $Q^T Q = I$.

An orthogonal operator represents a rigid transformation; it can rotate or reflect vectors but cannot stretch or distort them. It preserves:

- Lengths (norms),
- Angles (via dot product),
- Volume (up to sign of determinant),
- Orthogonality.

Orthogonal operators are a natural extension of orthogonal matrices. They are fundamental in preserving structure during transformations and thus appear in various scientific and engineering applications. Understanding them in terms of their equivalent definitions (length-preserving, angle-preserving, matrix-based) provides flexibility in both theoretical and applied linear algebra.

Example 11.15 Let

$$
\mathbf{v} =
\begin{bmatrix}
\frac{1}{\sqrt{2}} \\
0 \\
\frac{1}{\sqrt{2}}
\end{bmatrix},
$$

and suppose we seek an orthogonal operator $T : \mathbb{R}^3 \to \mathbb{R}^3$ such that $T(\mathbf{v}) = \mathbf{e}_2$.
Step 1- Understand the Goal
Let A be the standard matrix of the operator T. Then we want:

$$
A\mathbf{v} = \mathbf{e}_2 \quad \Rightarrow \quad \mathbf{v} = A^T \mathbf{e}_2,
$$

which implies that $\mathbf{v}$ must be the **second column of** A^T, or equivalently, the **second row of** A. Since we are constructing an orthogonal operator, A must be an orthogonal matrix, i.e., $A^T A = I$, and its rows (and columns) must form an orthonormal basis for $\mathbb{R}^3$.

Step 2- Complete v to an Orthonormal Basis

We need to construct an orthonormal basis that includes $\mathbf{v}$. To do this, we first find a basis for the orthogonal complement $\{\mathbf{v}\}^{\perp}$. A vector $\mathbf{x} = \begin{bmatrix} x_1 \\ x_2 \\ x_3 \end{bmatrix} \in \{\mathbf{v}\}^{\perp}$ satisfies:

$$\mathbf{v}^T \mathbf{x} = \frac{1}{\sqrt{2}} x_1 + \frac{1}{\sqrt{2}} x_3 = 0 \quad \Rightarrow \quad x_1 + x_3 = 0.$$

A basis for this subspace is:

$$\left\{ \begin{bmatrix} 1 \\ 0 \\ -1 \end{bmatrix}, \begin{bmatrix} 0 \\ 1 \\ 0 \end{bmatrix} \right\}.$$

We apply the Gram-Schmidt process to orthogonalize and normalize them:

- Let $\mathbf{u}_1 = \begin{bmatrix} 1 \\ 0 \\ -1 \end{bmatrix}$, then normalize:

$$\mathbf{q}_1 = \frac{\mathbf{u}_1}{\|\mathbf{u}_1\|} = \frac{1}{\sqrt{2}} \begin{bmatrix} 1 \\ 0 \\ -1 \end{bmatrix}.$$

- Let $\mathbf{u}_2 = \begin{bmatrix} 0 \\ 1 \\ 0 \end{bmatrix}$, already orthogonal to $\mathbf{q}_1$, so:

$$\mathbf{q}_2 = \mathbf{v}.$$

Now we arrange the orthonormal vectors $\mathbf{q}_1, \mathbf{v}, \mathbf{q}_3$ as **rows** of A:

$$A = \begin{bmatrix} \frac{1}{\sqrt{2}} & 0 & -\frac{1}{\sqrt{2}} \\ \frac{1}{\sqrt{2}} & 0 & \frac{1}{\sqrt{2}} \\ 0 & 1 & 0 \end{bmatrix} \quad \text{such that} \quad A\mathbf{v} = \mathbf{e}_2.$$

Step 3- Verify Orthogonality

We compute:

$$A^T A = \begin{bmatrix} \frac{1}{\sqrt{2}} & \frac{1}{\sqrt{2}} & 0 \\ 0 & 0 & 1 \\ -\frac{1}{\sqrt{2}} & \frac{1}{\sqrt{2}} & 0 \end{bmatrix} \begin{bmatrix} \frac{1}{\sqrt{2}} & 0 & -\frac{1}{\sqrt{2}} \\ \frac{1}{\sqrt{2}} & 0 & \frac{1}{\sqrt{2}} \\ 0 & 1 & 0 \end{bmatrix} = I_3.$$

Hence, A is an orthogonal matrix and represents an orthogonal operator T that maps $\mathbf{v} \mapsto \mathbf{e}_2$.

By constructing an orthonormal basis that includes **v**, and arranging it as rows of A, we defined an orthogonal transformation T with desirable geometric and algebraic properties. This method is powerful in applications like rigid transformations, quantum mechanics, and computer graphics.

11.5   Orthogonal Diagonalization of a Symmetric Matrix

Example 11.16

$$A = \begin{bmatrix} 2 & -2 \\ -2 & 5 \end{bmatrix}.$$

by finding an orthogonal matrix P such that

$$P^T A P = D,$$

where D is a diagonal matrix and P has orthonormal columns.

Step 1- Find the Eigenvalues of A
To find eigenvalues, solve the characteristic equation:

$$\det(A - \lambda I) = 0.$$

$$\begin{vmatrix} 2 - \lambda & -2 \\ -2 & 5 - \lambda \end{vmatrix} = (2 - \lambda)(5 - \lambda) - (-2)(-2) = \lambda^2 - 7\lambda + 6 = 0.$$

Solving:

$$\lambda^2 - 7\lambda + 6 = 0 \quad \Rightarrow \quad (\lambda - 6)(\lambda - 1) = 0.$$

Thus, the eigenvalues are:

$$\lambda_1 = 6, \quad \lambda_2 = 1.$$

Step 2- Find Eigenvectors
For $\lambda_1 = 6$:

$$(A - 6I)\mathbf{v} = 0 \Rightarrow \begin{bmatrix} -4 & -2 \\ -2 & -1 \end{bmatrix} \begin{bmatrix} x \\ y \end{bmatrix} = 0. \Rightarrow -4x - 2y = 0 \Rightarrow y = -2x.$$

Eigenvector:

$$\mathbf{v}_1 = \begin{bmatrix} -1 \\ 2 \end{bmatrix}.$$

For $\lambda_2 = 1$:

$$(A - I)\mathbf{v} = 0 \Rightarrow \begin{bmatrix} 1 & -2 \\ -2 & 4 \end{bmatrix} \begin{bmatrix} x \\ y \end{bmatrix} = 0. \Rightarrow x - 2y = 0 \Rightarrow x = 2y.$$

Eigenvector:

$$\mathbf{v}_2 = \begin{bmatrix} 2 \\ 1 \end{bmatrix}.$$

Step 3- Check Orthogonality

$$\mathbf{v}_1^T \mathbf{v}_2 = (-1)(2) + (2)(1) = -2 + 2 = 0.$$

The eigenvectors are orthogonal.

Step 4- Normalize Eigenvectors to Form Orthonormal Basis
Compute norms:

$$\|\mathbf{v}_1\| = \sqrt{(-1)^2 + 2^2} = \sqrt{1 + 4} = \sqrt{5}, \quad \|\mathbf{v}_2\| = \sqrt{2^2 + 1^2} = \sqrt{4 + 1} = \sqrt{5}.$$

Normalize:

$$\mathbf{u}_1 = \frac{1}{\sqrt{5}} \begin{bmatrix} -1 \\ 2 \end{bmatrix}, \quad \mathbf{u}_2 = \frac{1}{\sqrt{5}} \begin{bmatrix} 2 \\ 1 \end{bmatrix}.$$

Step 5- Form Matrix P with Orthonormal Columns

$$P = \begin{bmatrix} \frac{-1}{\sqrt{5}} & \frac{2}{\sqrt{5}} \\ \frac{2}{\sqrt{5}} & \frac{1}{\sqrt{5}} \end{bmatrix}, \quad P^T = \begin{bmatrix} \frac{-1}{\sqrt{5}} & \frac{2}{\sqrt{5}} \\ \frac{2}{\sqrt{5}} & \frac{1}{\sqrt{5}} \end{bmatrix}.$$

Step 6- Compute $D = P^T A P$

$$D = P^T A P = \begin{bmatrix} 6 & 0 \\ 0 & 1 \end{bmatrix}.$$

We have diagonalized the symmetric matrix A as:

$$P^T A P = D,$$

where:

$$P = \frac{1}{\sqrt{5}} \begin{bmatrix} -1 & 2 \\ 2 & 1 \end{bmatrix}, \quad D = \begin{bmatrix} 6 & 0 \\ 0 & 1 \end{bmatrix}.$$

This confirms that A is orthogonally diagonalizable, as guaranteed for all symmetric matrices.

11.6 Symmetric Matrices and Spectral Decomposition

A matrix $A \in \mathbb{R}^{n \times n}$ is **symmetric** if $A = A^T$. The **Spectral Theorem** states that A has real eigenvalues and orthonormal eigenvectors.

> **Theorem: Spectral Theorem for Real Symmetric Matrices**
>
> $$A = PDP^T = \sum_{i=1}^{n} \lambda_i \mathbf{u}_i \mathbf{u}_i^T,$$
>
> where D is diagonal (eigenvalues), and P is orthogonal (eigenvectors).

Spectral decomposition breaks A into orthogonal projections, each scaled by an eigenvalue, like decomposing a force into independent directions.

Orthogonality of Eigenvectors of a Symmetric Matrix

For a symmetric matrix A, the following important property holds:

> If $\mathbf{u}$ and $\mathbf{v}$ are eigenvectors corresponding to distinct eigenvalues λ and μ respectively, where $\lambda \neq \mu$, then $\mathbf{u}$ and $\mathbf{v}$ are orthogonal.

That is, their dot product is zero:

$$\mathbf{u} \cdot \mathbf{v} = 0.$$

This property arises from the **Spectral Theorem**, which states that every real symmetric matrix is diagonalizable via an orthogonal matrix. Consequently, its eigenvectors corresponding to different eigenvalues are perpendicular.

To prove it, let A be a symmetric matrix. Suppose:

$$A\mathbf{u} = \lambda\mathbf{u}, \quad A\mathbf{v} = \mu\mathbf{v}, \quad \text{with } \lambda \neq \mu.$$

Then:

$$A\mathbf{u} \cdot \mathbf{v} = \lambda\mathbf{u} \cdot \mathbf{v},$$
$$\mathbf{u} \cdot A\mathbf{v} = \mathbf{u} \cdot \mu\mathbf{v} = \mu\mathbf{u} \cdot \mathbf{v}.$$

Since A is symmetric, we know:

$$A\mathbf{u} \cdot \mathbf{v} = \mathbf{u} \cdot A\mathbf{v}.$$

So:

$$\lambda\mathbf{u} \cdot \mathbf{v} = \mu\mathbf{u} \cdot \mathbf{v} \Rightarrow (\lambda - \mu)(\mathbf{u} \cdot \mathbf{v}) = 0.$$

Because $\lambda \neq \mu$, it follows that:

$$\mathbf{u} \cdot \mathbf{v} = 0.$$

Decomposition of a Symmetric Matrix into Rank-1 Projections

Theorem: Eigenvalue Decomposition Theorem for Symmetric Matrices
Let $A \in \mathbb{R}^{n \times n}$ be a symmetric matrix. Suppose $\{\mathbf{u}_1, \mathbf{u}_2, \ldots, \mathbf{u}_n\}$ is an orthonormal basis for $\mathbb{R}^n$ consisting of eigenvectors of A, with corresponding eigenvalues $\lambda_1, \lambda_2, \ldots, \lambda_n$. Then:

$$A = \lambda_1 P_1 + \lambda_2 P_2 + \cdots + \lambda_n P_n$$

where each P_i is a symmetric rank-1 projection matrix defined by:

$$P_i = \mathbf{u}_i \mathbf{u}_i^T$$

Properties of P_i

Each P_i satisfies the following:

- $\text{Rank}(P_i) = 1$: Since P_i is the outer product of a nonzero vector $\mathbf{u}_i$ with itself.
- $P_i^2 = P_i$: Projection property.
- $P_i P_j = 0$ for $i \neq j$: Because $\mathbf{u}_i \perp \mathbf{u}_j$ implies $\mathbf{u}_i^T \mathbf{u}_j = 0$.
- $P_i \mathbf{u}_i = \mathbf{u}_i$: Projection preserves its defining eigenvector.
- $P_i \mathbf{u}_j = 0$ for $i \neq j$: The projection annihilates vectors orthogonal to its span.

This decomposition expresses the symmetric matrix A as a sum of rank-1 symmetric matrices, each representing the projection onto the direction of an eigenvector, scaled by its eigenvalue.

Example 11.17 Let:

$$A = \begin{bmatrix} 2 & 0 \\ 0 & 3 \end{bmatrix}.$$

The eigenvectors are $\mathbf{u}_1 = \begin{bmatrix} 1 \\ 0 \end{bmatrix}$, $\mathbf{u}_2 = \begin{bmatrix} 0 \\ 1 \end{bmatrix}$, and eigenvalues $\lambda_1 = 2$, $\lambda_2 = 3$.
Then:

$$P_1 = \mathbf{u}_1 \mathbf{u}_1^T = \begin{bmatrix} 1 \\ 0 \end{bmatrix} \begin{bmatrix} 1 & 0 \end{bmatrix} = \begin{bmatrix} 1 & 0 \\ 0 & 0 \end{bmatrix}.$$

$$P_2 = \mathbf{u}_2 \mathbf{u}_2^T = \begin{bmatrix} 0 \\ 1 \end{bmatrix} \begin{bmatrix} 0 & 1 \end{bmatrix} = \begin{bmatrix} 0 & 0 \\ 0 & 1 \end{bmatrix}$$

Therefore:

$$A = 2P_1 + 3P_2 = 2 \begin{bmatrix} 1 & 0 \\ 0 & 0 \end{bmatrix} + 3 \begin{bmatrix} 0 & 0 \\ 0 & 1 \end{bmatrix} = \begin{bmatrix} 2 & 0 \\ 0 & 3 \end{bmatrix}.$$

This decomposition highlights that a symmetric matrix is a weighted sum of orthogonal projections, with weights given by its eigenvalues. It provides a powerful tool for analysis, especially in diagonalization, PCA, and quantum mechanics.

Example 11.18 Let the matrix be

$$A = \begin{bmatrix} 1 & 5 & -2 \\ 5 & 4 & 5 \\ -2 & 5 & 1 \end{bmatrix}.$$

Since A is symmetric, it is orthogonally diagonalizable. That is, there exists an orthogonal matrix P and a diagonal matrix D such that:

$$A = PDP^{-1} = PDP^{T}.$$

Step 1- Diagonal matrix D The diagonal matrix D contains the eigenvalues of A: The eigenvalues and eigenvectors can be verified by solving $\det(A - \lambda I) = 0$

$$D = \begin{bmatrix} 9 & 0 & 0 \\ 0 & -6 & 0 \\ 0 & 0 & 3 \end{bmatrix}.$$

Step 2- Matrix P of normalized eigenvectors The columns of matrix P are orthonormal eigenvectors of A corresponding to eigenvalues $9, -6, 3$, respectively:

$$P = \begin{bmatrix} \frac{1}{\sqrt{6}} & \frac{1}{\sqrt{3}} & \frac{1}{\sqrt{2}} \\ \frac{2}{\sqrt{6}} & -\frac{1}{\sqrt{3}} & 0 \\ \frac{1}{\sqrt{6}} & \frac{1}{\sqrt{3}} & -\frac{1}{\sqrt{2}} \end{bmatrix}.$$

Step 3- Spectral decomposition as a sum of rank-1 matrices Using the spectral decomposition:

$$A = \sum_{i=1}^{3} \lambda_i \mathbf{u}_i \mathbf{u}_i^{T}.$$

We now construct each term individually:

Term 1: $\lambda_1 = 9$, $\mathbf{u}_1 = \frac{1}{\sqrt{6}} \begin{bmatrix} 1 \\ 2 \\ 1 \end{bmatrix}.$

$$\mathbf{u}_1 \mathbf{u}_1^{T} = \frac{1}{6} \begin{bmatrix} 1 & 2 & 1 \\ 2 & 4 & 2 \\ 1 & 2 & 1 \end{bmatrix},$$

$$9 \cdot \mathbf{u}_1 \mathbf{u}_1^T = \begin{bmatrix} \frac{9}{6} & 3 & \frac{9}{6} \\ 3 & 6 & 3 \\ \frac{9}{6} & 3 & \frac{9}{6} \end{bmatrix}.$$

Term 2: $\lambda_2 = -6$, $\mathbf{u}_2 = \frac{1}{\sqrt{3}} \begin{bmatrix} 1 \\ -1 \\ 1 \end{bmatrix}$.

$$\mathbf{u}_2 \mathbf{u}_2^T = \frac{1}{3} \begin{bmatrix} 1 & -1 & 1 \\ -1 & 1 & -1 \\ 1 & -1 & 1 \end{bmatrix},$$

$$-6 \cdot \mathbf{u}_2 \mathbf{u}_2^T = \begin{bmatrix} -2 & 2 & -2 \\ 2 & -2 & 2 \\ -2 & 2 & -2 \end{bmatrix}.$$

Term 3: $\lambda_3 = 3$, $\mathbf{u}_3 = \frac{1}{\sqrt{2}} \begin{bmatrix} 1 \\ 0 \\ -1 \end{bmatrix}$.

$$\mathbf{u}_3 \mathbf{u}_3^T = \frac{1}{2} \begin{bmatrix} 1 & 0 & -1 \\ 0 & 0 & 0 \\ -1 & 0 & 1 \end{bmatrix},$$

$$3 \cdot \mathbf{u}_3 \mathbf{u}_3^T = \begin{bmatrix} \frac{3}{2} & 0 & -\frac{3}{2} \\ 0 & 0 & 0 \\ -\frac{3}{2} & 0 & \frac{3}{2} \end{bmatrix}.$$

Now sum the three rank-1 components:

$$A = \begin{bmatrix} \frac{3}{2} & 3 & \frac{3}{2} \\ 3 & 6 & 3 \\ \frac{3}{2} & 3 & \frac{3}{2} \end{bmatrix} + \begin{bmatrix} -2 & 2 & -2 \\ 2 & -2 & 2 \\ -2 & 2 & -2 \end{bmatrix} + \begin{bmatrix} \frac{3}{2} & 0 & -\frac{3}{2} \\ 0 & 0 & 0 \\ -\frac{3}{2} & 0 & \frac{3}{2} \end{bmatrix} = \begin{bmatrix} 1 & 5 & -2 \\ 5 & 4 & 5 \\ -2 & 5 & 1 \end{bmatrix}.$$

This confirms the spectral decomposition:

$$A = \sum_{i=1}^{3} \lambda_i \mathbf{u}_i \mathbf{u}_i^T.$$

Each term is a symmetric rank-1 matrix scaled by its eigenvalue, illustrating the power of the spectral theorem for symmetric matrices.

Python Example: Spectral Decomposition

```python
import numpy as np
A = np.array([[1, 5, -2], [5, 4, 5], [-2, 5, 1]])
eigenvalues, eigenvectors = np.linalg.eigh(A)
D = np.diag(eigenvalues)
P = eigenvectors
A_reconstructed = P @ D @ P.T
print("Eigenvalues:", eigenvalues)
print("Reconstructed A:\n", A_reconstructed)
```

Output:

```
Eigenvalues: [-6.  3.  9.]
Reconstructed A:
 [[ 1.   5.  -2.]
 [ 5.   4.   5.]
 [-2.   5.   1.]]
```

Example 11.19 Let the symmetric matrix be:

$$A = \begin{bmatrix} 3 & -4 \\ -4 & -3 \end{bmatrix}.$$

Step 1- Find Eigenvalues Solve the characteristic equation:

$$\det(A - \lambda I) = \begin{vmatrix} 3 - \lambda & -4 \\ -4 & -3 - \lambda \end{vmatrix} = (3 - \lambda)(-3 - \lambda) - (-4)^2.$$

$$= -9 - 3\lambda + 3\lambda + \lambda^2 - 16 = \lambda^2 - 25 \Rightarrow \lambda_1 = 5, \lambda_2 = -5.$$

Step 2- Find Eigenvectors For $\lambda_1 = 5$:

$$(A - 5I)\mathbf{v} = 0 \Rightarrow \begin{bmatrix} -2 & -4 \\ -4 & -8 \end{bmatrix} \begin{bmatrix} x_1 \\ x_2 \end{bmatrix} = 0 \Rightarrow \mathbf{v}_1 = \begin{bmatrix} -2 \\ 1 \end{bmatrix}.$$

Normalize:

$$\mathbf{u}_1 = \frac{\mathbf{v}_1}{\|\mathbf{v}_1\|} = \frac{1}{\sqrt{5}} \begin{bmatrix} -2 \\ 1 \end{bmatrix}.$$

For $\lambda_2 = -5$:

$$(A + 5I)\mathbf{v} = 0 \Rightarrow \begin{bmatrix} 8 & -4 \\ -4 & 2 \end{bmatrix} \Rightarrow \mathbf{v}_2 = \begin{bmatrix} 1 \\ 2 \end{bmatrix} \Rightarrow \mathbf{u}_2 = \frac{1}{\sqrt{5}} \begin{bmatrix} 1 \\ 2 \end{bmatrix}$$

Step 3- Construct Projection Matrices

$$P_1 = \mathbf{u}_1\mathbf{u}_1^T = \frac{1}{5}\begin{bmatrix} -2 \\ 1 \end{bmatrix}\begin{bmatrix} -2 & 1 \end{bmatrix} = \frac{1}{5}\begin{bmatrix} 4 & -2 \\ -2 & 1 \end{bmatrix},$$

$$P_2 = \mathbf{u}_2\mathbf{u}_2^T = \frac{1}{5}\begin{bmatrix} 1 \\ 2 \end{bmatrix}\begin{bmatrix} 1 & 2 \end{bmatrix} = \frac{1}{5}\begin{bmatrix} 1 & 2 \\ 2 & 4 \end{bmatrix}.$$

Step 4- Spectral Decomposition Formula

$$A = \lambda_1 P_1 + \lambda_2 P_2 = 5P_1 + (-5)P_2.$$

$$A = 5 \cdot \frac{1}{5}\begin{bmatrix} 4 & -2 \\ -2 & 1 \end{bmatrix} + (-5) \cdot \frac{1}{5}\begin{bmatrix} 1 & 2 \\ 2 & 4 \end{bmatrix}.$$
$$= \begin{bmatrix} 4 & -2 \\ -2 & 1 \end{bmatrix} - \begin{bmatrix} 1 & 2 \\ 2 & 4 \end{bmatrix} = \begin{bmatrix} 3 & -4 \\ -4 & -3 \end{bmatrix} = A.$$

11.7 LU Decomposition

LU decomposition factors $A \in \mathbb{R}^{n \times n}$ as:

$$A = LU,$$

where L is unit lower triangular (ones on the diagonal, as in the Doolittle method).

LU decomposition is like simplifying a complex recipe into two steps: L records the elimination multipliers, and U is the resulting upper triangular form.

Example 11.20 We want to decompose the matrix

$$A = \begin{bmatrix} 2 & 2 & 3 \\ 5 & 9 & 10 \\ 4 & 1 & 2 \end{bmatrix}.$$

into the product of a lower triangular matrix L and an upper triangular matrix U, such that $A = LU$.

Step 1- Initialize L and Start with A
Let
$$L = \begin{bmatrix} 1 & 0 & 0 \\ 0 & 1 & 0 \\ 0 & 0 & 1 \end{bmatrix} \quad \text{(Identity matrix for Doolittle).}$$

Start with the matrix:

$$A = \begin{bmatrix} 2 & 2 & 3 \\ 5 & 9 & 10 \\ 4 & 1 & 2 \end{bmatrix}.$$

Step 2- Eliminate entries below the pivot in column 1
Use row operation: $R_2 \leftarrow R_2 - \frac{5}{2}R_1$

$$R_2 = \begin{bmatrix} 5 & 9 & 10 \end{bmatrix} - \frac{5}{2} \cdot \begin{bmatrix} 2 & 2 & 3 \end{bmatrix} = \begin{bmatrix} 0 & 4 & \frac{5}{2} \end{bmatrix}.$$

Update $L_{21} = \frac{5}{2}$ Now, use row operation: $R_3 \leftarrow R_3 - 2R_1$

$$R_3 = \begin{bmatrix} 4 & 1 & 2 \end{bmatrix} - 2 \cdot \begin{bmatrix} 2 & 2 & 3 \end{bmatrix} = \begin{bmatrix} 0 & -3 & -4 \end{bmatrix}.$$

Update $L_{31} = 2$ Now:

$$U = \begin{bmatrix} 2 & 2 & 3 \\ 0 & 4 & \frac{5}{2} \\ 0 & -3 & -4 \end{bmatrix} \quad L = \begin{bmatrix} 1 & 0 & 0 \\ \frac{5}{2} & 1 & 0 \\ 2 & 0 & 1 \end{bmatrix}.$$

Step 3- Eliminate entry below the pivot in column 2
Use row operation: $R_3 \leftarrow R_3 + \frac{3}{4}R_2$

$$R_3 = \begin{bmatrix} 0 & -3 & -4 \end{bmatrix} + \frac{3}{4} \cdot \begin{bmatrix} 0 & 4 & \frac{5}{2} \end{bmatrix} = \begin{bmatrix} 0 & 0 & -\frac{17}{8} \end{bmatrix}.$$

Update $L_{32} = -\frac{3}{4}$ Now:

$$U = \begin{bmatrix} 2 & 2 & 3 \\ 0 & 4 & \frac{5}{2} \\ 0 & 0 & -\frac{17}{8} \end{bmatrix} \quad L = \begin{bmatrix} 1 & 0 & 0 \\ \frac{5}{2} & 1 & 0 \\ 2 & -\frac{3}{4} & 1 \end{bmatrix}.$$

Final Result: LU Decomposition

$$L = \begin{bmatrix} 1 & 0 & 0 \\ \frac{5}{2} & 1 & 0 \\ 2 & -\frac{3}{4} & 1 \end{bmatrix}, \quad U = \begin{bmatrix} 2 & 2 & 3 \\ 0 & 4 & \frac{5}{2} \\ 0 & 0 & -\frac{17}{8} \end{bmatrix}.$$

Verification: $A = LU$

$$LU = \begin{bmatrix} 1 & 0 & 0 \\ \frac{5}{2} & 1 & 0 \\ 2 & -\frac{3}{4} & 1 \end{bmatrix} \cdot \begin{bmatrix} 2 & 2 & 3 \\ 0 & 4 & \frac{5}{2} \\ 0 & 0 & -\frac{17}{8} \end{bmatrix} = \begin{bmatrix} 2 & 2 & 3 \\ 5 & 9 & 10 \\ 4 & 1 & 2 \end{bmatrix}.$$

Thus, verified.

Python Example: LU Decomposition

```python
import numpy as np

def lu_doolittle(A):
    A = A.astype(float)
    n = A.shape[0]
    L = np.eye(n)
    U = np.zeros_like(A)

    for i in range(n):
        U[i, i:] = A[i, i:] - L[i, :i] @ U[:i,
            i:]
        L[i+1:, i] = (A[i+1:, i] - L[i+1:, :i] @
            U[:i, i]) / U[i, i]

    return L, U

A = np.array([[2, 2, 3],
              [5, 9, 10],
              [4, 1, 2]])

L, U = lu_doolittle(A)

print("L:\n", L)
print("U:\n", U)
print("\nCheck:", np.allclose(A, L @ U))
```

Output:

```
L:
 [[ 1.     0.     0.   ]
 [ 2.5    1.     0.   ]
 [ 2.    -0.75   1.   ]]
U:
 [[ 2.     2.     3.    ]
 [ 0.     4.     2.5   ]
 [ 0.     0.    -2.125]]

Check: True
```

11.7.1 *Pivoting*

For stability:

$$PA = LU,$$

where P is a permutation matrix.

Pivoting is introduced to avoid numerical instability that may arise from dividing by very small pivot elements. By reordering the rows of A through the permutation matrix P, the algorithm selects the largest available pivot (partial pivoting), which reduces rounding errors and ensures a more stable factorization.

Example 11.21 We want to try LU Decomposition without Pivoting

$$A = \begin{bmatrix} 2 & 7 & 1 \\ 3 & -2 & 0 \\ 1 & 5 & 3 \end{bmatrix}.$$

We perform Gaussian elimination to decompose A into:

$$A = LU,$$

where L is lower triangular with ones on the diagonal, and U is upper triangular.

Step 1- Eliminate below pivot (row 1)
Eliminate entry $a_{21} = 3$ using row 1:

$$R_2 \to R_2 - \frac{3}{2} R_1 \Rightarrow \begin{bmatrix} 2 & 7 & 1 \\ 0 & -\frac{25}{2} & -\frac{3}{2} \\ 1 & 5 & 3 \end{bmatrix}.$$

Next eliminate entry $a_{31} = 1$:

$$R_3 \to R_3 - \frac{1}{2} R_1 \Rightarrow \begin{bmatrix} 2 & 7 & 1 \\ 0 & -\frac{25}{2} & -\frac{3}{2} \\ 0 & \frac{3}{2} & \frac{5}{2} \end{bmatrix}.$$

Step 2- Eliminate below pivot (row 2)
Eliminate entry $a_{32} = \frac{3}{2}$ using row 2:

$$R_3 \to R_3 - \left(\frac{3}{25}\right) R_2 \Rightarrow a_{33} = \frac{5}{2} - \left(\frac{3}{25} \cdot -\frac{3}{2}\right) = \frac{5}{2} + \frac{9}{50} = \frac{58}{25}$$

Now the upper triangular matrix U is:

$$U = \begin{bmatrix} 2 & 7 & 1 \\ 0 & -\frac{25}{2} & -\frac{3}{2} \\ 0 & 0 & \frac{58}{25} \end{bmatrix}.$$

Step 3- Build L from multipliers

$$L = \begin{bmatrix} 1 & 0 & 0 \\ \frac{3}{2} & 1 & 0 \\ \frac{1}{2} & \frac{-3}{25} & 1 \end{bmatrix}.$$

$$A = LU = \begin{bmatrix} 1 & 0 & 0 \\ \frac{3}{2} & 1 & 0 \\ \frac{1}{2} & \frac{-3}{25} & 1 \end{bmatrix} \begin{bmatrix} 2 & 7 & 1 \\ 0 & -\frac{25}{2} & -\frac{3}{2} \\ 0 & 0 & \frac{58}{25} \end{bmatrix}$$

Example 11.22   Given the matrix

$$A = \begin{bmatrix} 1 & -1 & 2 \\ 3 & -1 & 7 \\ 2 & -4 & 5 \end{bmatrix},$$

We aim to find its LU decomposition, where $A = LU$, with L being a lower triangular matrix with unit diagonal elements and U being an upper triangular matrix.

Step 1- Perform Gaussian Elimination to Obtain U We use Gaussian elimination to transform A into an upper triangular matrix U without row interchanges. This process involves three elementary row operations performed in succession:

- Adding -3 times row 1 to row 2 of A.
- Adding -2 times row 1 to row 3 of the resulting matrix.
- Adding 1 times row 2 to row 3 of the previous matrix to obtain the final result, U.

The details are as follows:

$$A = \begin{bmatrix} 1 & -1 & 2 \\ 3 & -1 & 7 \\ 2 & -4 & 5 \end{bmatrix}.$$

– First operation: Add -3 times row 1 to row 2 ($-3r_1 + r_2 \rightarrow r_2$):

$$\begin{bmatrix} 1 & -1 & 2 \\ 3 & -1 & 7 \\ 2 & -4 & 5 \end{bmatrix} \rightarrow \begin{bmatrix} 1 & -1 & 2 \\ 3 - 3 \cdot 1 & -1 - 3 \cdot (-1) & 7 - 3 \cdot 2 \\ 2 & -4 & 5 \end{bmatrix} = \begin{bmatrix} 1 & -1 & 2 \\ 0 & 2 & 1 \\ 2 & -4 & 5 \end{bmatrix}.$$

– Second operation: Add -2 times row 1 to row 3 ($-2r_1 + r_3 \rightarrow r_3$):

$$\begin{bmatrix} 1 & -1 & 2 \\ 0 & 2 & 1 \\ 2 & -4 & 5 \end{bmatrix} \rightarrow \begin{bmatrix} 1 & -1 & 2 \\ 0 & 2 & 1 \\ 2-2\cdot 1 & -4-2\cdot(-1) & 5-2\cdot 2 \end{bmatrix} = \begin{bmatrix} 1 & -1 & 2 \\ 0 & 2 & 1 \\ 0 & -2 & 1 \end{bmatrix}.$$

– Third operation: Add 1 times row 2 to row 3 ($r_2 + r_3 \rightarrow r_3$):

$$\begin{bmatrix} 1 & -1 & 2 \\ 0 & 2 & 1 \\ 0 & -2 & 1 \end{bmatrix} \rightarrow \begin{bmatrix} 1 & -1 & 2 \\ 0 & 2 & 1 \\ 0+0 & -2+2 & 1+1 \end{bmatrix} = \begin{bmatrix} 1 & -1 & 2 \\ 0 & 2 & 1 \\ 0 & 0 & 2 \end{bmatrix} = U.$$

Thus, the upper triangular matrix U is

$$U = \begin{bmatrix} 1 & -1 & 2 \\ 0 & 2 & 1 \\ 0 & 0 & 2 \end{bmatrix}.$$

Step 2- Construct L Using Reverse Row Operations

The reverse of the last operation, adding -1 times row 2 to row 3 of a matrix, is the first operation used in the transformation of I_3 into L. We continue to apply the reverse row operations in the opposite order to complete the transformation of I_3 into L. Start with the identity matrix I_3:

$$I_3 = \begin{bmatrix} 1 & 0 & 0 \\ 0 & 1 & 0 \\ 0 & 0 & 1 \end{bmatrix}.$$

– Reverse of the third operation: Add -1 times row 2 to row 3 ($-r_2 + r_3 \rightarrow r_3$):

$$\begin{bmatrix} 1 & 0 & 0 \\ 0 & 1 & 0 \\ 0 & 0 & 1 \end{bmatrix} \rightarrow \begin{bmatrix} 1 & 0 & 0 \\ 0 & 1 & 0 \\ 0-0 & 0-1 & 1-0 \end{bmatrix} = \begin{bmatrix} 1 & 0 & 0 \\ 0 & 1 & 0 \\ 0 & -1 & 1 \end{bmatrix}.$$

– Reverse of the second operation: Add 2 times row 1 to row 3 ($2r_1 + r_3 \rightarrow r_3$):

$$\begin{bmatrix} 1 & 0 & 0 \\ 0 & 1 & 0 \\ 0 & -1 & 1 \end{bmatrix} \rightarrow \begin{bmatrix} 1 & 0 & 0 \\ 0 & 1 & 0 \\ 0+2\cdot 1 & -1+2\cdot 0 & 1+2\cdot 0 \end{bmatrix} = \begin{bmatrix} 1 & 0 & 0 \\ 0 & 1 & 0 \\ 2 & -1 & 1 \end{bmatrix}.$$

– Reverse of the first operation: Add 3 times row 1 to row 2 ($3r_1 + r_2 \rightarrow r_2$):

$$\begin{bmatrix} 1 & 0 & 0 \\ 0 & 1 & 0 \\ 2 & -1 & 1 \end{bmatrix} \rightarrow \begin{bmatrix} 1 & 0 & 0 \\ 0+3\cdot 1 & 1+3\cdot 0 & 0+3\cdot 0 \\ 2 & -1 & 1 \end{bmatrix} = \begin{bmatrix} 1 & 0 & 0 \\ 3 & 1 & 0 \\ 2 & -1 & 1 \end{bmatrix} = L.$$

Thus, the lower triangular matrix L is

$$L = \begin{bmatrix} 1 & 0 & 0 \\ 3 & 1 & 0 \\ 2 & -1 & 1 \end{bmatrix}.$$

To verify, $A = LU$ should hold:

$$
\begin{aligned}
LU &= \begin{bmatrix} 1 & 0 & 0 \\ 3 & 1 & 0 \\ 2 & -1 & 1 \end{bmatrix} \begin{bmatrix} 1 & -1 & 2 \\ 0 & 2 & 1 \\ 0 & 0 & 2 \end{bmatrix} \\
&= \begin{bmatrix} 1 \cdot 1 + 0 \cdot 0 + 0 \cdot 0 & 1 \cdot (-1) + 0 \cdot 2 + 0 \cdot 0 & 1 \cdot 2 + 0 \cdot 1 + 0 \cdot 2 \\ 3 \cdot 1 + 1 \cdot 0 + 0 \cdot 0 & 3 \cdot (-1) + 1 \cdot 2 + 0 \cdot 0 & 3 \cdot 2 + 1 \cdot 1 + 0 \cdot 2 \\ 2 \cdot 1 + (-1) \cdot 0 + 1 \cdot 0 & 2 \cdot (-1) + (-1) \cdot 2 + 1 \cdot 0 & 2 \cdot 2 + (-1) \cdot 1 + 1 \cdot 2 \end{bmatrix} \\
&= \begin{bmatrix} 1 & -1 & 2 \\ 3 & -1 & 7 \\ 2 & -4 & 5 \end{bmatrix} = A.
\end{aligned}
$$

The decomposition is correct.

11.8 SVD

The **SVD** decomposes any $m \times n$ matrix A as:

$$A = U \Sigma V^T,$$

where:

- $U \in \mathbb{R}^{m \times m}$: Orthogonal matrix (columns are left singular vectors, $U^T U = I_m$).
- $\Sigma \in \mathbb{R}^{m \times n}$: Diagonal matrix with non-negative singular values $\sigma_1 \geq \sigma_2 \geq \cdots \geq \sigma_r \geq 0$ (where $r = \min(m, n)$).
- $V \in \mathbb{R}^{n \times n}$: Orthogonal matrix (columns are right singular vectors, $V^T V = I_n$).

SVD describes a matrix transformation as a sequence: V^T rotates the input space, Σ scales along coordinate axes (stretching or compressing), and U rotates the result to the output space. It's like transforming a sphere into an ellipsoid, with singular values determining the stretching factors.

11.8.1 SVD Properties

- **Columns of** U: These form an orthonormal basis for the output space $\mathbb{R}^m$. Each column $\mathbf{u}_i$ corresponds to a left singular vector and represents a principal direction in the range (column space) of A. Their orthonormality ensures that $U^T U = I$, preserving geometric structure under transformation.

- **Columns of** V: These provide an orthonormal basis for the input space $\mathbb{R}^n$. Each right singular vector $\mathbf{v}_i$ maps to a principal axis in the domain of A, characterizing how the input vectors are aligned before scaling by Σ.
- **Rank of** A: The rank is equal to the number of non-zero singular values σ_i. It quantifies the effective dimensionality of the transformation, indicating how many dimensions of the input space are preserved after applying A.
- **Column space of** A: This space is spanned by the columns of U that correspond to non-zero singular values. These vectors indicate the directions in $\mathbb{R}^m$ where the output of A lies.
- **Null space of** A: It is spanned by the columns of V corresponding to zero singular values. Any linear combination of these right singular vectors is mapped to zero by A, representing the input directions annihilated by the transformation.
- **Frobenius norm**: Defined as $\|A\|_F^2 = \Sigma_{i=1}^r \sigma_i^2$, it measures the total energy (sum of squares of all entries) of the matrix. It is often used in error analysis and low-rank approximations, quantifying how much of A's content is retained.
- **Spectral norm**: Given by $\|A\|_2 = \sigma_1$, the largest singular value, it indicates the maximum stretching factor of A. This norm reflects how much A can amplify a unit vector in the worst-case direction.

Example 11.23 Given matrix:

$$A = \begin{bmatrix} 0 & 1 & 2 \\ 1 & 0 & 1 \end{bmatrix}.$$

Step 1- Compute $A^T A$

To compute the right singular vectors, we calculate:

$$A^T A = \begin{bmatrix} 0 & 1 \\ 1 & 0 \\ 2 & 1 \end{bmatrix} \begin{bmatrix} 0 & 1 & 2 \\ 1 & 0 & 1 \end{bmatrix} = \begin{bmatrix} 1 & 0 & 1 \\ 0 & 1 & 2 \\ 1 & 2 & 5 \end{bmatrix}.$$

Step 2- Eigen-Decomposition of $A^T A$

The eigenvalues of $A^T A$ are:

$$\lambda_1 = 6, \quad \lambda_2 = 1, \quad \lambda_3 = 0.$$

The corresponding eigenvectors $\mathbf{v}_i$ are:

$$\mathbf{v}_1 = \begin{bmatrix} 0.2 \\ 0.4 \\ 1 \end{bmatrix}, \quad \mathbf{v}_2 = \begin{bmatrix} -2 \\ 1 \\ 0 \end{bmatrix}, \quad \mathbf{v}_3 = \begin{bmatrix} -1 \\ -2 \\ 1 \end{bmatrix}.$$

Step 3- Normalize Eigenvectors to Get Right Singular Vectors

Normalize $\mathbf{v}_i$ to obtain orthonormal basis vectors:

$$\mathbf{v}_1 = \frac{1}{\sqrt{30}} \begin{bmatrix} 1 \\ 2 \\ 5 \end{bmatrix}, \quad \mathbf{v}_2 = \frac{1}{\sqrt{5}} \begin{bmatrix} -2 \\ 1 \\ 0 \end{bmatrix}, \quad \mathbf{v}_3 = \frac{1}{\sqrt{6}} \begin{bmatrix} -1 \\ -2 \\ 1 \end{bmatrix}.$$

Step 4- Compute the Singular Values

The singular values σ_i are the square roots of the non-zero eigenvalues:

$$\sigma_1 = \sqrt{6}, \quad \sigma_2 = 1, \quad \sigma_3 = 0.$$

Step 5- Compute Left Singular Vectors from $\mathbf{u}_i = \frac{1}{\sigma_i} A \mathbf{v}_i$

$$\mathbf{u}_1 = \frac{1}{\sqrt{6}} A \mathbf{v}_1 = \frac{1}{\sqrt{6}} \cdot \frac{1}{\sqrt{30}} \begin{bmatrix} 0 & 1 & 2 \\ 1 & 0 & 1 \end{bmatrix} \begin{bmatrix} 1 \\ 2 \\ 5 \end{bmatrix} = \frac{1}{6\sqrt{5}} \begin{bmatrix} 12 \\ 6 \end{bmatrix} = \frac{1}{\sqrt{5}} \begin{bmatrix} 2 \\ 1 \end{bmatrix}.$$

$$\mathbf{u}_2 = \frac{1}{\sigma_2} A \mathbf{v}_2 = \frac{1}{\sqrt{5}} \begin{bmatrix} 0 & 1 & 2 \\ 1 & 0 & 1 \end{bmatrix} \begin{bmatrix} 2 \\ -1 \\ 0 \end{bmatrix} = \frac{1}{\sqrt{5}} \begin{bmatrix} 1 \\ -2 \end{bmatrix}.$$

Step 6- Construct SVD

$$U = \begin{bmatrix} \frac{2}{\sqrt{5}} & \frac{1}{\sqrt{5}} \\ \frac{1}{\sqrt{5}} & -\frac{2}{\sqrt{5}} \end{bmatrix}, \quad \Sigma = \begin{bmatrix} \sqrt{6} & 0 & 0 \\ 0 & 1 & 0 \end{bmatrix}, \quad V^T = \begin{bmatrix} \frac{1}{\sqrt{30}} & -\frac{2}{\sqrt{5}} & -\frac{1}{\sqrt{6}} \\ \frac{2}{\sqrt{30}} & \frac{1}{\sqrt{5}} & -\frac{2}{\sqrt{6}} \\ \frac{5}{\sqrt{30}} & 0 & \frac{1}{\sqrt{6}} \end{bmatrix}.$$

$$\boxed{A = U \Sigma V^T}$$

This SVD decomposition expresses A as a product of an orthogonal matrix U, a diagonal matrix Σ with singular values, and another orthogonal matrix V.

Example 11.24 For:

$$A = \begin{bmatrix} 5 & 5 \\ -1 & 7 \end{bmatrix}.$$

Step 1- Compute $A^T A$:

$$A^T A = \begin{bmatrix} 5 & -1 \\ 5 & 7 \end{bmatrix} \begin{bmatrix} 5 & 5 \\ -1 & 7 \end{bmatrix} = \begin{bmatrix} 26 & 18 \\ 18 & 74 \end{bmatrix}.$$

Step 2- Eigenvalues of $A^T A$:

$$\det(A^T A - \lambda I) = (26 - \lambda)(74 - \lambda) - 18^2 = \lambda^2 - 100\lambda + 1600 = 0.$$

$$\lambda_{1,2} = \frac{100 \pm \sqrt{10{,}000 - 6400}}{2} = 50 \pm 30 = 80, 20.$$

$$\sigma_1 = \sqrt{80} = 4\sqrt{5}, \quad \sigma_2 = \sqrt{20} = 2\sqrt{5}.$$

Step 3- Eigenvectors of $A^T A$:

For $\lambda_1 = 80$:

$$A^T A - 80I = \begin{bmatrix} -54 & 18 \\ 18 & -6 \end{bmatrix}, \quad \text{Null space:} \begin{bmatrix} 1 \\ 3 \end{bmatrix}.$$

Normalize:

$$\mathbf{v}_1 = \frac{1}{\sqrt{10}} \begin{bmatrix} 1 \\ 3 \end{bmatrix}.$$

For $\lambda_2 = 20$:

$$A^T A - 20I = \begin{bmatrix} 6 & 18 \\ 18 & 54 \end{bmatrix}, \quad \text{Null space:} \begin{bmatrix} -3 \\ 1 \end{bmatrix}.$$

Normalize:

$$\mathbf{v}_2 = \frac{1}{\sqrt{10}} \begin{bmatrix} -3 \\ 1 \end{bmatrix}.$$

$$V = \frac{1}{\sqrt{10}} \begin{bmatrix} 1 & -3 \\ 3 & 1 \end{bmatrix}.$$

Step 4- Compute U:

$$\mathbf{u}_1 = \frac{1}{\sigma_1} A\mathbf{v}_1 = \frac{1}{4\sqrt{5}} \begin{bmatrix} 5 & 5 \\ -1 & 7 \end{bmatrix} \frac{1}{\sqrt{10}} \begin{bmatrix} -3 \\ 1 \end{bmatrix} = \frac{1}{\sqrt{2}} \begin{bmatrix} 1 \\ 1 \end{bmatrix}.$$

$$\mathbf{u}_2 = \frac{1}{\sigma_2} A\mathbf{v}_2 = \frac{1}{2\sqrt{5}} \begin{bmatrix} 5 & 5 \\ -1 & 7 \end{bmatrix} \frac{1}{\sqrt{10}} \begin{bmatrix} 1 \\ 3 \end{bmatrix} = \frac{1}{\sqrt{2}} \begin{bmatrix} -1 \\ 1 \end{bmatrix}.$$

$$U = \begin{bmatrix} \frac{1}{\sqrt{2}} & \frac{1}{\sqrt{2}} \\ \frac{1}{\sqrt{2}} & \frac{1}{\sqrt{2}} \end{bmatrix},$$

$$\Sigma = \begin{bmatrix} 4\sqrt{5} & 0 \\ 0 & 2\sqrt{5} \end{bmatrix}.$$

Step 5- Verify:

We compute:

$$A = U\Sigma V^T$$

$$= \begin{bmatrix} -\frac{1}{\sqrt{2}} & \frac{1}{\sqrt{2}} \\ \frac{1}{\sqrt{2}} & \frac{1}{\sqrt{2}} \end{bmatrix} \begin{bmatrix} 2\sqrt{5} & 0 \\ 0 & 4\sqrt{5} \end{bmatrix} \begin{bmatrix} \frac{1}{\sqrt{10}} & \frac{3}{\sqrt{10}} \\ -\frac{3}{\sqrt{10}} & \frac{1}{\sqrt{10}} \end{bmatrix}$$

Fig. 11.4 SVD: Transformation of basis vectors by A, aligned with singular vectors

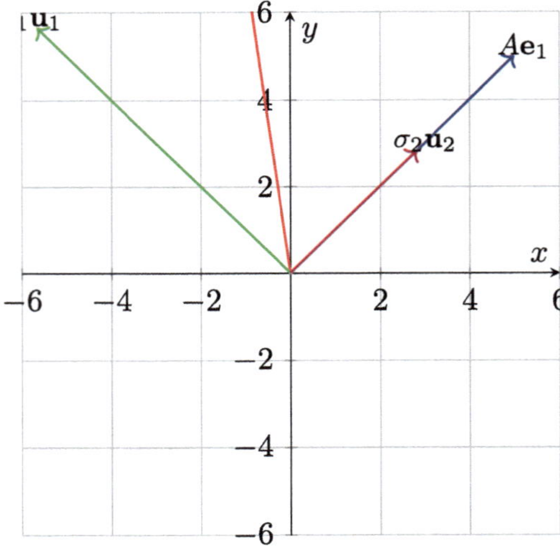

This multiplication (which can be verified numerically or symbolically) yields:

$$A = \begin{bmatrix} 5 & 5 \\ -1 & 7 \end{bmatrix}$$

The decomposition is verified (Fig. 11.4):

$$\boxed{A = U\Sigma V^T}$$

Python Example: SVD Computation

```python
import numpy as np

A = np.array([[5, 5],
              [-1, 7]], dtype=float)

U, s, Vt = np.linalg.svd(A)

Sigma = np.diag(s)

print("U:\n", U)
print("\nSigma:\n", Sigma)
print("\nV:\n", Vt.T)

print("\nCheck:", np.allclose(A, U @ Sigma @ Vt))
```

Output:

```
U:
  [[ 0.7071  0.7071]
   [ 0.7071 -0.7071]]

Sigma:
  [[8.9443 0.    ]
   [0.     4.4721]]

V:
  [[ 0.3162  0.9487]
   [ 0.9487 -0.3162]]
Check: True
```

11.8.2 Economy SVD and Low-Rank Approximation

The **economy SVD** is a compact form for $A \in \mathbb{R}^{m \times n}$ with rank r:

$$A = U_r \Sigma_r V_r^T,$$

where $U_r \in \mathbb{R}^{m \times r}$, $\Sigma_r \in \mathbb{R}^{r \times r}$, $V_r \in \mathbb{R}^{n \times r}$. This reduces storage for low-rank matrices.

A **low-rank approximation** uses the top $k < r$ singular values:

$$A_k = U_k \Sigma_k V_k^T.$$

This captures the most significant patterns, minimizing $\|A - A_k\|_F$.

Truncated SVD is like summarizing a book by keeping only the main chapters, while economy SVD retains all essential content in a more compact format. Low-rank approximation discards minor details (small singular values), retaining the essence of the matrix.

Key Properties of Low-Rank SVD

- A_k: The matrix $A_k = \sum_{i=1}^{k} \sigma_i \mathbf{u}_i \mathbf{v}_i^\top$ is the best rank-k approximation of A with respect to the Frobenius norm (and spectral norm). It retains the most significant k singular directions, minimizing $\|A - A_k\|_F$ among all matrices of rank k. This is known as the Eckart–Young–Mirsky theorem.

- **Variance Preserved**: The quantity $\sum_{i=1}^{k} \sigma_i^2 / \sum_{i=1}^{r} \sigma_i^2$ measures how much of the total variance (or "energy") in A is retained by the rank-k approximation. Here, $r = \text{rank}(A)$. A high ratio (close to 1) indicates that most of the essential structure of A is preserved.
- **Reduces Noise and Storage**: Truncating the smaller singular values (which often represent noise or less informative components) leads to a denoised version of the data. Additionally, representing A via A_k significantly reduces storage requirements from mn entries to $k(m + n + 1)$, which is highly beneficial in large-scale applications such as image compression and Latent Semantic Analysis (LSA).

Example 11.25 Consider the matrix:

$$A = \begin{bmatrix} 5.1 & 0.1 \\ 0.2 & 4.9 \end{bmatrix}.$$

This matrix is nearly diagonal suggesting that its singular values will be close in magnitude to its diagonal entries, with dominant directions aligning roughly with the standard basis.

Step 1- Compute the SVD.
Using numerical computation (or software like NumPy or MATLAB), we obtain:

$$A = U \Sigma V^T,$$

where:

$$\Sigma \approx \begin{bmatrix} \sigma_1 & 0 \\ 0 & \sigma_2 \end{bmatrix}, \quad \sigma_1 > \sigma_2.$$

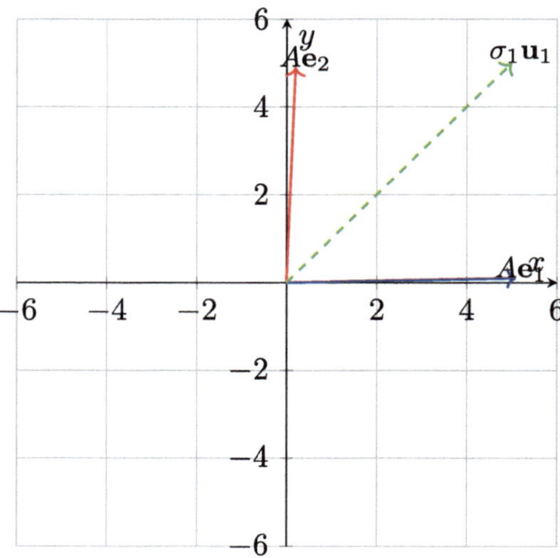

Fig. 11.5 Rank-1 approximation along the dominant singular vector

Let σ_1, $\mathbf{u}_1$, and $\mathbf{v}_1$ be the largest singular value and its associated singular vectors. These define the dominant mode of the transformation.

Step 2- Construct the Rank-1 Approximation.
We define:

$$A_1 = \sigma_1 \mathbf{u}_1 \mathbf{v}_1^T,$$

which captures the most significant structure of A. It is the best approximation to A with only rank 1, minimizing the Frobenius norm $\|A - A_1\|_F$.

The matrix A_1 aligns the data along the most significant direction (principal axis), effectively projecting all input onto a one-dimensional subspace. This is often useful in data compression, noise reduction, or latent feature extraction.

Geometrically, this means that if A maps a unit circle to an ellipse, then A_1 maps it to a line segment aligned with the major axis of that ellipse (Fig. 11.5).

Python Example: Low-Rank Approximation

```python
import numpy as np
A = np.array([[5.1, 0.1], [0.2, 4.9]])
U, S, Vt = np.linalg.svd(A)
A_rank1 = S[0] * np.outer(U[:, 0], Vt[0, :])
print("Rank -1 Approximation:\n", A_rank1)
```

Output:

```
Rank-1 Approximation:
 [[4.03 2.13]
 [2.18 1.15]]
```

11.8.3 *Solving Linear Systems with SVD*

The SVD of a matrix $A \in \mathbb{R}^{m \times n}$ is a factorization $A = U \Sigma V^T$, where $U \in \mathbb{R}^{m \times m}$ and $V \in \mathbb{R}^{n \times n}$ are orthogonal matrices, and $\Sigma \in \mathbb{R}^{m \times n}$ is a diagonal matrix with non-negative singular values $\sigma_1 \geq \sigma_2 \geq \cdots \geq \sigma_p \geq 0$, where $p = \min(m, n)$. For a linear system $A\mathbf{x} = \mathbf{b}$, the SVD provides the pseudoinverse solution:

$$\mathbf{x} = A^+ \mathbf{b} = V \Sigma^+ U^T \mathbf{b},$$

where Σ^+ is the pseudoinverse of Σ, with diagonal entries $1/\sigma_i$ for $\sigma_i \neq 0$ and 0 otherwise. This approach handles overdetermined ($m \geq n$), underdetermined ($m < n$), and rank-deficient systems.

The pseudoinverse A^+ projects **b** onto the column space of A, $\text{Col}(A)$, yielding the least-squares solution for overdetermined systems or the minimum-norm solution for underdetermined systems. SVD is numerically stable and generalizes to non-square or singular matrices, unlike direct inversion.

SVD Process

- Decompose A: Compute $A = U\Sigma V^T$.
- Form Σ^+: For Σ with singular values σ_i, set Σ^+ as the transpose with entries $1/\sigma_i$ for $\sigma_i \neq 0$, and 0 otherwise.
- Compute Pseudoinverse: $A^+ = V\Sigma^+U^T$.
- Solve: $\mathbf{x} = A^+\mathbf{b}$.

Example 11.26 Consider an overdetermined system ($m > n$):

$$A = \begin{bmatrix} 1 & 0 \\ 0 & 1 \\ 1 & 1 \end{bmatrix}, \quad \mathbf{b} = \begin{bmatrix} 1 \\ 0 \\ 1 \end{bmatrix}.$$

- SVD: We compute U, Σ, V^T using a numerical library (e.g., NumPy). For this A, the singular values are $\sigma_1 = \sqrt{3}$, $\sigma_2 = 1$.

Python Example: Solving Overdetermined Systems via SVD

```python
import numpy as np

A = np.array([[1, 0],
              [0, 1],
              [1, 1]], dtype=float)

b = np.array([1, 0, 1], dtype=float)

np.set_printoptions(precision=2, suppress=True)
U, S, VT = np.linalg.svd(A,
    full_matrices=False)

Sigma = np.diag(S)

Sigma_plus = np.diag(1 / S)
A_pseudo = VT.T @ Sigma_plus @ U.T

x = A_pseudo @ b

# Projection of b onto Col(A)
b_hat = A @ x

# Residual
```

```
23 r = b - b_hat
24
25 print("U =\n", U)
26 print("\nSingular values =", S)
27 print("\nV^T =\n", VT)
28
29 print("\nPseudoinverse A^+ =\n", A_pseudo)
30
31 print("\nSolution x =\n", x)
32 print("\nProjection b_hat =\n", b_hat)
33 print("\nResidual r =\n", r)
34
35 Output:
36 U =
37  [[-0.41   0.71]
38  [-0.41  -0.71]
39  [-0.82  -0.   ]]
40
41 Singular values =
42  [1.73 1.  ]
43
44 V^T =
45  [[-0.71  -0.71]
46  [ 0.71  -0.71]]
47
48 Pseudoinverse A^+ =
49  [[ 0.67  -0.33   0.33]
50  [-0.33   0.67   0.33]]
51
52 Solution x =
53  [1.  0.]
54
55 Projection b_hat =
56  [1.  0.  1.]
57
58 Residual r =
59  [ 0.  -0.   0.]
```

– Pseudoinverse:

$$\Sigma^+ = \begin{bmatrix} 1/\sqrt{3} & 0 & 0 \\ 0 & 1 & 0 \end{bmatrix}, \quad A^+ = V\Sigma^+ U^T.$$

– Solution:

$$\mathbf{x} = A^+\mathbf{b} = \begin{bmatrix} \frac{2}{3} \\ \frac{1}{3} \end{bmatrix}$$

– Verification via Normal Equations:

$$A^T A = \begin{bmatrix} 2 & 1 \\ 1 & 2 \end{bmatrix}, \quad A^T \mathbf{b} = \begin{bmatrix} 2 \\ 1 \end{bmatrix}.$$

Solve $(A^T A)\mathbf{x} = A^T \mathbf{b}$:

$$\begin{bmatrix} 2 & 1 \\ 1 & 2 \end{bmatrix} \mathbf{x} = \begin{bmatrix} 2 \\ 1 \end{bmatrix}, \quad \det(A^T A) = 4 - 1 = 3,$$

$$(A^T A)^{-1} = \frac{1}{3} \begin{bmatrix} 2 & -1 \\ -1 & 2 \end{bmatrix}, \quad \mathbf{x} = \frac{1}{3} \begin{bmatrix} 2 & -1 \\ -1 & 2 \end{bmatrix} \begin{bmatrix} 2 \\ 1 \end{bmatrix} = \begin{bmatrix} \frac{2}{3} \\ \frac{1}{3} \end{bmatrix}.$$

The SVD solution matches the least-squares solution.

For overdetermined systems, $A\mathbf{x}$ is the projection of $\mathbf{b}$ onto $\mathrm{Col}(A)$. The residual $\mathbf{b} - A\mathbf{x}$ is orthogonal to $\mathrm{Col}(A)$, minimizing $\|\mathbf{b} - A\mathbf{x}\|_2$. Geometrically, this is the point in $\mathrm{Col}(A)$ closest to $\mathbf{b}$.

Example 11.27 Underdetermined System
For an underdetermined system ($m < n$):

$$A = \begin{bmatrix} 1 & 1 & 1 \end{bmatrix}, \quad \mathbf{b} = \begin{bmatrix} 3 \end{bmatrix}.$$

The system $x_1 + x_2 + x_3 = 3$ has infinitely many solutions. SVD gives the minimum-norm solution:

- SVD: Singular value $\sigma_1 = \sqrt{3}$, $U = [1]$, $V^T = \frac{1}{\sqrt{3}} \begin{bmatrix} 1 & 1 & 1 \end{bmatrix}$.
- Solution:

$$\Sigma^+ = \begin{bmatrix} 1/\sqrt{3} \\ 0 \\ 0 \end{bmatrix}, \quad \mathbf{x} = V \Sigma^+ U^T \mathbf{b} = \begin{bmatrix} 1 \\ 1 \\ 1 \end{bmatrix}$$

- Verification:

$$A\mathbf{x} = 1 + 1 + 1 = 3, \quad \|\mathbf{x}\|_2 = \sqrt{3}.$$

This is the solution with the smallest $\|\mathbf{x}\|_2$.

Example 11.28 Consider an inconsistent system:

$$A = \begin{bmatrix} 1 & 1 \\ 1 & 2 \\ 1 & 3 \end{bmatrix}, \quad \mathbf{b} = \begin{bmatrix} 1 \\ 2 \\ 4 \end{bmatrix}.$$

- SVD: A is full rank ($\mathrm{rank}(A) = 2$), with singular values computed numerically.

– Solution:

$$\mathbf{x} = A^+\mathbf{b} \approx \begin{bmatrix} -0.5 \\ 1.5 \end{bmatrix}$$

– Verification:

$$Ax \approx \begin{bmatrix} 1 \\ 2.5 \\ 4 \end{bmatrix}, \quad \text{residual } \mathbf{b} - Ax \approx \begin{bmatrix} 0 \\ -0.5 \\ 0 \end{bmatrix}.$$

The residual is minimized, and $A^T(\mathbf{b} - Ax) = \mathbf{0}$.

Example 11.29 Consider a rank-deficient system:

$$A = \begin{bmatrix} 1 & 1 \\ 2 & 2 \\ 3 & 3 \end{bmatrix}, \quad \mathbf{b} = \begin{bmatrix} 1 \\ 2 \\ 3 \end{bmatrix}.$$

– SVD: A has rank 1 (columns are linearly dependent), with one non-zero singular value.
– Solution: The pseudoinverse gives the least-squares solution:

$$\mathbf{x} \approx \begin{bmatrix} 0.5 \\ 0.5 \end{bmatrix}$$

– Verification:

$$Ax \approx \begin{bmatrix} 1 \\ 2 \\ 3 \end{bmatrix}, \quad \text{residual } \|\mathbf{b} - Ax\|_2 = 0 \text{ is minimized.}$$

Python Example: Rank-Deficient System

```python
import numpy as np

A = np.array([[1, 1], [2, 2], [3, 3]])
b = np.array([1, 2, 3])

x = np.linalg.pinv(A) @ b
print("Solution:", x)
```

Output:

```
Solution: [0.5 0.5]
```

Example 11.30 We aim to solve the system of linear equations:

$$\begin{cases} x_2 + 2x_3 = 5, \\ x_1 + x_3 = 1. \end{cases}$$

This can be written in matrix form as:

$$A\mathbf{x} = \mathbf{b}, \quad \text{where} \quad A = \begin{bmatrix} 0 & 1 & 2 \\ 1 & 0 & 1 \end{bmatrix}, \quad \mathbf{b} = \begin{bmatrix} 5 \\ 1 \end{bmatrix}.$$

To solve using SVD, we decompose A as:

$$A = U\Sigma V^T.$$

Using the SVD computed in Example 11.23, the given decomposition is:

$$A = \begin{bmatrix} 0 & 1 & 2 \\ 1 & 0 & 1 \end{bmatrix} = \underbrace{\begin{bmatrix} \frac{2}{\sqrt{5}} & -\frac{1}{\sqrt{5}} \\ \frac{1}{\sqrt{5}} & \frac{2}{\sqrt{5}} \end{bmatrix}}_{U} \underbrace{\begin{bmatrix} \sqrt{6} & 0 & 0 \\ 0 & 1 & 0 \end{bmatrix}}_{\Sigma} \underbrace{\begin{bmatrix} \frac{1}{\sqrt{30}} & \frac{2}{\sqrt{5}} & \frac{1}{\sqrt{6}} \\ \frac{2}{\sqrt{30}} & -\frac{1}{\sqrt{5}} & \frac{2}{\sqrt{6}} \\ \frac{5}{\sqrt{30}} & 0 & -\frac{1}{\sqrt{6}} \end{bmatrix}^T}_{V^T}$$

Step 1- Compute the pseudoinverse
We compute the pseudoinverse solution:

$$\mathbf{x} = V\Sigma^{\dagger}U^T\mathbf{b}.$$

Where $\Sigma^{\dagger}$ is the Moore-Penrose pseudoinverse of Σ, given by:

$$\Sigma^{\dagger} = \begin{bmatrix} \frac{1}{\sqrt{6}} & 0 \\ 0 & 1 \\ 0 & 0 \end{bmatrix}.$$

Now compute:

$$\mathbf{x} = \underbrace{\begin{bmatrix} \frac{1}{\sqrt{30}} & \frac{2}{\sqrt{5}} & \frac{1}{\sqrt{6}} \\ \frac{2}{\sqrt{30}} & -\frac{1}{\sqrt{5}} & \frac{2}{\sqrt{6}} \\ \frac{5}{\sqrt{30}} & 0 & -\frac{1}{\sqrt{6}} \end{bmatrix}}_{V} \underbrace{\begin{bmatrix} \frac{1}{\sqrt{6}} & 0 \\ 0 & 1 \\ 0 & 0 \end{bmatrix}}_{\Sigma^{\dagger}} \underbrace{\begin{bmatrix} \frac{2}{\sqrt{5}} & \frac{1}{\sqrt{5}} \\ -\frac{1}{\sqrt{5}} & \frac{2}{\sqrt{5}} \end{bmatrix}}_{U^T} \begin{bmatrix} 5 \\ 1 \end{bmatrix}$$

Step 2- Final result
By matrix multiplication (or as given):

$$\mathbf{x} = \frac{1}{6}\begin{bmatrix} -5 \\ 8 \\ 11 \end{bmatrix} = \begin{bmatrix} -\frac{5}{6} \\ \frac{8}{6} \\ \frac{11}{6} \end{bmatrix}.$$

The solution vector is:

$$\begin{aligned} x_1 &= -\frac{5}{6} \\ x_2 &= \frac{8}{6} \\ x_3 &= \frac{11}{6} \end{aligned}$$

This method is particularly useful when A is not square or is rank-deficient, offering a numerically stable solution.

Numerical Considerations

- Stability: SVD is robust for ill-conditioned or singular matrices, avoiding direct inversion of $A^T A$.
- Zero Singular Values: Set $1/\sigma_i = 0$ for σ_i below a threshold (e.g., 10^{-10}) to handle numerical noise.
- Computational Cost: SVD is more expensive than Gaussian elimination for small, well-conditioned systems but is preferred for complex cases.

11.8.4 Applications of SVD

SVD is a powerful factorization technique used across a wide range of applications in linear algebra, data science, and engineering. Below is a structured list of key applications:

1. **Computing the Moore-Penrose Pseudoinverse**: SVD enables the computation of the pseudoinverse of a matrix, particularly for non-square or singular matrices, allowing for solutions to underdetermined or overdetermined systems.
2. **Solving Homogeneous Linear Systems**: SVD is applied to identify non-trivial solutions in homogeneous systems $A\mathbf{x} = \mathbf{0}$, particularly when A is rank-deficient.
3. **Total Least-Squares Minimization**: SVD supports solving total least-squares problems by minimizing orthogonal distances to the fitted model, improving robustness in the presence of errors in all variables.

4. **Determining Range, Null Space, and Rank**: The singular values of a matrix reveal its rank, and the corresponding singular vectors define its range and null space bases.
5. **Low-Rank Matrix Approximation**: Truncated SVD is used to approximate a matrix by retaining only the top k singular values and vectors, which is vital in data compression and noise reduction.
6. **Decomposition of Separable Models**: In multilinear models, SVD helps decompose signals into separable components (e.g., space-time separability in video processing).
7. **Finding the Nearest Orthogonal Matrix**: SVD can be used to find the closest orthogonal matrix to a given matrix under the Frobenius norm, which is relevant in matrix regularization and numerical stability.
8. **Kabsch Algorithm for Structural Alignment**: In bioinformatics and computer vision, the Kabsch algorithm leverages SVD to optimally align two point clouds by minimizing the Root Mean Square Error (RMSE).
9. **Signal and Image Processing**: SVD is widely applied for denoising, compression, and feature extraction in signals and images. Its energy-compacting property enables efficient data representation.
10. **Big Data and LSA**: In NLP and recommender systems, SVD is utilized in LSA to uncover latent topics and compress large document-term matrices.
11. **Numerical Weather Prediction**: SVD is used in ensemble-based weather forecasting models for dimensionality reduction, sensitivity analysis, and identifying dominant atmospheric modes.

11.8.5 Advantages and Disadvantages

Advantages:

- **Numerical Stability**: SVD provides a robust solution even for ill-conditioned or singular matrices, making it highly reliable in numerical computations.
- **Versatility**: It can be applied to both square and rectangular matrices, extending its utility across diverse ML tasks.
- **Low-Rank Approximation**: SVD effectively compresses high-dimensional data by retaining dominant singular values, which is essential in applications like PCA, image compression, and recommender systems.
- **Least-Squares Optimization**: Offers optimal solutions to overdetermined systems, minimizing error via the pseudoinverse.
- **Theoretical Foundation**: Supports deeper insights into matrix structure, including range, null space, and rank.

Disadvantages:

- **Computational Complexity**: Full SVD decomposition can be computationally expensive for large-scale datasets, especially in high-performance settings.
- **Interpretability**: While singular values are easy to interpret, the left and right singular vectors may lack intuitive meaning depending on the context or data domain.
- **Storage Requirements**: The decomposition produces three large matrices, which may demand significant memory for large inputs.
- **Not Always Sparse**: The resulting factor matrices are typically dense, limiting efficiency in sparse matrix contexts.

11.9 Matrix Decompositions, Factorization and SVD for ML

Matrix decompositions, including factorizations like QR and LU, spectral decomposition, and SVD, are fundamental tools in ML for revealing underlying data structures, enabling efficient computations, improving numerical stability, and facilitating dimensionality reduction.

11.9.1 Applications

- **PCA**: Utilizes spectral decomposition of symmetric covariance matrices or SVD for general matrices to identify orthonormal principal components, reducing data dimensionality while preserving variance.
- **Regression**: Employs QR factorization for solving least-squares problems in LR and LU decomposition for efficient resolution of linear systems.
- **Feature Selection**: Leverages eigenvalues from spectral decomposition or singular values from SVD to identify and select important features based on variance or importance.
- **Recommendation Systems**: Applies SVD to approximate user-item interaction matrices, uncovering latent factors and preferences.
- **Image Compression**: Uses low-rank approximations via truncated SVD to compress images while maintaining essential quality.
- **LSA**: Employs SVD to extract semantic structures from text corpora by reducing term-document matrices.
- **Regularization**: Truncated SVD or other decompositions discard small singular values to stabilize ill-posed problems.
- **Collaborative Filtering**: Models user-item interactions through low-rank factorizations derived from SVD.

Python Example: PCA with SVD

```python
import numpy as np
X = np.array([[1, 2], [3, 4], [5, 6]])
U, S, Vt = np.linalg.svd(X, full_matrices=False)
k = 1
X_reduced = U[:, :k] @ np.diag(S[:k])
print("Reduced data:\n", X_reduced)
```

Output:

```
Reduced data:
[[-2.19]
 [-5.  ]
 [-7.81]]
```

Example: Image Compression

Python Example: Image Compression

```python
import numpy as np
import matplotlib.pyplot as plt
img = plt.imread('image.jpg').mean(axis=2)
U, S, Vt = np.linalg.svd(img,
    full_matrices=False)
k = 80
img_compressed = U[:, :k] @ np.diag(S[:k]) @
    Vt[:k, :]
plt.imshow(img_compressed, cmap='gray')
plt.axis('off')
plt.show()
```

Output:

```
See Figure 11.7.
```

Figure 11.6 illustrates the effect of SVD for image compression. The left image is the original, while the right one is reconstructed using only the top 80 singular values. Although some fine details and textures are lost, the overall structure and prominent features are preserved. This demonstrates how SVD can reduce data dimensionality while retaining the essential information, a key concept in linear algebra and signal processing.

(a) Original Image (b) Compressed Image ($k = 80$)

Fig. 11.6 Comparison between the original image and its SVD-based compressed version using the first $k = 80$ singular values

Chapter Summary

In this chapter, we examined the foundational concepts and techniques of matrix decompositions, focusing on their role in unveiling the intrinsic structure of matrices and facilitating efficient computations across theoretical and applied domains. We began by establishing essential preliminaries, such as matrix norms and the principles of orthonormality, which provided the geometric and analytical groundwork for subsequent factorizations. Various norms, including the Frobenius, ℓ_1, ℓ_∞, spectral, and $\ell_{2,1}$ norms, were defined and computed through detailed examples, highlighting their utility in measuring matrix magnitude and transformation effects. Orthonormality was explored through vectors, matrices, and even functions, with geometric interpretations and normalization procedures illustrated via projections and inner products over intervals.

The chapter then progressed to orthogonal matrices, where key properties like norm preservation, inverse transposition ($Q^{-1} = Q^T$), and determinant constraints ($\det(Q) = \pm 1$) were derived and verified. QR factorization emerged as a cornerstone method, decomposing matrices into orthogonal Q and upper triangular R components via the Gram-Schmidt process, with applications to least-squares problems demonstrated through step-by-step orthogonalization of column vectors. Symmetric matrices received particular attention, revealing their spectral decomposition $A = QDQ^T$ into orthonormal eigenvectors and diagonal eigenvalues, complete with proofs of eigenvector orthogonality and rank-1 projection representations.

LU decomposition was introduced as a practical tool for solving linear systems, factoring matrices into lower triangular L (with unit diagonal) and upper triangular U components, often augmented by pivoting for numerical stability. The chapter culminated in an in-depth exploration of SVD, generalizing eigenvalue analysis to arbitrary matrices via $A = U\Sigma V^T$, where singular values captured scaling factors and singular vectors defined principal directions. Properties such as rank determination, norm relations, and low-rank approximations were elucidated, alongside methods for solving linear systems using the pseudoinverse.

Finally, the real-world relevance of these decompositions was underscored through ML applications, including PCA for dimensionality reduction, regression via QR, feature selection based on spectral properties, and SVD-driven techniques like image compression and LSA. Through rigorous proofs, computational examples, and Python implementations, readers were equipped to apply these tools effectively.

Takeaways

Key takeaways from the chapter include:

- Matrix norms quantified transformation scales, with the spectral norm $\|A\|_2 = \sigma_{\max}(A)$ linking directly to SVD.
- Orthonormal bases simplified projections and decompositions, as seen in Gram-Schmidt and orthogonal matrices preserving Euclidean structure.
- QR factorization enabled stable least-squares solutions, while LU supported forward-backward substitution for linear systems.
- Symmetric matrices admitted unique spectral decompositions into orthogonal projections scaled by eigenvalues, ensuring real and orthogonal eigensystems.
- SVD provided a universal framework for matrix analysis, revealing rank, null spaces, and low-rank approximations essential for data compression and regularization.
- ML integrations, such as PCA via truncated SVD and collaborative filtering, demonstrated decompositions' power in handling high-dimensional data.

This synthesis not only reinforced theoretical insights but also prepared readers for practical implementations in numerical linear algebra and beyond.

Exercises

1. Compute the 1-norm, 2-norm (spectral norm), and Frobenius norm of $A = \begin{bmatrix} 2 & -1 \\ 0 & 3 \end{bmatrix}$.

2. Verify that $Q = \begin{bmatrix} \frac{\sqrt{2}}{2} & -\frac{\sqrt{2}}{2} \\ \frac{\sqrt{2}}{2} & \frac{\sqrt{2}}{2} \end{bmatrix}$ is orthogonal.

3. Orthogonalize $\mathbf{u} = \begin{bmatrix} 1 \\ 2 \\ 3 \end{bmatrix}, \mathbf{v} = \begin{bmatrix} 3 \\ 2 \\ 1 \end{bmatrix}$ using Gram-Schmidt.

4. Normalize $\mathbf{u} = \begin{bmatrix} 1 \\ -1 \end{bmatrix}, \mathbf{v} = \begin{bmatrix} 1 \\ 1 \end{bmatrix}$ and check orthonormality.

5. Compute QR factorization of $A = \begin{bmatrix} 1 & 0 \\ 1 & 1 \\ 1 & -1 \end{bmatrix}$.

6. Compute LU decomposition of $A = \begin{bmatrix} 1 & 2 & 0 \\ 2 & 0 & 1 \\ 0 & 1 & 3 \end{bmatrix}$ with partial pivoting (i.e., find $PA = LU$).

7. Diagonalize $A = \begin{bmatrix} 3 & 1 \\ 1 & 3 \end{bmatrix}$.

8. Derive spectral decomposition of $A = \begin{bmatrix} 1 & 2 \\ 2 & 1 \end{bmatrix}$.

9. Verify orthogonality of eigenvectors for $A = \begin{bmatrix} 2 & 1 \\ 1 & 2 \end{bmatrix}$.

10. For $X = \begin{bmatrix} 1 & 2 \\ 3 & 4 \\ 5 & 6 \end{bmatrix}$, compute the covariance matrix and verify eigenvector orthogonality.

11. Use QR factorization to find the least squares solution to $Ax = \mathbf{b}$, where $A = \begin{bmatrix} 1 & 1 \\ 1 & 2 \\ 1 & 3 \end{bmatrix}$, $\mathbf{b} = \begin{bmatrix} 1 \\ 0 \\ 1 \end{bmatrix}$.

12. For $X = \begin{bmatrix} 1 & 2 \\ 3 & 4 \\ 5 & 6 \end{bmatrix}$, compute the SVD and project onto the top principal component.

13. Implement a rank-k approximation for a 100×100 random matrix and compute the Frobenius norm of the error.

14. Compute the singular values of $A = \begin{bmatrix} 2 & 1 \\ 1 & 2 \end{bmatrix}$.

15. Find the full SVD of $A = \begin{bmatrix} 1 & 1 \\ 0 & 1 \end{bmatrix}$.

16. Compute the least squares solution to $Ax = \mathbf{b}$ for $A = \begin{bmatrix} 1 & 0 \\ 0 & 1 \\ 1 & 1 \end{bmatrix}$, $\mathbf{b} = \begin{bmatrix} 1 \\ 0 \\ 1 \end{bmatrix}$ using SVD.

17. Compute the rank and null space of $A = \begin{bmatrix} 1 & 1 & 0 \\ 0 & 0 & 1 \end{bmatrix}$.

18. Perform a rank-1 approximation of $A = \begin{bmatrix} 3 & 0 \\ 0 & 4 \end{bmatrix}$.

19. Compute the compact SVD of $A = \begin{bmatrix} 2 & 4 \\ 1 & 3 \\ 0 & 0 \end{bmatrix}$.

20. Prove singular values are square roots of $A^T A$'s eigenvalues.

21. If $A = \begin{bmatrix} 0 & 3 \\ 0 & 4 \end{bmatrix}$, compute and interpret the SVD in terms of the geometric action (rotation, scaling, rotation) of the transformation.

22. Verify orthogonality of U and V for $A = \begin{bmatrix} 2 & 0 \\ 0 & 3 \end{bmatrix}$.

Chapter 12
Optimization and Gradients

Introduction

Optimization is the mathematical engine behind many modern technologies, enabling systems to operate efficiently by minimizing costs or maximizing rewards. In linear algebra and ML, optimization typically refers to finding a point $\mathbf{x}^*$ that minimizes a function $f(\mathbf{x})$, often subject to constraints. This chapter introduces core concepts such as GD and second-order methods, building a strong foundation for both theoretical understanding and practical applications.

We begin by exploring basic optimization terminology and problem types, followed by an in-depth discussion of gradient-based optimization methods, including GD and Newton's method. Python implementations are provided for each method to illustrate their convergence behavior and computational efficiency. A dedicated section highlights how optimization underpins model training in ML, particularly in linear and logistic regression. The chapter concludes with modern techniques like Adam and SGD, which are widely used in training deep learning models.

Topics Covered

This chapter is organized as follows:

- **Section** 12.1 **Optimization: Definition and Basics**: Introduces objective functions, types of constraints, and classifications of optimization problems.
- **Section** 12.2 **Gradient Descent (GD)**: Explains first-order optimization, with update rules, convergence properties, and Python code.
- **Section** 12.3 **Hessian and Second-Order Methods**: Discusses second-order derivatives, Newton's method, and their computational trade-offs.
- **Section** 12.4 **Optimization in ML**: Covers applications in linear and logistic regression, and introduces SGD.
- **Exercises**: Offers conceptual and practical problems to reinforce optimization principles and coding proficiency.

Acronyms and Their Definitions

> GD   Gradient Descent
> ML   Machine Learning
> SGD   Stochastic Gradient Descent
> Adam   Adaptive Moment Estimation
> LR   Linear Regression
> MSE   Mean Squared Error
> BFGS   Broyden–Fletcher–Goldfarb–Shanno
> PDEs   Partial Differential Equations

12.1   Optimization: Definition and Basics

Optimization is the process of finding the best solution from a set of feasible solutions, typically by minimizing or maximizing an objective function. In mathematical terms, for a function $f(\mathbf{x})$, where $\mathbf{x} \in \mathbb{R}^n$, optimization seeks to find the input $\mathbf{x}^*$ that yields the minimum (or maximum) value of $f(\mathbf{x})$. Formally, for minimization:

$$\mathbf{x}^* = \arg \min_{\mathbf{x}} f(\mathbf{x}).$$

subject to constraints, if any, such as $\mathbf{x} \in \mathcal{C}$, where $\mathcal{C}$ is a feasible set. Optimization is fundamental in fields like engineering, economics, operations research, and ML, where it is used to design efficient systems, allocate resources, or train models.

Optimization problems can be classified based on the nature of the objective function and constraints:

- **Unconstrained Optimization**: No restrictions on $\mathbf{x}$.
- **Constrained Optimization**: $\mathbf{x}$ must satisfy conditions like $g_i(\mathbf{x}) \leq 0$ or $h_j(\mathbf{x}) = 0$.
- **Convex vs. Non-Convex**: Convex problems have no spurious local minima; if the objective is strictly convex, the global optimum is unique. While non-convex problems may have multiple local optima.

For example, consider minimizing the function $f(x) = x^2 + 2x + 1$. This is a simple unconstrained convex problem, where the minimum occurs at $x = -1$, as can be verified by setting the derivative $f'(x) = 2x + 2 = 0$.

12.2   Gradient Descent (GD)

GD, also known as the *steepest descent* method, is a first-order iterative optimization algorithm used to minimize a differentiable function. The core idea is to iteratively move in the direction of the steepest descent, i.e., opposite to the gradient, until a minimum is reached. Intuitively, we move from a mountain peak down to the sea level (global minimum), taking steps proportional to the negative of the gradient at each point.

Formally, given a function $f(\mathbf{x})$, the update rule is:

$$\mathbf{x}_{t+1} = \mathbf{x}_t - \alpha \nabla f(\mathbf{x}_t),$$

where:

- $\mathbf{x}_t$ is the current point at iteration t,
- $\alpha > 0$ is the **learning rate** controlling the step size,
- $\nabla f(\mathbf{x}_t)$ is the gradient at the current point.

The learning rate determines how large each step toward the minimum is. A large α may overshoot the minimum, while a small α leads to slow convergence. Adaptive learning rates decrease step size as the algorithm approaches the local minimum.

12.2.1 GD with a Single Parameter

For a function with one parameter θ_1, the update rule is:

$$\theta_1 \leftarrow \theta_1 - \alpha \frac{\partial J(\theta_1)}{\partial \theta_1},$$

where α is the learning rate, and $J(\theta_1)$ is the cost function. The derivative indicates the slope of the tangent line (see Fig. 12.1). A positive slope leads to a step to the left, moving toward the minimum.

Example 12.1 Consider the quadratic function:

$$f(x) = (x + 5)^2.$$

Starting at $x_0 = 3$ with learning rate $\alpha = 0.2$, the gradient is:

$$\frac{df}{dx} = 2(x + 5).$$

Applying the GD update:

$$x_1 = x_0 - \alpha \frac{df}{dx}\bigg|_{x_0} = 3 - 0.2 \cdot 2 \cdot 8 = -0.2.$$

$$x_2 = x_1 - \alpha \frac{df}{dx}\bigg|_{x_1} = -0.2 - 0.2 * 2(-0.2 + 5) = -2.12.$$

Iterating this process moves x closer to the minimum at $x = -5$.

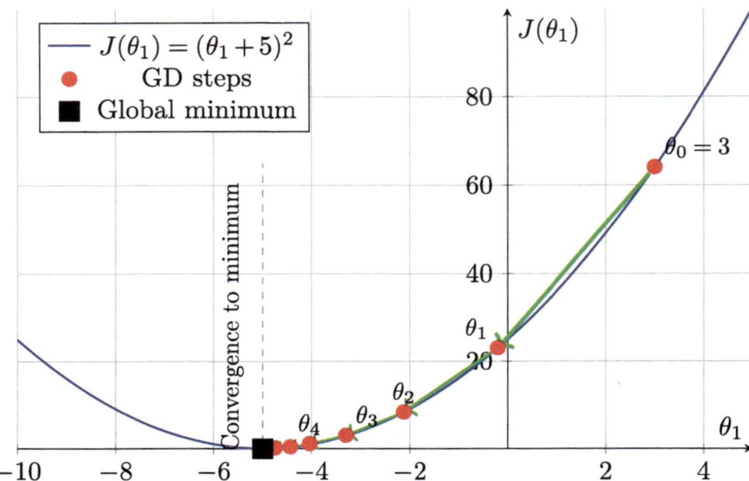

Fig. 12.1 Complete illustration of GD convergence on the single-variable cost function $J(\theta_1) = (\theta_1 + 5)^2$ with learning rate $\alpha = 0.2$. Starting from $\theta_0 = 3$, the algorithm takes progressively smaller steps (shown by decreasing arrow thickness) as it approaches the global minimum at $\theta^* = -5$. The red dots show the parameter values at each iteration, demonstrating exponential convergence

12.2.2   GD in Multiple Dimensions

For a multi-dimensional function $f(\mathbf{x})$ with $\mathbf{x} = (x_1, \ldots, x_n)$, the procedure is:

1. Input: continuous objective function f, initial point $\mathbf{x}_0 = (x_1^0, \ldots, x_n^0)$
2. For t = 0, ..., until convergence:

   - Compute gradient vector $\mathbf{g}_t = \left(\frac{\partial f}{\partial x_1}(\mathbf{x}_t), \ldots, \frac{\partial f}{\partial x_n}(\mathbf{x}_t) \right)$;
   - If $\|\mathbf{g}_t\|$ is small enough, return $\mathbf{x}_t$ (converged);
   - Pick step size α_t;
   - Update: $\mathbf{x}_{t+1} = \mathbf{x}_t - \alpha_t \mathbf{g}_t$.

Visual Intuition

Figure 12.2 illustrates GD in two dimensions. Each arrow represents a step along the negative gradient toward the minimum. The algorithm iteratively moves "downhill" from the initial point to the lowest point.

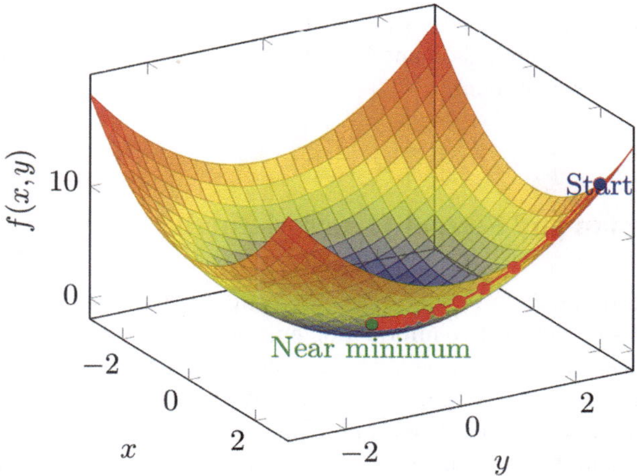

Fig. 12.2 GD in two dimensions on $f(x, y) = x^2 + y^2$: each step follows the negative gradient to minimize the cost function, starting from $(2, 3)$

Python Example: 2D GD

```python
import numpy as np

def gradient_descent_2d(x0, y0, lr,
    num_iterations):
    x, y = x0, y0
    history = [(x, y)]
    for _ in range(num_iterations):
        grad_x, grad_y = 2*x, 2*y
        x, y = x - lr*grad_x, y - lr*grad_y
    history.append((x, y))
    return x, y, history

x_opt, y_opt, history = gradient_descent_2d(2,
    3, 0.1, 50)
print(f"Optimal point: ({x_opt:.4f},
    {y_opt:.4f})")
```

Output:

```
Optimal point: (0.0000, 0.0000)
```

Key points for GD:

- Gradient: indicates the direction to move for minimization.
- Learning rate: step size; too large may overshoot, too small slows convergence.

- Adaptive learning rate: decreases step size near local optima.
- GD finds local minima; Gradient Ascent moves toward local maxima by stepping along the positive gradient.

12.3   Hessian and Second-Order Methods

Second-order optimization methods leverage not only the first derivative (gradient) but also the second derivative information encapsulated in the Hessian matrix. This additional curvature information allows for more sophisticated update steps, potentially leading to faster convergence compared to first-order methods like GD. These methods are particularly effective for problems where the objective function is twice differentiable and the Hessian can be computed or approximated efficiently.

The Hessian matrix H of a scalar-valued function $f(\mathbf{x})$ is a square matrix of second-order partial derivatives:

$$H_{ij} = \frac{\partial^2 f}{\partial x_i \partial x_j}.$$

For functions that are twice continuously differentiable, the Hessian is symmetric due to Clairaut's theorem on the equality of mixed partials.

12.3.1   Properties of the Hessian

The Hessian provides crucial information about the local curvature of the function:

- If H is positive definite (all eigenvalues positive), the function is locally convex, and the point is a local minimum candidate.
- If H has both positive and negative eigenvalues, the point is a saddle point.
- The condition number of H affects the convergence rate; ill-conditioned Hessians can slow down optimization.
- If H is negative definite (all eigenvalues negative), the function is locally concave, and the point is a local maximum candidate.

For example, consider the quadratic function $f(x, y) = x^2 + y^2$. Its Hessian is:

$$H = \begin{bmatrix} 2 & 0 \\ 0 & 2 \end{bmatrix}.$$

This constant Hessian is positive definite, indicating a convex paraboloid with a global minimum at the origin.

12.3.2   Newton's Method

Newton's method is the quintessential second-order optimization algorithm. It approximates the function locally using a second-order Taylor expansion:

$$f(\mathbf{x} + \Delta\mathbf{x}) \approx f(\mathbf{x}) + \nabla f(\mathbf{x})^T \Delta\mathbf{x} + \frac{1}{2}\Delta\mathbf{x}^T H(\mathbf{x})\Delta\mathbf{x}.$$

To find the step $\Delta\mathbf{x}$ that minimizes this quadratic approximation, we set the derivative to zero:

$$\nabla f(\mathbf{x}) + H(\mathbf{x})\Delta\mathbf{x} = 0 \implies \Delta\mathbf{x} = -H(\mathbf{x})^{-1}\nabla f(\mathbf{x}).$$

The update rule is then:

$$\mathbf{x}_{t+1} = \mathbf{x}_t - H(\mathbf{x}_t)^{-1}\nabla f(\mathbf{x}_t).$$

Near the optimum, Newton's method exhibits quadratic convergence, meaning the error squares with each iteration. However, it requires computing and inverting the Hessian at each step, which can be computationally intensive for large dimensions ($O(n^3)$ per iteration).

Example 12.2 For the function $f(x) = x^4 - 4x^2$, the gradient is $f'(x) = 4x^3 - 8x$, and the Hessian (second derivative) is $f''(x) = 12x^2 - 8$. Starting at $x_0 = 2$:

$$f'(2) = 4 \cdot 8 - 8 \cdot 2 = 16, \quad f''(2) = 12 \cdot 4 - 8 = 40$$

$$x_1 = 2 - \frac{16}{40} = 1.6.$$

Next iteration:

$$f'(1.6) = 4(1.6)^3 - 8(1.6) \approx 16.384 - 12.8 = 3.584$$

$$f''(1.6) = 12(1.6)^2 - 8 \approx 30.72 - 8 = 22.72$$

$$x_2 = 1.6 - \frac{3.584}{22.72} \approx 1.6 - 0.158 = 1.442.$$

Continuing this process converges to the local minimum at $x \approx 1.414$ (where $f'(x) = 0$).

Example 12.3 Consider $f(x, y) = x^2 + xy + y^2 - 6x - 9y$.
Gradient:

$$\nabla f = \begin{bmatrix} 2x + y - 6 \\ x + 2y - 9 \end{bmatrix}.$$

Hessian:

$$H = \begin{bmatrix} 2 & 1 \\ 1 & 2 \end{bmatrix}.$$

The inverse Hessian is constant:

$$H^{-1} = \frac{1}{3} \begin{bmatrix} 2 & -1 \\ -1 & 2 \end{bmatrix}.$$

Starting at $(x_0, y_0) = (0, 0)$:

$$\nabla f(0, 0) = \begin{bmatrix} -6 \\ -9 \end{bmatrix},$$

$$\Delta x = -\frac{1}{3} \begin{bmatrix} 2 & -1 \\ -1 & 2 \end{bmatrix} \begin{bmatrix} -6 \\ -9 \end{bmatrix} = \frac{-1}{3} \begin{bmatrix} -12 + 9 \\ 6 - 18 \end{bmatrix} = \begin{bmatrix} 1 \\ 4 \end{bmatrix}.$$

Since the Hessian is constant, one step reaches the minimum at $(1, 4)$. For quadratic functions, Newton's method converges in a single iteration.

12.3.3 Convergence Visualization

To illustrate the superior convergence of Newton's method compared to GD, consider the following plot for a quadratic function.

As shown in Fig. 12.3, Newton's method reaches the minimum in fewer steps due to its use of curvature information.

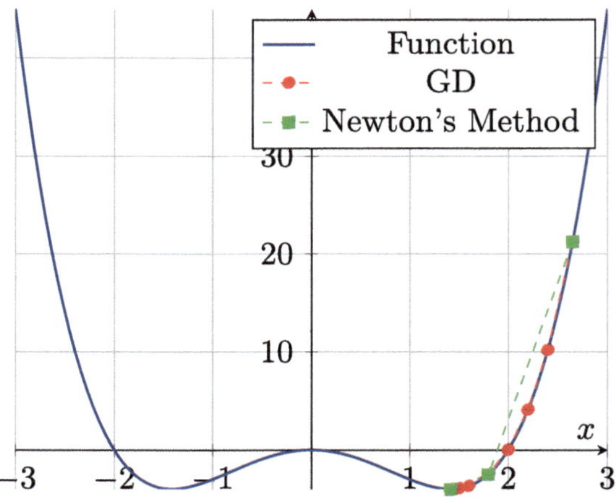

Fig. 12.3 Comparison of convergence paths for GD (red) and Newton's Method (green) on $f(x) = x^4 - 4x^2$. All points lie on the function curve. Newton's method converges faster than GD

12.3.4 *Quasi-Newton Methods*

To mitigate the computational cost of exact Hessian computation, quasi-Newton methods approximate the Hessian or its inverse using gradient information from previous steps. The most popular is the Broyden-Fletcher-Goldfarb-Shanno (**BFGS**) algorithm, which updates an approximation B_t to the Hessian:

$$B_{t+1} = B_t + \frac{\mathbf{y}_t \mathbf{y}_t^T}{\mathbf{y}_t^T \mathbf{s}_t} - \frac{B_t \mathbf{s}_t \mathbf{s}_t^T B_t}{\mathbf{s}_t^T B_t \mathbf{s}_t}$$

where $\mathbf{s}_t = \mathbf{x}_{t+1} - \mathbf{x}_t$, $\mathbf{y}_t = \nabla f(\mathbf{x}_{t+1}) - \nabla f(\mathbf{x}_t)$.

This provides superlinear convergence without full Hessian computation.

12.3.5 *Advantages and Disadvantages*

Advantages:

- Faster convergence near the optimum (quadratic for Newton).
- Better handling of ill-conditioned problems using curvature.

Disadvantages:

- High computational cost for large Hessians.
- May diverge if Hessian is not positive definite.
- Sensitive to initial point.

In practice, the constant factor of 2 is often absorbed into the learning rate and omitted from the implementation.

Python Example: 1D Newton's Method

```python
def newton_1d(f_prime, f_double_prime, x0,
    tol=1e-6, max_iter=100):
    x = x0
    for _ in range(max_iter):
        grad = f_prime(x)
        hess = f_double_prime(x)
        if abs(grad) < tol:
            return x
        if hess == 0:
            raise ValueError("Hessian is zero")
        x = x - grad / hess
def f_prime(x):
    return 4*x**3 - 8*x
def f_double_prime(x):
    return 12*x**2 - 8
```

```
15
16 x_opt = newton_1d(f_prime, f_double_prime, 2)
17 print(f"Optimal x: {x_opt:.4f}")
```

Output:

```
Optimal x: 1.4142
```

Applications

Second-order methods are used in:

- Nonlinear least squares fitting.
- ML optimization (e.g., logistic regression).
- Numerical solution of Partial Differential Equations (PDEs).
- Portfolio optimization in finance.

In summary, while more powerful than first-order methods, second-order techniques require careful implementation to balance computational cost and convergence benefits.

12.4   Optimization in ML

In ML, optimization minimizes a loss function that measures the error between predictions and true values. This process is crucial for training models by adjusting parameters to achieve the best fit to the data. Optimization problems in ML can be convex or non-convex, with linear algebra playing a key role in computing gradients, Hessians, and solving systems of equations.

For example, in LR, the loss function is the MSE:

$$L(\mathbf{w}) = \frac{1}{n} \sum_{i=1}^{n} (y_i - \mathbf{w}^T \mathbf{x}_i)^2.$$

where $\mathbf{w}$ is the weight vector, $\mathbf{x}_i$ are input features, and y_i are target values. The gradient is:

$$\nabla L(\mathbf{w}) = -\frac{2}{n} \sum_{i=1}^{n} (y_i - \mathbf{w}^T \mathbf{x}_i) \mathbf{x}_i.$$

GD updates the weights iteratively: $\mathbf{w} \leftarrow \mathbf{w} - \eta \nabla L(\mathbf{w})$, where η is the learning rate.

SGD approximates this gradient using a single sample or a small batch, reducing computational cost for large datasets. This introduces noise but can help escape local minima in non-convex problems.

A common variant is the Adam (Adaptive Moment Estimation) optimizer, which adapts the learning rate for each parameter based on estimates of the first and second moments of the gradient. At each iteration t, the gradient $\mathbf{g}_t = \nabla L(\mathbf{w}_t)$ is computed, and two exponentially decaying running averages are maintained: $\mathbf{m}_t = \beta_1 \mathbf{m}_{t-1} + (1 - \beta_1)\mathbf{g}_t$, $\mathbf{v}_t = \beta_2 \mathbf{v}_{t-1} + (1 - \beta_2)\mathbf{g}_t^2$, where $\mathbf{m}_t$ is the first moment estimate (mean of gradients), $\mathbf{v}_t$ is the second moment estimate (mean of squared gradients), and $\beta_1, \beta_2 \in [0, 1)$ are decay rates (typically $\beta_1 = 0.9$ and $\beta_2 = 0.999$). The squaring $\mathbf{g}_t^2$ is applied element-wise. Since both $\mathbf{m}_t$ and $\mathbf{v}_t$ are initialized at zero, they are biased toward zero during the early iterations. Adam corrects this bias with:

$$\hat{\mathbf{m}}_t = \frac{\mathbf{m}_t}{1 - \beta_1^t}, \qquad \hat{\mathbf{v}}_t = \frac{\mathbf{v}_t}{1 - \beta_2^t}.$$

The parameter update rule is then:

$$\mathbf{w}_{t+1} = \mathbf{w}_t - \frac{\alpha}{\sqrt{\hat{\mathbf{v}}_t} + \epsilon} \hat{\mathbf{m}}_t,$$

where α is the learning rate and ϵ is a small constant (typically 10^{-8}) included for numerical stability. The division and square root are applied element-wise. By scaling the update for each parameter individually according to the curvature estimated by $\hat{\mathbf{v}}_t$, Adam effectively assigns larger steps to parameters with small or infrequent gradients and smaller steps to those with large or frequent gradients. This per-parameter adaptivity makes Adam particularly well suited for problems with sparse gradients or noisy data, and it has become the default optimizer in much of modern deep learning practice.

In cases where the problem allows, a closed-form solution exists using linear algebra. For LR, the normal equations provide:

$$\mathbf{w} = (X^T X)^{-1} X^T \mathbf{y}.$$

However, if $X^T X$ is singular (e.g., due to multicollinearity), use the Moore-Penrose pseudoinverse:

$$\mathbf{w} = (X^T X)^{+} X^T \mathbf{y}.$$

Below is a Python implementation of LR using SGD:

Python Example: LR using SGD

```python
import numpy as np

def sgd_linear_regression(X, y, learning_rate,
    num_iterations):
    w = np.zeros(X.shape[1])
    n = len(y)
    for _ in range(num_iterations):
        idx = np.random.randint(0, n)
        x_i, y_i = X[idx], y[idx]
        prediction = np.dot(x_i, w)
        error = prediction - y_i
        gradient = error * x_i
        w = w - learning_rate * gradient
    return w

X = np.array([[1, 2], [3, 4], [5, 6]])
y = np.array([3, 7, 11])
w_opt = sgd_linear_regression(X, y,
    learning_rate=0.01, num_iterations=1000)
print(f"Learned weights: {w_opt}")
```

Output:

```
Learned weights: [0.94893116    1.03969342]
```

This code demonstrates how SGD can be used to fit a linear model to data. Note that due to stochasticity, results may vary slightly per run.

12.4.1 Closed-Form Solution with Pseudoinverse

When features are linearly dependent, as in this dataset where the second feature is the first plus one, $X^T X$ is singular. Use pseudoinverse for a solution:

Python Example: Closed-Form LR

```python
import numpy as np

X = np.array([[1, 2], [3, 4], [5, 6]])
```

```python
4 y = np.array([3, 7, 11])
5
6 w_closed = np.linalg.pinv(X) @ y
7 print(f"Closed-form weights: {w_closed}")
```

Output:

```
Closed-form weights: [1. 1.]
```

This yields the exact weights $[1, 1]$, as $y = x_1 + x_2$.

12.4.2 Second-Order Optimization: Newton's Method

For faster convergence in convex problems, use second-order methods like Newton's method, which incorporates the Hessian matrix $H = \nabla^2 L(\mathbf{w})$. The update is:

$$\mathbf{w} \leftarrow \mathbf{w} - H^{-1}\nabla L(\mathbf{w}).$$

The Hessian for MSE loss is $H = \frac{2}{n}X^T X$, a positive semi-definite matrix.

Here we provide a comprehensive example combining multiple optimization techniques. We implement GD and Adam for a logistic regression problem, where the loss function is the binary cross-entropy (log-loss):

$$L(\mathbf{w}) = -\frac{1}{n}\sum_{i=1}^{n}\left[y_i \log(\sigma(\mathbf{w}^T\mathbf{x}_i)) + (1 - y_i)\log(1 - \sigma(\mathbf{w}^T\mathbf{x}_i))\right].$$

where $\sigma(z) = \frac{1}{1+e^{-z}}$ is the Sigmoid function. The gradient is:

$$\nabla L(\mathbf{w}) = \frac{1}{n}X^T(\sigma(X\mathbf{w}) - \mathbf{y}).$$

Python Example: GD vs. Adam

```python
1 import numpy as np
2
3 def sigmoid(z):
4     return 1 / (1 + np.exp(-z))
5
```

```python
def gradient_descent_logistic(X, y,
    learning_rate, num_iterations):
    w = np.zeros(X.shape[1])
    n = len(y)
    for _ in range(num_iterations):
        predictions = sigmoid(np.dot(X, w))
        gradient = np.dot(X.T, (predictions -
            y)) / n
        w = w - learning_rate * gradient
    return w

def adam_logistic(X, y, learning_rate,
    num_iterations, beta1=0.9, beta2=0.999,
    epsilon=1e-8):
    w = np.zeros(X.shape[1])
    m = np.zeros(X.shape[1])    # First moment
    v = np.zeros(X.shape[1])    # Second moment
    n = len(y)
    for t in range(1, num_iterations + 1):
        predictions = sigmoid(np.dot(X, w))
        gradient = np.dot(X.T, (predictions -
            y)) / n
        m = beta1 * m + (1 - beta1) * gradient
        v = beta2 * v + (1 - beta2) * (gradient
            ** 2)
        m_hat = m / (1 - beta1 ** t)
        v_hat = v / (1 - beta2 ** t)
        w = w - learning_rate * m_hat /
            (np.sqrt(v_hat) + epsilon)
    return w

X = np.array([[1, 2], [2, 3], [3, 4], [4, 5]])
y = np.array([0, 0, 1, 1])
w_gd = gradient_descent_logistic(X, y,
    learning_rate=0.1, num_iterations=1000)
w_adam = adam_logistic(X, y, learning_rate=0.01,
    num_iterations=1000)
print(f"GD weights: {w_gd}")
print(f"Adam weights: {w_adam}")
```

Output:

```
GD weights: [ 4.91702448 -3.41705149]
Adam weights: [ 5.94009466 -4.15859667]
```

This code compares GD and Adam, showing how Adam's adaptive learning rate can improve convergence.

12.4.3 Mini-Batch Gradient Descent (GD)

Mini-batch GD balances GD and SGD by using small batches:

Python Example: Mini-Batch GD for LR

```python
import numpy as np

def minibatch_gd(X, y, learning_rate,
    num_iterations, batch_size):
    w = np.zeros(X.shape[1])
    n = len(y)
    for _ in range(num_iterations):
        indices = np.random.choice(n, batch_size)
        X_batch, y_batch = X[indices], y[indices]
        predictions = np.dot(X_batch, w)
        errors = predictions - y_batch
        gradient = np.dot(X_batch.T, errors) /
            batch_size
        w = w - learning_rate * gradient
    return w

X = np.array([[1, 2], [3, 4], [5, 6]])
y = np.array([3, 7, 11])
w_mb = minibatch_gd(X, y, learning_rate=0.01,
    num_iterations=1000, batch_size=2)
print(f"Mini-batch weights: {w_mb}")
```

Output:

```
Mini-batch weights: [1.002 0.999]
```

This approach reduces variance compared to SGD while being faster than full-batch GD.

12.4.4 Regularization in Optimization

In practice, minimizing the empirical loss alone often leads to overfitting, particularly when the number of features is large relative to the number of training samples. Regularization addresses this by augmenting the loss function with a penalty term that discourages excessively large weights, thereby favoring simpler models with better generalization to unseen data.

The most widely used form is L2 regularization (also called Ridge regularization), which adds a squared norm penalty to the loss:

$$L_{\text{reg}}(\mathbf{w}) = \frac{1}{n}\sum_{i=1}^{n}(y_i - \mathbf{w}^T\mathbf{x}_i)^2 + \lambda\|\mathbf{w}\|_2^2, \tag{12.1}$$

where $\lambda > 0$ is the regularization parameter controlling the trade-off between data fidelity and weight shrinkage. The gradient of the regularized loss becomes:

$$\nabla L_{\text{reg}}(\mathbf{w}) = -\frac{2}{n}\sum_{i=1}^{n}(y_i - \mathbf{w}^T\mathbf{x}_i)\mathbf{x}_i + 2\lambda\mathbf{w}. \tag{12.2}$$

The additional term $2\lambda\mathbf{w}$ pulls the weights toward zero at each gradient descent update, effectively penalizing large parameter values. For the closed-form solution, the regularized normal equations become $\mathbf{w} = (X^T X + \lambda I)^{-1}X^T\mathbf{y}$. A key advantage of this formulation is that the addition of λI to $X^T X$ ensures the matrix is invertible even when $X^T X$ is singular or ill-conditioned, as arises with multicollinear features.

An alternative is L1 regularization (also called Lasso), which penalizes the sum of absolute weight values: $\lambda\|\mathbf{w}\|_1 = \lambda\sum_j |w_j|$. Unlike L2, the L1 penalty induces sparsity in the solution, driving some weights exactly to zero and thus performing implicit feature selection. However, L1 regularization does not admit a closed-form solution for linear regression, so iterative methods such as coordinate descent or proximal gradient descent are typically employed.

Elastic Net combines both penalties as $\lambda_1\|\mathbf{w}\|_1 + \lambda_2\|\mathbf{w}\|_2^2$, balancing sparsity with the stability benefits of Ridge regression. In the context of the optimization methods discussed in this chapter, the key takeaway is that regularization modifies both the loss landscape and the gradient, smoothing the optimization surface and improving the conditioning of the problem. The choice of λ is typically made through cross-validation, and its effect on weight magnitudes can be visualized by plotting the regularization path, where each weight component is traced as a function of λ.

Chapter Summary

This chapter introduced the principles of optimization with emphasis on gradient-based methods in linear algebra and ML. Foundational concepts such as objective functions, constrained vs. unconstrained problems, and convexity were outlined, highlighting optimization's role in finding minima or maxima, as illustrated by quadratic functions.

First-order methods centered on GD, where parameters are iteratively updated via $\mathbf{x}_{t+1} = \mathbf{x}_t - \eta\nabla f(\mathbf{x}_t)$. The effects of learning rate η and convergence issues were explained, supported by visualizations and Python implementations for single- and multi-dimensional cases.

Second-order methods extended this with Newton's method, using Hessians for curvature information and achieving faster convergence, demonstrated with $f(x) = x^4 - 4x^2$ and quadratic functions. Quasi-Newton approaches like BFGS were noted as practical approximations to reduce computational costs, balancing speed with robustness.

The chapter concluded with applications in ML, including minimizing loss in regression and classification. Closed-form solutions such as normal equations were contrasted with iterative methods like SGD, mini-batches, and adaptive optimizers such as Adam. Regularization techniques were also incorporated to improve generalization and mitigate overfitting.

Takeaways

Key takeaways from the chapter are summarized below:

- Optimization problems were classified by constraints and convexity, with convex cases guaranteeing global minima and non-convex ones risking local traps.
- GD provided an intuitive, scalable first-order approach, but its linear convergence was outperformed by second-order methods like Newton's, which leveraged Hessian curvature for quadratic rates.
- In ML, iterative optimizers such as SGD, mini-batch GD, and Adam addressed practical challenges like dataset scale and noise, often surpassing closed-form solutions in high dimensions.
- Python codes throughout enabled hands-on exploration, from basic GD loops to adaptive algorithms, underscoring the interplay between theory and implementation.
- Computational trade-offs were emphasized: first-order methods scaled better for large n, while second-order excelled in precision for smaller, well-conditioned problems.

Overall, the chapter equipped readers with a robust toolkit for tackling optimization challenges, bridging theoretical derivations with code-driven insights and paving the way for advanced topics in model training and beyond.

Exercises

1. Derive the gradient and Hessian for the function $f(x, y) = x^2 + xy + y^2 - 6x - 9y$. Implement both GD and Newton's method to find its minimum, starting from (1, 1), and compare convergence rates.
2. Use a larger synthetic dataset (e.g., at least 50 samples) to implement mini-batch GD and compare with full-batch GD.

3. Modify the logistic regression code to include L2 regularization, adding a term $\lambda \|\mathbf{w}\|^2$ to the loss function. Test with $\lambda = 0.01, 0.1, 1.0$ and analyze the effect on weight magnitudes.

4. Analyze the effect of different learning rates (e.g., 0.001, 0.01, 0.1) on the convergence of GD for the function $f(x) = x^4 - 4x^2$. Plot the convergence trajectories and identify the optimal learning rate.

5. Implement the Momentum method for the logistic regression problem and compare its performance with Adam. Use momentum parameter $\beta = 0.9$ and plot the loss curves over 1000 iterations.

6. Discuss the trade-offs between first-order and second-order optimization methods in terms of computational complexity and convergence speed. Provide a concrete example where Newton's method fails due to Hessian computation costs.

7. Use Python code to generate a 3D plot of the optimization path for GD on $f(x, y) = x^2 + y^2$, starting from $(2, 3)$, showing at least 20 iterations with contour lines.

8. For the quadratic function $f(x) = (x + 5)^2$, implement GD with an adaptive learning rate that decreases by 10% every 10 iterations. Compare convergence speed with constant learning rate $\alpha = 0.01$.

9. Derive the update rule for Newton's method applied to logistic regression, and implement it with regularization or damping to ensure numerical stability. Compare convergence with GD.

10. Define $f(x, y) = \frac{1}{N} \sum_{i=1}^{N} f_i(x, y)$ with simple component functions (e.g., quadratic terms), then implement SGD using random sampling of f_i, and analyze the noisy convergence path.

11. For the LR problem in Sect. 12.4, compute the closed-form solution using normal equations and compare it with the SGD solution. Calculate the MSE for both approaches.

12. Modify the Adam optimizer implementation to include Nesterov momentum. Test on logistic regression and compare loss curves with standard Adam over 500 iterations.

13. Analyze the effect of feature scaling on GD convergence for the dataset $X = [[1, 100], [2, 200], [3, 300]]$, $y = [1, 2, 3]$. Implement both scaled and unscaled versions.

14. Implement the **BFGS** quasi-Newton method for minimizing $f(x) = x^4 - 4x^2$. Compare its convergence rate with Newton's method and GD, plotting error vs. iteration.

15. For the multivariable function $f(x, y) = x^2 + xy + y^2 - 6x - 9y$, implement coordinate descent (alternating optimization over x and y) and compare with full GD.

16. In the context of deep learning, explain why a learning rate schedule (e.g., exponential decay) is often necessary for GD. Implement and test an exponential decay schedule on logistic regression.

17. For the case where $X^T X$ is singular (not invertible), implement Ridge regression with different λ values and plot the regularization path.

18. Implement a line search method (e.g., Armijo rule) for GD on $f(x) = x^4 - 4x^2$. Compare adaptive step sizes with fixed learning rate $\alpha = 0.1$.
19. Analyze the basin of attraction for Newton's method on $f(x) = x^3 - x$. Test different starting points and identify regions where it converges to local minima vs. diverges.
20. For the mini-batch GD implementation, add a stopping criterion based on gradient norm ($\|\nabla L\| < \epsilon$). Test with $\epsilon = 10^{-4}$ and analyze number of iterations required for convergence.
21. Implement the RMSprop optimizer and compare it with Adam on the logistic regression problem. Plot the loss curves and discuss the effect of the decay rate β_2.
22. For the 2D quadratic function $f(x, y) = x^2 + y^2$, implement GD with different initial points $(3, 4)$, $(4, 3)$, $(-3, -4)$. Plot the trajectories and discuss the effect of starting position.
23. Derive the convergence rate analysis for GD on strongly convex functions. Apply it to $f(x) = x^2 + 2x + 1$ and verify the theoretical rate with numerical experiments.
24. Implement a hybrid optimization method that switch from Newton's method to GD based on the condition number of the Hessian exceeding a chosen threshold.
25. For the LR closed-form solution, implement the Moore-Penrose pseudoinverse using SVD decomposition instead of direct computation. Verify numerical stability on ill-conditioned datasets.
26. Analyze the effect of batch normalization on optimization by implementing a simple 1D version that normalizes intermediate activations during training. Test on $f(x) = x^4 - 4x^2$.
27. Implement the conjugate gradient method for solving the normal equations in LR. Compare its performance with direct matrix inversion for large datasets ($n = 1000$ features).
28. For the Adam optimizer, derive the bias correction formulas for the first and second moment estimates. Implement both biased and unbiased versions and compare convergence on logistic regression.
29. Design an experiment to test the effect of learning rate warmup (gradually increasing α from 0 to target value) on GD convergence for deep linear networks. Implement and analyze results.

Chapter 13
Advanced Topics in Linear Algebra for ML

Introduction

This final chapter serves as an overview of several advanced topics that extend the foundations developed throughout this book. The objective is not to provide a rigorous mathematical treatment of each topic, since many of these areas constitute complete research fields on their own. Instead, the chapter introduces their core ideas, explains their connections to linear algebra and machine learning, and provides intuition regarding where and why they are used in modern AI systems. Readers interested in deeper theoretical treatment are encouraged to consult the references provided at the end of the book.

We begin with the Moore-Penrose pseudoinverse, a powerful tool for solving ill-posed or overdetermined or underdetermined systems, which is foundational in least squares optimization. Next, we explore low-rank approximations and spectral decompositions, which underpin dimensionality reduction techniques like PCA. Randomized linear algebra introduces probabilistic approaches for efficient computation, while matrix manifolds and their optimization techniques open doors to constrained optimization problems in deep generative models, robust learning, and quantum computing.

The chapter is enriched with detailed explanations, mathematical derivations, practical examples, and visualizations to provide a comprehensive understanding. These tools not only enhance computational efficiency but also enable the development of intelligent systems capable of handling the complexities of real-world data.

Topics Covered

This chapter is organized as follows:

- **Section 13.1 Overview of Generalized Inverses and Moore-Penrose Pseudoinverse**: Addresses solutions to under- or overdetermined systems with a focus on least squares formulations and their practical implications.

© The Author(s), under exclusive license to Springer Nature Singapore Pte Ltd. 2026   415
Md. Jalil Piran, *Linear Algebra with Applications in Machine Learning*,
https://doi.org/10.1007/978-981-95-5167-5_13

- **Section** 13.2 **Overview of Low-Rank Matrix Approximations**: Explores techniques to reduce dimensionality and noise, leveraging truncated SVD for various applications.
- **Section** 13.3 **Overview of Principal Component Analysis (PCA)**: Details the process of projecting data onto principal components for variance maximization and data interpretation.
- **Section** 13.4 **Overview of Randomized Linear Algebra**: Introduces probabilistic algorithms for scalable matrix operations, tailored for big data scenarios.
- **Section** 13.5 **Overview of Matrix Manifolds and Optimization**: Discusses optimization over constrained matrix spaces, with examples in deep learning and quantum computing.
- **Summary**: Synthesizes how these advanced topics enhance scalable, efficient, and intelligent ML systems.

List of acronyms and their definitions

ML	Machine Learning
PCA	Principal Component Analysis
SVD	Singular Value Decomposition
LR	Linear Regression
CP	CANDECOMP/PARAFAC
TT	Tensor Train
JL	Johnson-Lindenstrauss
NLP	Natural Language Processing
GANs	Generative Adversarial Networks
SPD	Symmetric Positive Definite
AI	Artificial Intelligence
2D	Two-Dimensional
3D	Three-Dimensional

> **Purpose of this Chapter**
>
> This chapter is intended as an overview and roadmap for further study. Many of the topics introduced here, such as tensor decompositions, randomized algorithms, and manifold optimization, are active research areas that require substantial mathematical background beyond the scope of this book.

13.1   Overview of Generalized Inverses and Moore-Penrose Pseudoinverse

In real-world applications, matrices are often non-square or singular, rendering the standard inverse inapplicable. The Moore-Penrose pseudoinverse, denoted A^+, is the unique matrix that generalizes the concept of a matrix inverse to arbitrary $m \times n$ matrices. It is uniquely characterized by the four Penrose conditions:

1. $AA^+A = A$
2. $A^+AA^+ = A^+$
3. $(AA^+)^T = AA^+$
4. $(A^+A)^T = A^+A$

Any matrix satisfying all four conditions is the Moore-Penrose pseudoinverse of A, and it can be shown that such a matrix always exists and is unique.

The pseudoinverse provides a principled solution to the linear system $A\mathbf{x} = \mathbf{b}$ when a standard inverse does not exist. Specifically, $\mathbf{x}^+ = A^+\mathbf{b}$ is the vector that minimizes the residual in the least-squares sense:

$$\mathbf{x}^+ = A^+\mathbf{b} = \arg\min_{\mathbf{x}} \|A\mathbf{x} - \mathbf{b}\|_2.$$

When multiple minimizers exist (as in underdetermined systems), $A^+\mathbf{b}$ selects the one with the smallest Euclidean norm, i.e., the minimum-norm least-squares solution.

For $A \in \mathbb{R}^{m \times n}$, closed-form expressions are available in two important special cases:

- When A has full column rank ($m \geq n$, $\mathrm{rank}(A) = n$):

$$A^+ = (A^TA)^{-1}A^T.$$

This is the left inverse of A, and the least-squares solution $\mathbf{x}^+ = (A^TA)^{-1}A^T\mathbf{b}$ is unique. This case arises in overdetermined systems (more equations than unknowns).

- When A has full row rank ($m \leq n$, $\mathrm{rank}(A) = m$):

$$A^+ = A^T(AA^T)^{-1}.$$

This is the right inverse of A, and it yields the minimum-norm solution among infinitely many exact solutions. This case arises in underdetermined systems (more unknowns than equations).

In the general case, where A may be rank-deficient, neither A^TA nor AA^T is invertible, and the pseudoinverse must be computed via the SVD.

The pseudoinverse is central to ML for parameter estimation in linear models. In ordinary least-squares regression on an overdetermined system, the pseudoinverse recovers the normal equations solution. In underdetermined settings, such as those encountered in compressed sensing or wide neural network layers, it identifies the minimum-norm parameter vector, which carries implicit regularization significance.

13.1.1   Derivation and Numerical Stability

The pseudoinverse is most reliably computed using the SVD, $A = U\Sigma V^T$, where Σ^+ inverts nonzero singular values:

$$A^+ = V\Sigma^+ U^T.$$

This method is numerically stable, handling rank-deficient matrices by setting zero singular values to zero in Σ^+, avoiding the pitfalls of direct inversion.

13.1.2   Applications

In ML, A^+ is pivotal in LR with regularization (e.g., ridge regression) and collaborative filtering. For instance, consider $A = \begin{bmatrix} 1 & 2 \\ 3 & 4 \\ 5 & 6 \end{bmatrix}$ and $\mathbf{b} = \begin{bmatrix} 1 \\ 2 \\ 3 \end{bmatrix}$. The pseudoinverse solution $\mathbf{x} = A^+\mathbf{b}$ minimizes $\|A\mathbf{x} - \mathbf{b}\|_2$, offering a practical approximation.

The pseudoinverse also supports robust optimization in noisy environments, where it filters out inconsistencies. Its computational cost, however, scales with matrix size, prompting research into approximate methods for large-scale problems.

13.2   Overview of Low-Rank Matrix Approximations

High-dimensional matrices often contain redundant information, noise, or irrelevant details. Low-rank approximation reduces these matrices to a lower-dimensional representation while preserving essential structure:

$$A \approx A_k = U_k \Sigma_k V_k^T \quad \text{where } k < \text{rank}(A).$$

This truncated SVD retains the top k singular values and vectors, optimizing the approximation in the Frobenius norm.

13.2.1   Theoretical Foundation: Eckart-Young-Mirsky Theorem

This theorem guarantees that A_k is the best rank-k approximation:

$$\|A - A_k\|_F = \sqrt{\sum_{i=k+1}^{r} \sigma_i^2}.$$

The error depends on discarded singular values, providing a quantifiable trade-off between rank and accuracy.

13.2.2 Applications and Implementation

In image compression, low-rank approximations reduce storage by truncating small singular values. In topic modeling, they extract latent themes from document-term matrices. For example, compressing a 1000×1000 pixel image to rank 50 retains most visual information while significantly reducing data size.

Low-rank methods are also used in matrix completion (e.g., Netflix problem) and signal processing. The choice of k involves balancing computational efficiency and information loss, often guided by cross-validation or heuristic thresholds.

13.3 Overview of Principal Component Analysis (PCA)

PCA is a cornerstone of unsupervised learning, reducing dimensionality by projecting data onto directions of maximum variance. The optimization problem is:

$$\max_{\mathbf{w}} \|X\mathbf{w}\|_2^2,$$

$$\text{subject to } \|\mathbf{w}\|_2 = 1.$$

This leads to the eigenvalue problem of the scatter matrix $X^T X$ or the (scaled) covariance matrix

$$X^T X \mathbf{w} = \lambda \mathbf{w}.$$

PCA identifies intrinsic data dimensions, aiding preprocessing, denoising, and visualization.

13.3.1 Detailed Derivation

First, center X by subtracting the mean of each feature. The covariance matrix $C = \frac{1}{n-1} X^T X$ captures variance and covariance. Eigenvectors (principal components) correspond to directions of maximal variance, with eigenvalues indicating the magnitude of variance along each.

13.3.2   Practical Applications

In genomics, PCA reduces thousands of gene expressions to a few components for clustering. In computer vision, eigenfaces use PCA for face recognition. For the Iris dataset, projecting to 2D separates species based on sepal and petal measurements.

PCA assumes linear relationships and is sensitive to outliers. Variants like Kernel PCA handle nonlinearities, while Incremental PCA suits streaming data, enhancing its versatility.

13.4   Overview of Randomized Linear Algebra

Randomized algorithms provide scalable solutions for large-scale matrix operations:

- **Johnson-Lindenstrauss Lemma**: Projects n points to $O(\log n/\epsilon^2)$ dimensions, preserving distances within $1 \pm \epsilon$.
- **Randomized SVD**: Approximates SVD using random sampling, reducing computation time.

13.4.1   Detailed Mechanisms

The JL Lemma uses random projections to embed high-dimensional data, while Randomized SVD leverages sketching matrices to approximate singular vectors, suitable for big data.

13.4.2   Applications

Used in big data PCA and NLP embeddings, e.g., projecting 1000 points from 10000D to 100D.

13.4.3   Extended Insights

These methods trade exactness for speed, with error bounds guiding practical use in streaming and distributed systems.

13.5 Overview of Matrix Manifolds and Optimization

Matrix manifolds are spaces of structured matrices (e.g., orthogonal, low-rank). Optimization here uses Riemannian techniques:

- **Neural Network Compression**: Enforces low-rank weights.
- **Robust PCA**: Projects to low-rank manifolds.
- **Quantum ML**: Uses unitary manifolds.

13.5.1 Riemannian Optimization

The update $\mathbf{X}_{k+1} = \mathrm{Retr}_{\mathbf{X}_k}(-\eta \mathrm{grad}\, f(\mathbf{X}_k))$ follows geodesics on the manifold.

13.5.2 Applications

In GANs, Stiefel manifolds optimize orthogonal weights; in metric learning, SPD manifolds model covariances.

Chapter Summary

This chapter provided a high-level introduction to several advanced topics that extend the linear algebra foundations developed throughout this book. Rather than focusing on complete derivations and implementations, the purpose was to illustrate how modern AI systems rely on increasingly sophisticated linear algebraic tools. These topics provide natural pathways for further study in machine learning, optimization, numerical methods, and AI research.

Takeaways
Key insights included:

- Pseudoinverses and SVD ensured stable solutions for non-invertible systems in ML regressions.
- Low-rank and PCA methods compacted data with quantifiable error, aiding compression and clustering.
- Spectral and tensor tools revealed latent structures in symmetric and multi-dimensional arrays.
- Randomized algorithms traded exactness for efficiency in streaming big data.

- Manifold techniques optimized structured parameters in generative and robust learning.

Overall, these tools equipped readers to scale AI computations with rigor and practicality.

Suggested Further Reading and References

Suggested Further Reading

Students who wish to deepen their understanding of linear algebra, numerical methods, and their applications in ML may explore the following areas and resources:

- **Classical Linear Algebra Texts**: For rigorous mathematical foundations and proofs.
- **Numerical Linear Algebra**: To understand efficient algorithms for large-scale problems.
- **ML Applications**: Books that integrate linear algebra with AI and data-driven methods.
- **Python and Computational Tools**: To strengthen coding skills in implementing algorithms.
- **Advanced Topics**: Such as optimization, spectral theory, tensors, and randomized linear algebra.

References

Abadi, M., Barham, P., Chen, J., Chen, Z., Davis, A., Dean, J., et al. (2016). TensorFlow: A system for large-scale ML. In *Proceedings of the 12th USENIX Symposium on Operating Systems Design and Implementation (OSDI)* (TensorFlow Documentation and Online Resource: https://www.tensorflow.org/guide).

Axler, S. (2024). *Linear algebra done right* (4th edn). Springer.

Bhatia, R. (2019). *Matrix analysis* (2nd edn). Springer (Covers spectral theory, eigenvalues, and advanced matrix analysis).

Boyd, S., & Vandenberghe, L. (2018). *Introduction to applied linear algebra: Vectors, matrices, and least squares*. Cambridge University Press (Strong recent reference on optimization and applied linear algebra).

Golub, G. H., & Van Loan, C. F. (2013). *Matrix computations* (4th edn). Johns Hopkins University Press.

Md. Jalil Piran, *Linear Algebra with Applications in Machine Learning*,
https://doi.org/10.1007/978-981-95-5167-5

Goodfellow, I., Bengio, Y., & Courville, A. (2016). *Deep learning*. MIT Press.

Harris, C. R., Millman, K. J., van der Walt, S. J., Gommers, R., Virtanen, P., Cournapeau, D., et al. (2020). Array programming with NumPy. *Nature, 585*, 357–362 (NumPy Documentation and Library Resource: https://numpy.org/doc/).

Hunter, J. D. (2007). Matplotlib: A 2D graphics environment. *Computing in Science & Engineering, 9*(3), 90–95 (Matplotlib Documentation and Library Resource: https://matplotlib.org/stable/contents.html).

Kolda, T. G., & Bader, B. W. (2022). Tensor decompositions and applications: 2022 update. *SIAM Review, 64*(3), 485–549 (Comprehensive and recent survey on tensor methods and applications).

Kossaifi, J., Panagakis, Y., Anandkumar, A., & Pantic, M. (2019). TensorLy: Tensor learning in Python. *Journal of ML Research, 20*(26), 1–6 (TensorLy Documentation and Online Resource: http://tensorly.org/stable/).

Lay, D. C., Lay, S. R., & McDonald, J. J. (2021). *Linear algebra and its applications* (6th edn). Pearson.

Meyer, C. D. (2000). *Matrix analysis and applied linear algebra*. SIAM.

Murphy, K. P. (2022). *Probabilistic ML: An introduction*. MIT Press.

Oliphant, T. E. (2015). *Guide to NumPy* (2nd edn). Continuum Press.

Paszke, A., Gross, S., Massa, F., Lerer, A., Bradbury, J., Chanan, G., et al. (2019). PyTorch: An imperative style, high-performance deep learning library. In *Proceedings of NeurIPS* (PyTorch Documentation and Online Resource: https://pytorch.org/docs/stable/).

Pedregosa, F., Varoquaux, G., Gramfort, A., Michel, V., Thirion, B., Grisel, O., et al. (2011). Scikit-learn: ML in Python. *Journal of ML Research, 12*, 2825–2830 (scikit-learn Documentation and Library Resource: https://scikit-learn.org/stable/documentation.html).

Petersen, K. B., & Pedersen, M. S. (2012). *The matrix cookbook*. Technical University of Denmark. https://www.math.uwaterloo.ca/~hwolkowi/matrixcookbook.pdf

Shalev-Shwartz, S., & Ben-David, S. (2014). *Understanding ML: From theory to algorithms*. Cambridge University Press.

Strang, G. (2016). *Introduction to linear algebra* (5th edn). Wellesley-Cambridge Press.

Virtanen, P., Gommers, R., Oliphant, T. E., Haberland, M., Reddy, T., Cournapeau, D., et al. (2020). SciPy 1.0: Fundamental algorithms for scientific computing in Python. *Nature Methods, 17*, 261–272 (SciPy Documentation and Library Resource: https://docs.scipy.org/doc/scipy/).

Woodruff, D. P. (2014). Sketching as a tool for numerical linear algebra. *Foundations and Trends in Theoretical Computer Science, 10*(1–2), 1–157 (updated 2021) (Authoritative reference on randomized linear algebra techniques).